T0187469

CRC Series in
CONTEMPORARY FOOD SCIENCE

MODELING MICROBIAL
RESPONSES in FOOD

CRC Series in
CONTEMPORARY FOOD SCIENCE

Fergus M. Clydesdale, Series Editor
University of Massachusetts, Amherst

Published Titles:

America's Foods Health Messages and Claims:
Scientific, Regulatory, and Legal Issues
James E. Tillotson

New Food Product Development: From Concept to Marketplace
Gordon W. Fuller

Food Properties Handbook
Shafiur Rahman

Aseptic Processing and Packaging of Foods:
Food Industry Perspectives
Jarius David, V. R. Carlson, and Ralph Graves

The Food Chemistry Laboratory: A Manual for Experimental Foods,
Dietetics, and Food Scientists, Second Edition
Connie M. Weaver and James R. Daniel

Handbook of Food Spoilage Yeasts
Tibor Deak and Larry R. Beauchat

Food Emulsions: Principles, Practice, and Techniques
David Julian McClements

Getting the Most Out of Your Consultant: A Guide
to Selection Through Implementation
Gordon W. Fuller

Antioxidant Status, Diet, Nutrition, and Health
Andreas M. Papas

Food Shelf Life Stability
N.A. Michael Eskin and David S. Robinson

Bread Staling
Pavinee Chinachoti and Yael Vodovotz

Interdisciplinary Food Safety Research
Neal M. Hooker and Elsa A. Murano

Automation for Food Engineering: Food Quality Quantization
and Process Control
Yanbo Huang, A. Dale Whittaker, and Ronald E. Lacey

Modeling Microbial Responses in Food
Robin C. McKellar and Xuewen Lu

CRC Series in
CONTEMPORARY FOOD SCIENCE

MODELING MICROBIAL RESPONSES in FOOD

Edited by
Robin C. McKellar
Xuewen Lu

CRC Press
Taylor & Francis Group
Boca Raton London New York

CRC Press is an imprint of the
Taylor & Francis Group, an **informa** business

CRC Press
Taylor & Francis Group
6000 Broken Sound Parkway NW, Suite 300
Boca Raton, FL 33487-2742

First issued in paperback 2019

ISBN-13: 978-0-8493-1237-3 (hbk)
ISBN-13: 978-0-367-39465-3 (pbk)

Library of Congress Cataloging-in-Publication Data

Modeling microbial responses in foods / edited by Robin C. McKellar and Xuewen Lu.
     p. cm. — (CRC series in contemporary food science)
  Includes bibliographical references and index.
  ISBN 0-8493-1237-X (alk. paper)
  1. Food—Microbiology. I. McKellar, Robin C., 1949–II. Lu, Xuewen. III. Series.

QR115.M575 2003
664'.001'579—dc22   2003055715

Library of Congress Card Number 2003055715

Visit the Taylor & Francis Web site at
http://www.taylorandfrancis.com

and the CRC Press Web site at
http://www.crcpress.com

Preface

The field of food microbiology is a broad one, encompassing the study of microorganisms that have both beneficial and deleterious effects on the quality and safety of raw and processed foods. The microbiologist's primary objective is to identify and quantify food-borne microorganisms; however, the inherent inaccuracies in the enumeration process and the natural variation found in all bacterial populations complicate the microbiologist's job. Accumulating sufficient data on the behavior of microorganisms in foods requires an extensive amount of work and is costly. In addition, while data can describe a microorganism's response in food, they provide little insight into the relationship between physiological processes and growth or survival. One way this link can be made is through the use of mathematical models.

In its simplest form, a mathematical model is a simple mathematical description of a process. Models have been used extensively in all scientific disciplines. They were first used in food microbiology in the early 20th century to describe the inactivation kinetics of food-borne pathogens during thermal processing of foods. Since then, with the advent of personal computers and more powerful statistical software packages, the use of modeling in food microbiology has grown to the point of being recognized as a distinct discipline of food microbiology, termed predictive microbiology. This concept was introduced and extensively discussed (with particular reference to growth of food-borne pathogens) by McMeekin and his colleagues at the University of Tasmania.[1]

Predictive microbiology has application in both microbial safety and quality of foods; indeed, early development of the concept was based on seafood spoilage. Extensive research in recent years has shown, however, that the most important application of predictive microbiology is in support of food safety initiatives. Microbial growth and survival models are now sufficiently detailed and accurate to make important contributions — scientists and regulators can make reasonable predictions of the relative risk posed by a particular food or food process. It has been argued, however, that predictive microbiology is a misnomer, for predictive microbiology does not actually make predictions at all. To examine this further, we need to first look at some definitions of modeling.

In a general sense, a model simplifies a system by using a combination of descriptions, mathematical functions or equations, and specific starting conditions. There are two general classes of models: descriptive and explanatory, the latter being composed of analytical and numerical models. Descriptive (i.e., observational, empirical, "black box," or inductive) models are data-driven — approaches such as polynomial functions, artificial neural nets, and principal component analysis are used to classify the data. True predictions with this class of models are difficult to make because models cannot be extrapolated beyond the data used to build the model. In spite of this, descriptive models have been used extensively with considerable success in predictive microbiology.

Explanatory (i.e., mechanistic, "white box," or deductive) models aim to relate the given data to fundamental scientific principles, or at least to measurable physiological processes. Many predictive microbiology models have parameters that are related to observed phenomena and are therefore considered mechanistic. Analytical models — explicit equations that can be fit to data — are the most common form of mechanistic models. To be truly mechanistic, however, a model should raise new questions and hypotheses that can be tested. This is not always easy with explicit functions, however, because it is difficult to extend such functions to dynamic situations or add additional steps to the model. Numerical approaches are designed specifically to allow further development of the model. These models are hierarchical, containing submodels at least one level deeper than the response being described. They have been extensively developed for complex ecological systems, but have not been applied to any great extent in predictive microbiology. Because their use requires extensive programming skills, numerical models have been difficult for nonmathematicians to apply. New software platforms and concepts, such as object-oriented programming, have provided new tools, allowing microbiologists to expand on traditional modeling approaches.

The models discussed so far (and the majority of existing predictive microbiology models) have all been deterministic. In a deterministic model, knowledge of the starting conditions, combined with a mathematical function describing the behavior of the system over time, is sufficient to predict the state of the system at any point in time. Bacteria, however, are not so cooperative. While it may be possible to define a function, the starting conditions are less clear — particularly when dealing with individual bacterial cells. Models that recognize and account for uncertainty or variability in an experimental system are called stochastic or probabilistic models. Probability models have been used extensively in the past to predict the probability of germination of pathogenic spore-forming bacteria. Recently, the behavior of individual bacteria has been likened to that of atoms, in a concept referred to as quantal microbiology, analogous in some ways to quantum mechanics.[2] In addition, the effect of environmental stress on microorganisms leads to interpopulation diversity, where individual cells may be phenotypically but not genetically different from each other.[3] Recent advances in the use of software that allows the development of probability models have helped to make these approaches more accessible and provided more support for the development of risk assessment procedures.

McMeekin's monograph set the stage for an explosive increase in predictive microbiology research in the last decade of the 20th century. As we begin the new millennium, there is a need for a definitive work on the subject, above and beyond the many comprehensive and stimulating reviews that have appeared. This book is intended to serve many purposes. First, we believe it will be a primer for many who are not familiar with the field. Chapter 1, "Experimental Design and Data Collection"; Chapter 2, "Primary Models"; and Chapter 3, "Secondary Models" are designed in part to give the uninitiated sufficient information to start developing their own models. Other chapters address more complex issues such as the difficulties in fitting models (Chapter 4, "Model Fitting and Uncertainty") and the relevance of models to the real world (Chapter 5, "Challenge of Food and the Environment"). Extensive applications of predictive microbiology are covered in Chapter 6, "Software Programs to Increase the Utility of Predictive Microbiology Information," and in Chapter 7, "Modeling Microbial Dynamics under Time-Varying Conditions." The important contribution made by predictive microbiology to quantitative risk assessment is described in Chapter 8, "Predictive Microbiology in Quantitative Risk Assessment"; and the further complication of individual cell behavior and intercell variability is addressed in Chapter 9, "Modeling the History Effect on Microbial Growth and Survival: Deterministic and Stochastic Approaches." The future of predictive microbiology is the subject of Chapter 10, "Models — What Comes after the Next Generation?" Application of predictive microbiology concepts to the study of fungi is dealt with in Chapter 11, "Predictive Mycology"; and in Chapter 12, "An Essay on the Unrealized Potential of Predictive Microbiology," Tom McMeekin discusses the contribution made by predictive modeling to the field of food microbiology.

We have attempted to cover the basics of predictive microbiology, as well as the more up-to-date and challenging aspects of the field. There are extensive references to earlier work, as well as to recent publications. It is anticipated that this book will reflect the extensive research that has been instrumental in placing predictive microbiology at the forefront of food microbiology, and that it will stimulate future discussion and research in this exciting field.

REFERENCES

1. McMeekin, T.A., Olley, J.N., Ross, T., and Ratkowsky, D.A., *Predictive Microbiology: Theory and Application,* John Wiley & Sons, New York, 1993.
2. Bridson, E.Y. and Gould, G.W., Quantal microbiology, *Lett. Appl. Microbiol.,* 30, 95, 2000.
3. Booth, I.R., Stress and the single cell: intrapopulation diversity is a mechanism to ensure survival upon exposure to stress, *Int. J. Food Microbiol.,* 78, 19, 2002.

About the Editors

Robin C. McKellar is a senior research scientist who obtained a B.Sc. in biology and chemistry, an M.Sc. in microbiology from the University of Waterloo, and a Ph.D. in microbiology from the University of Ottawa. Dr. McKellar joined the Food Research Institute in Ottawa in 1979 to study the problem of psychrotrophic bacteria in milk. After the formation of the Centre for Food and Animal Research, he served as team leader of the Food Safety Team and initiated a research program on the control of food-borne pathogens. He relocated to Guelph in 1996, and now serves as program science advisor for Theme 410 (Food Safety), and research leader of the Food Preservation Technologies section at the Food Research Program. He has been actively involved in research in such areas as quality of dairy products; enzymatic and microbiological methods development; characterization of the virulence factors of food-borne pathogens; control of pathogens using antimicrobial agents; use of the electronic nose to monitor quality of foods and beverages; and mathematical modeling of pathogen survival and growth in foods. He is currently an adjunct professor with the Department of Food Science, University of Guelph, and has cosupervised several graduate students.

Xuewen Lu is an assistant professor of statistics in the Department of Mathematics and Statistics at the University of Calgary, Canada. Before he joined the university, he was a research scientist in the Atlantic Food and Horticulture Research Centre, Agriculture and Agri-Food Canada from 1997 to 1998, a biostatistician at the Food Research Program, Agriculture and Agri-Food Canada, and a Special Graduate Faculty member at the University of Guelph from 1998 to 2002. He is a member of the Statistical Society of Canada and the Canadian Research Institute for Food Safety, University of Guelph. His main research areas are functional data analysis, survival analysis, and predictive microbiology. He has cosupervised several graduate students. He received his B.Sc. degree (1987) in mathematics from Hunan Normal University, China, M.Sc. degree (1990) in statistics from Peking University, China, and Ph.D. degree (1997) in statistics from the University of Guelph.

Contributors

József Baranyi
Institute of Food Research
Norwich Research Park
Colney, Norwich, United Kingdom

Kristel Bernaerts
BioTeC-Bioprocess Technology and
   Control
Department of Chemical
   Engineering
Katholieke Universiteit Leuven
Leuven, Belgium

Tim Brocklehurst
Institute of Food Research
Norwich Research Park
Colney, Norwich, United Kingdom

Paw Dalgaard
Department of Seafood Research
Danish Institute for Fisheries
   Research
Ministry of Food, Agriculture, and
   Fisheries
Lyngby, Denmark

Philippe Dantigny
Laboratoire de Microbiologie
Université de Bourgogne
Dijon, France

Johan Debevere
Laboratory for Food Microbiology and
   Food Preservation
Department of Food Technology and
   Nutrition
Ghent University, Belgium

Els Dens
BioTeC-Bioprocess Technology and
   Control
Department of Chemical Engineering
Katholieke Universiteit Leuven
Leuven, Belgium

Frank Devlieghere
Laboratory for Food Microbiology and
   Food Preservation
Department of Food Technology and
   Nutrition
Ghent University
Ghent, Belgium

Annemie Geeraerd
BioTeC-Bioprocess Technology and
   Control
Department of Chemical Engineering
Katholieke Universiteit Leuven
Leuven, Belgium

Anna M. Lammerding
Health Canada
Laboratory for Foodborne Zoonoses
Guelph, Ontario, Canada

Xuewen Lu
Department of Mathematics and
   Statistics
University of Calgary
Calgary, Alberta, Canada

Robin C. McKellar
Food Research Program
Agriculture and Agri-Food Canada
Guelph, Ontario, Canada

Tom McMeekin
Centre for Food Safety and Quality
School of Agricultural Science
University of Tasmania
Hobart, Tasmania, Australia

Greg Paoli
Decisionalysis Risk Consultants
Ottawa, Canada

Carmen Pin
Institute of Food Research
Norwich Research Park
Colney, Norwich, United Kingdom

Maria Rasch
Department of Seafood Research
Danish Institute for Fisheries Research
Ministry of Food, Agriculture, and
  Fisheries
Lyngby, Denmark

David A. Ratkowsky
Centre for Food Safety and Quality
School of Agricultural Science
University of Tasmania
Hobart, Tasmania, Australia

Thomas Ross
Centre for Food Safety and Quality
School of Agricultural Science
University of Tasmania
Hobart, Tasmania, Australia

Donald W. Schaffner
Cook College
Rutgers University
New Brunswick, New Jersey

Mark Tamplin
Microbial Food Safety Research
  Unit
United States Department of
  Agriculture
Agricultural Research Service
Eastern Regional Research Center
Wyndmoor, Pennsylvania

Jan F. Van Impe
BioTeC-Bioprocess Technology and
  Control
Department of Chemical Engineering
Katholieke Universiteit Leuven
Leuven, Belgium

Karen Vereecken
BioTeC-Bioprocess Technology and
  Control
Department of Chemical Engineering
Katholieke Universiteit Leuven
Leuven, Belgium

Contents

1 Experimental Design and Data Collection

Maria Rasch

CONTENTS

1.1 EXPERIMENTAL DESIGN

Assessments of the impact of environmental factors on the response of food-borne microorganisms are the primary sources of data for the development of predictive models. When investigating the influence of more than one factor and accurately describing how those factors interact, it is important to consider how to design the experiment. Unfortunately, in modeling bacterial growth in foods, the design is commonly not accounted for, or it is chosen based on habit rather than the experiment's specific purpose. But carefully considering the experiment's design is vital to extracting the desired information (e.g., interactions) and to avoiding excessive experimental work. Furthermore, researchers should be aware of the experimental design in order to avoid extrapolation.[1-3] The following sections describe the most common experimental designs used in modeling of microbial responses in food.

1.1.1 COMPLETE FACTORIAL DESIGN

A complete factorial design is one in which all combinations of the different factors are investigated (Figure 1.1). This allows straightforward modeling of interactions between, for example, environmental factors influencing growth or inactivation of microorganisms. The experimental design is simple, easy to set up, and easy to handle statistically. The main disadvantage is the large increase in number of experiments for every new factor/level added to the experiment. A simple example of a complete $3 \times 3 \times 3$ factorial design was applied by Chhabra et al.[4] for investigating thermal inactivation of *Listeria monocytogenes* in milk. The factors were milk fat content, pH, and heating temperature; the experiment was performed in triplicate, resulting in $3^4 = 81$ experiments. A complete factorial experimental design was also used by Uljas et al.[5] for modeling the combined effect of different processing steps on the reduction of *Escherichia coli* O157:H7 in apple cider. The response variable measured was binary (whether a 5-$\log_{10}$-unit reduction was obtained or not), resulting in a logistic model. Three class variables (cider from three different cider plants, a freeze–thaw treatment, and the preservation agents potassium sorbate and sodium benzoate) and four continuous variables (cider pH, storage temperature, storage time, and preservation concentration) were investigated.[5] This resulted in 1,596 treatments for each of the three types of cider. As one type of cider was tested in duplicate and the other two in triplicate, the total number of experiments was 12,768, which very clearly illustrates the major drawback of complete factorial designs, namely, the very large number of experiments required. However, complete factorial designs are still widely used within predictive modeling of microorganisms, and have been used for different purposes such as the effect of inoculum size, pH, and NaCl on the time-to-detection (TTD) of *Clostridium botulinum*;[6] the effect of pH, NaCl, and temperature on coculture growth of *L. monocytogenes* and *Pseudomonas fluorescens*;[7] the effect of temperature, NaCl, and pH on the inhibitory effect of the antimicrobial compound reuterin on *E. coli*;[8] and the growth of *L. monocytogenes* under combined chilling processes.[9]

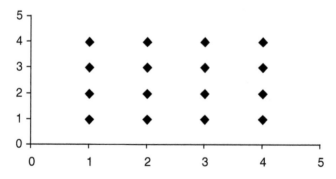

FIGURE 1.1 Example of a complete factorial design for the simple case with two variables ($k = 2$), e.g., temperature and pH, each at four levels.

1.1.2 FRACTIONAL FACTORIAL DESIGN

In order to reduce the number of experiments, several different alternative experimental designs can be applied. Among these are fractional factorial designs, described in this section. In contrast to the complete factorial designs, the fractional factorial designs are not as easy to construct, and thus different software packages are often used for determining which combinations of the different parameters to include in the experimental setup. Examples of software programs used for fractional factorial designs are the Screening Design procedure from STATGRAPHICS Plus (Manugistics, Rockport, MD)[10] and Modde (Umetri, Umeå, Sweden).[11] Farber et al.[10] used a fractional factorial design for modeling the growth of *L. monocytogenes* on liver pâté. The factors investigated were temperature, salt, nitrite, erythrobate, and spice, each at two different levels. The fractional factorial design yielded a total of 16 experiments ($= 2^{5-1}$), where a complete factorial design would have resulted in 32 different experiments. Juneja and Eblen[12] also obtained a large reduction in number of experiments (compared to the number of experiments in a complete factorial design) when they modeled thermal inactivation of *L. monocytogenes*. They investigated 47 combinations of four different environmental factors (temperature, NaCl, sodium pyrophosphate, and pH, each at five levels), where a complete factorial design would have resulted in $5^4 = 625$ experiments. Fractional factorial designs have also been applied for investigating the heat resistance of *E. coli* O157:H7 in beef gravy.[13]

A particular class of fractional factorial designs has been widely used for modeling of bacterial growth, namely, the Box–Behnken designs. These designs are formed by combining two-level factorial designs with balanced incomplete block designs (Figure 1.2).[14] Often, more than one experiment is performed at the central point of the experimental design in order to evaluate the repeatability of the model. A Box–Behnken design was applied for three studies of spoilage of cold-filled ready-to-drink beverages investigating the bacteria *Acinetobacter calcoaceticus* and *Gluconobacter oxydans*,[15] the molds *Aspergillus niger* and *Penicillium spinulosum*,[16] and the yeasts *Saccharomyces cerevisiae*, *Zygosaccharomyces bailii*, and *Candida*

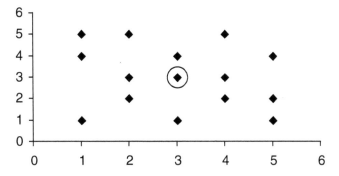

FIGURE 1.2 Example of a Box–Behnken design for the simple case with two variables ($k = 2$), e.g., temperature and pH. The circle denotes the central point.

lipolytica.[17] In each study, the effect of pH, titratable acidity, sugar content, and concentrations of the preservatives sodium benzoate and potassium sorbate were tested at three different levels each.[15–17] The Box–Behnken design was constructed by using the JMP software (SAS Institute, Cary, NC), with two points at the center of the design, resulting in 42 experiments.[15–17] A Box–Behnken design has also been used to show that the CO_2 concentration in the water phase of a model food system was the most important factor when describing modified atmosphere packaging and its inhibitory effect towards microorganisms.[18] The CO_2 concentration in the water phase was investigated as a function of gas/product ratio, initial CO_2 concentration in the gas phase, temperature, pH, and lard content.[18]

1.1.3 CENTRAL COMPOSITE DESIGN

A central composite design consists of a complete (or fraction of a) 2^k factorial design, n_0 center points, and two axial points on the axis of each design variable at a distance of α from the design center (Figure 1.3). The number of experiments for k variables is $2^k + 2k + n_0$, where n_0 denotes the number of experiments at the central point ($n_0 \geq 1$).[14] For $k = 2$ and 3 and $n_0 = 2$, this results in 9 and 16 experiments, respectively.

In a validation study by Walls and Scott,[19] the effect of temperature, pH, and NaCl on the growth of *L. monocytogenes* was described by the use of a central composite design. The experiment was repeated six times at the design center in order to estimate the experimental variance. Guerzoni et al.[20] used central composite design to optimize the composition of an egg-based product in order to prevent survival and growth of *Salmonella enteritidis*. The factors studied were pH, NaCl, and pressure treatment. Lebert et al.[21] used a central composite design to study the growth of *L. monocytogenes* in meat broth. Three variables were studied: pH, a_w,

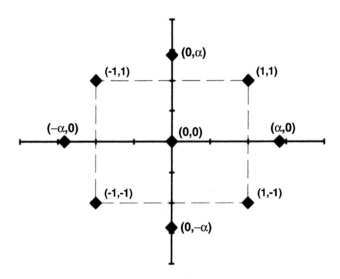

FIGURE 1.3 Example of a central composite design for the simple case with two variables ($k = 2$), e.g., temperature and pH.

and temperature. Two experiments were performed at the central point, resulting in 16 experiments. Later, a similar design was used to study the mixed growth of *Pseudomonas* spp. and *Listeria* in meat, where the three variables were NaCl, temperature, and pH.[22]

A combination of two central composite designs and a factorial design has been applied to study the effect of osmotic and acid/alkaline stresses on *L. monocytogenes*. The two central composite designs were set up in the acid and alkaline pH range, i.e., one covered the pH range from 5.6 to 7 and the other from 7 to 9.5.[23,24] As pointed out by Pin et al.,[2] the risk of extrapolation can be very high when using central composite design as the vertices of the nominal variable space (the unit cube) are far from the interpolation region (the minimal convex polyhedron). The shape of the minimal convex polyhedron is determined by a convex linear combination of the environmental factors at which the experiments were performed for the model development. If a prediction is made randomly in the unit cube, the risk of extrapolation is as high as 75%.[2]

1.1.4 DOEHLERT MATRIX

The Doehlert matrix is another form of experimental design that to some extent resembles the central composite design. The Doehlert matrices consist of points uniformly spaced on concentric spherical shells, and are therefore also called uniform shell designs (Figure 1.4).[14] The number of experiments for k variables is $k^2 + k + n_0$, i.e., for $n_0 = 1$ this gives 13 experiments for $k = 3$ and 21 experiments for $k = 4$. The experiment performed at the center of the experimental domain (n_0) can be repeated several times in order to estimate residual variance. An advantage of

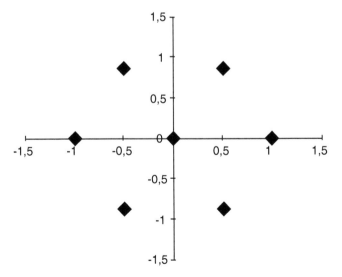

FIGURE 1.4 Example of a Doehlert matrix design for the simple case with two variables ($k = 2$), e.g., temperature and pH.

Doehlert matrices is that they are easy to expand. Expansion can be done both by investigating new variables (provided that these variables were set to their central point during the first experiments) and by enlarging the range of the parameters tested without having to repeat the former experiments.[25]

The Doehlert matrix has been used by Terebiznik et al.[26] to investigate the effect of combinations of nisin and pulsed electric fields on the inactivation of *E. coli*. The experiments were designed with three variables, namely, nisin concentration, electric field strength, and number of pulses. In a later study by the same group, the effect of water activity in combination with nisin and electric field strength was studied.[27] Bouttefroy et al.[28,29] used Doehlert design to study the inhibition of *L. monocytogenes* at different combinations of NaCl, pH, incubation time, and the inhibitory effect of the bacteriocins nisin and curvaticin 13, respectively. Doehlert design has also been applied to investigate the conidial germination of *Penicillium chrysogenum* at different combinations of temperature, water activity, and pH.[30]

1.1.5 OPTIMAL EXPERIMENTAL DESIGN

A new approach within mathematical modeling of growth or inactivation of microorganisms is optimal experimental design. The basic idea is to optimize the experimental conditions with respect to parameter estimation by the use of an established methodology from bioreactor engineering.[31] Ideally, the optimal design of dynamic experiments will result in increased information content from each experiment, and thereby to more accurate parameter estimates from a smaller number of experiments. In the approach by Bernaerts et al.,[32,33] the growth data are modeled directly by the square root model of Ratkowsky et al.[34] (see Chapter 3) integrated into the dynamic model of Baranyi and Roberts[35] (see Chapter 2). Thus, the so-called secondary model parameters are estimated directly from the population density data. Optimal dynamic experimental conditions are then obtained by a stepwise change in temperature, which is first shown with a one-step change[32] and later with three smaller temperature increments in order to avoid an intermediate lag phase.[33] The optimization process was performed by designing the optimal step-temperature profile in order to minimize the standard deviation of the parameter estimates and the correlation between parameters.[32,33] Grijspeerdt and Vanrolleghem[36] used optimal experimental design for optimizing sampling times for the Baranyi growth model. This resulted in lower error on the parameter estimates and decreased correlation between them.[36]

1.2 DATA COLLECTION

1.2.1 STRAIN SELECTION

There are several different approaches that one can use when choosing which strain to use for model building purposes. Furthermore, there is the choice between using a single strain or a mixture of different strains (i.e., cocktail). Before choosing which strain to use it is important to clarify the intended use of the model: is the model going to be used for prediction of possible growth of one particular pathogenic species, or is it a model of the spoilage flora of a specific food product? Using a strain (type strain or other) that has previously been used for several studies or maybe even modeling purposes gives the benefit of the previously accumulated knowledge on the particular strain. On the other hand, a strain isolated from the particular food product which is the goal for the application of the model gives the advantage of relevance for the product, and being able to grow the strain at the environmental conditions investigated.

A well-known strategy is to choose the fastest growing strain at the environmental conditions investigated, as it is the fastest growing strain that will dominate the growth in, e.g., food products. McMeekin et al.[37] recommended independent modeling of several different strains before choosing the strain that grows fastest under the environmental conditions of most interest. This strain then simulates a worst-case scenario.[37] This strategy was followed by Neumeyer et al.[38] who, after an initial screening of different *Pseudomonas* strains, chose the fastest growing strain for modeling, and later during the validation stage confirmed that the chosen strain was the fastest strain.[39] For modeling the growth of *Bacillus cereus* in boiled rice, three different strains were examined and the fastest growing of the three chosen for the modeling studies.[40] Miles et al.[41] examined four different strains of *Vibrio parahaemolyticus* and found that one strain was the most resistant at all conditions of temperature and water activity tested, and hence the growth data of this strain were used for model development. A different method was employed by Lebert et al.[22] who modeled the growth of three different strains, one fast and one slow growing strain of *Pseudomonas fragi* and one slow growing strain of *P. fluorescens*. Any growth was then assumed to be within a zone delimited by the predicted growth curves of these three different organisms.[22] A similar approach was followed by Benito et al.,[42] who initially investigated the resistance of six different strains to high hydrostatic pressure and heat before choosing one pressure- and heat-resistant strain and one pressure- and heat-sensitive strain for further analyses.

The strains used for model development can also be isolated from the food that is under investigation. For modeling the spoilage of ready-to-drink beverages, strains of *Saccharomyces cerevisiae*, *Z. bailii*, and *C. lipolytica* were used.[17] These strains were all isolated from spoiled ready-to-drink beverages.[17] Oscar[43] chose a specific strain of *Salmonella typhimurium* as it exhibits the same growth kinetics as *Salmonella* strains commonly found on chicken in the U.S. The use of strains related to the food in question was also recommended by Hudson,[44] who used strains isolated from smoked mussels and sliced smoked salmon to investigate the growth of *L. monocytogenes*.

The importance of using more than one strain of a species in order to assess the influence of strain variation has also been stressed.[44–46] According to Whiting and Golden,[46] the between-strain variation should be equal to or smaller than the experimental and statistical variation. However, when investigating the growth, survival, thermal inactivation, and toxin production by 17 different strains of *E. coli*, they found that the variations among the strains were larger than the uncertainties related to the experimental error.[46] The variation among 58 strains of *L. monocytogenes* and 8 strains of *Listeria innocua* was examined by Begot et al.[45] Most of the strains had been isolated from meat, meat products, and related industrial sites, and four additional strains that had been involved in outbreaks were also included. Large variations in lag times were found between the strains, whereas the variations in generation times were less pronounced.[45] The opposite conclusion was reached by Oscar[47] when studying 11 different strains of *Salmonella*. He found that the mean coefficients of variation for four repetitions with the same strain were 11.7 and 6.7% for the lag time (λ) and the specific growth rate (μ), respectively, whereas the mean coefficients of variation among the different strains were 9.4 and 5.7% for λ and μ, respectively.[47]

Salter et al.[48] compared the growth of the nonpathogenic *E. coli* M23 with the growth of different pathogenic strains of *E. coli* and found only little difference in the growth responses of the different strains. They also found that the model based on *E. coli* M23 was able to describe the growth of pathogenic strains of *E. coli*, including *E. coli* O157:H7.[48] This result has practical value, as many research groups do not have access to laboratory facilities suitable for work with *E. coli* O157:H7. However, the general suitability of nonpathogenic strains as models for the growth or survival of pathogenic strains would have to be confirmed for each species.

Mixtures of different strains, so-called cocktails, have also been widely used. The main arguments for using cocktails are as follows: first, that a mixture of several different strains is more representative of the situation found in foods, where a flora of strains is likely to be present. Second, it is not necessarily the same strain that shows the fastest growth under all the investigated growth conditions, i.e., a strain with a high salt tolerance might be the fastest growing at high salt concentrations and high pH, but not necessarily at low salt and low pH conditions. For building the Food Micromodel, which is a database software system for predicting growth and survival of microorganisms in foods (see Chapter 6), it was decided to use a cocktail of strains for the growth experiments, but a single strain for thermal inactivation studies, as a cocktail of strains for the latter procedure could produce thermal inactivation kinetics data that would be difficult to interpret.[49]

A cocktail of five strains of *Staphylococcus aureus* was used for the determination of growth/no growth boundaries by measuring turbidity in microtiter plates[50] and Uljas et al.[5] used a mixture of three different strains to characterize the effect of different preservation methods on the survival of *E. coli* in apple cider.

1.2.2 VIABLE COUNT

Viable count determinations by spreading on agar plates are still a very common method for enumeration of microorganisms and it remains the method of reference. To a certain extent it has been possible to automate viable count plating by the use of automated platers such as the spiral plater and automatic colony readers.

Vast numbers of modeling studies have been based on viable counts. A few studies have, however, observed problems with the viable count method compared with other methods. As described in Section 1.2.3.2, enumeration of *Brochothrix thermosphacta* by flow cytometry gave a more accurate result than with viable counts when both were compared to manual counting by microscopy.[51]

1.2.3 NOVEL METHODS

Construction of models using viable count data is time-consuming and expensive, and several alternative, more rapid methods for accumulating sufficient data for modeling have been explored. A novel method for data capture should either be faster, cheaper, and less labor intensive or be able to provide more information on the cells than do viable counts, e.g., physiological status or expression of different phenotypic traits. In the following sections, four of these novel methods will be described, namely, turbidity, flow cytometry, microscopy, and impedance. A number of other methods have been used to indirectly model bacterial growth, but extensive development of these approaches has not been attempted. Some of these include headspace measurements of evolved CO_2 by gas chromatography[52,53] and bioluminescence.[54,55]

1.2.3.1 Turbidity

One of the simplest methods for data collection is the use of optical density (OD), where growth can be related to the increase in turbidity of a bacterial culture. OD, or absorbance, is a measure of the amount of light that is absorbed or scattered by a solution of bacteria. The bacteria absorb or scatter light depending on their concentration, size, and shape. According to Beer's law, absorbance is proportional to concentration, and is related to the percent transmitted light (%T) by the following equation:

$$OD = 2 - \log_{10}(\%T)$$

Some of the fundamentals of this approach have been discussed by McMeekin et al.[37] There are some limitations associated with this approach to data collection. Deviations from responses predicted by Beer's law occur at high cell densities, requiring that dilutions be made to OD < 0.3 before accurate absorbance measurements can be taken.[56] In addition, OD methods are comparative only, and cannot be used to predict viable counts unless some attempt at calibration is made. Detectable absorbance changes occur at a minimum bacterial concentration of 10^6 cfu ml^{-1}, depending on the sensitivity of the instrument,[56] and a linear relationship between OD and viable count exists only between the detection limit and approximately $10^{7.5}$ cfu ml^{-1}. With the maximum cell density in most growth media limited to approximately 10^9 cfu ml^{-1}, the μ measured using OD will represent the rate towards the end of the growth phase, and this will be less than the maximum specific growth rate (μ_{max}) experienced during the midexponential phase of growth. Another drawback is the inability to distinguish between dead and living cells, which can lead to an overestimation of the cell concentration. Furthermore, bacterial cultures that change cell morphology under different environmental conditions, e.g., elongated cells of *L. monocytogenes* at high salt concentrations, again lead to an overestimation of the cell number.[57] Hudson and Mott[58] showed that the cell length of *P. fragi* increased during lag phase, and consequently models based on OD measurements underestimated λ, unless a conversion equation was applied.[58] This method lends itself particularly well to automation, and a number of studies have used automated turbidimetric instruments such as the Bioscreen.[57,59,60]

   A number of attempts have been made to calibrate OD data. McClure et al.[57] used a simple quadratic equation to relate OD to viable counts. Dalgaard et al.[56]

used two equivalent methods for calibration: one in which stationary-phase cells were diluted to the appropriate OD, and the other in which samples for OD and viable count were taken during growth. Predicted generation times were lower with viable count data,[56] and this factor has been taken into account in later studies.[41] Similar methods have been used to relate turbidimetric and viable count data.[58,61,62]

In some studies, the Gompertz equation was fitted directly to OD data; however, no data were available at below the minimum detectable OD (ca. 10^6 cfu ml^{-1}) and thus the estimates for μ and μ_{max} should be questioned.[44,58] A form of calibration was achieved by relating λ determined using OD measurements to that determined with viable counts by a regression equation.[58] McMeekin et al. have discussed the correct way to fit the Gompertz function to % transmittance data (Appendix 2A.9 of their book[37]), and this method has been used to calculate generation times.[38]

Other studies have been carried out without any apparent calibration.[59] λ values have been estimated from OD data by extrapolation of the exponential portion of the curve back to the initial cell numbers;[63] however, this method may be inaccurate since the μ estimated from the OD data may be lower than that obtained during the period of maximum growth.[37] Lebert et al.[21] estimated λ values of *L. monocytogenes* with OD data, but the inoculation level during these experiments was kept at 10^7 cfu ml^{-1}, i.e., above the detection limit. This procedure, however, gives only a very small dynamic range of growth of about 2 log units.

Interestingly, the TTD approach has not been used to any great extent. The TTD for a turbidimetric instrument can be defined as the time required for a detectable increase in OD. The difference between TTD for serial twofold dilutions gives the doubling time, from which μ can be determined.[60,64] λ can be calculated subsequently by the difference between the predicted TTD based on λ, and the observed TTD.[60,64] This method was used to estimate λ for individual cells.[65] This method was also used by Augustin et al.[61] for estimating μ_{max} of 10 different strains of *L. monocytogenes*. They, however, observed large variations in the time separating the two successive growth curves (i.e., doubling time).

In spite of the problems associated with the use of turbidimetric data for modeling, there appears to be some value in this approach. Models based on viable counts were compared with those obtained using either OD or transmittance data, and it was concluded that turbidimetric methods may be used for reliably estimating μ_{max}.[56]

OD measurements have been used extensively for modeling purposes. This includes modeling of the growth boundaries of *S. aureus* at different levels of relative humidity, pH, potassium sorbate, and calcium propionate[50] and modeling the effect of the antimicrobial compound reuterin on the growth of *E. coli* at different combinations of temperature, pH, and NaCl.[8] OD has also been used to determine 5-log$_{10}$-unit reductions of *E. coli* in apple cider (see also Section 1.1.1).[5] Cider inoculated with 10^7 cfu ml^{-1} was exposed to the different treatments investigated, after which a 10-μl sample of the cider was transferred to a microtiter well containing Tryptic Soy Broth and incubated. If a 5-log$_{10}$-unit reduction occurred during the treatment, the 10-μl sample would contain <1 cfu and therefore no growth would be observed in the well.[5]

OD data have also been used for the determination of growth boundaries, i.e., the growth/no growth models (see Chapter 3). The growth boundaries of the spoilage organism *Z. bailii* were investigated at different combinations of salt, sugar, acetic acid, and pH at a constant temperature of 30°C. Growth was measured by a Bioscreen analyzer, but between measurements the Bioscreen plates were incubated in closed containers in an incubator.[66] Masana and Baranyi[67] studied the growth boundaries of *B. thermosphacta* in multi-well plates, but inspected the wells visually for growth. The interface between survival and death of *E. coli* O157:H7 in a mayonnaise model system was studied by McKellar et al.[68] A cocktail of five strains of *E. coli* O157:H7 was inoculated at a level of 10^7 cfu ml^{-1} into 5-ml tubes under different environmental conditions, and growth was observed visually. In the case of no growth, the samples were diluted 100-fold into Tryptic Soy Broth and incubated again. Continued absence of growth was interpreted as a >5.7 log reduction in viable cell numbers under the test conditions. Survival was hence defined as a <5.7 log reduction in viable cell numbers.[68]

1.2.3.2 Flow Cytometry

Flow cytometry is a rapid technique for measurement of single cells in suspension. Individual cells confined within a rapidly flowing jet of water pass a measuring window, in which several parameters can be simultaneously measured for several thousand cells per second with high precision.[69] Light scattering reflects cellular size and structure, while fluorescence measurements can determine the cellular content of any constituent that can be labeled with a fluorescent dye.[70] In this way flow cytometry combines the advantages of being a single cell technique with the power of being able to measure a very large number of cells in a very short time. The resulting data are not a mere average of the measured cells but a distribution of the measured parameters for the cells. The possibility of measuring the distribution gives an estimate of the heterogeneity of the microbial population and thereby also the possibility to detect subpopulations that, e.g., are resistant to a treatment under investigation. With a flow cytometer equipped with a cell sorter it is furthermore possible to sort cells out on the basis of the parameters measured. These cells can then be sorted into microtiter wells and be used for new growth experiments to monitor, e.g., λ for the single cells as shown by Smelt et al.[71] In general a good correlation between the number of cells determined by plate counting and by flow cytometry has been found for both bacteria[72] and yeast,[73] with detection limits of approximately 10^4 and 10^2 cells ml^{-1} determined for *L. monocytogenes* and *Debaryomyces hansenii*, respectively.

The use of flow cytometry for predictive microbiology is still very limited. Sørensen and Jakobsen[73] used flow cytometry to enumerate viable cells of *D. hansenii* at different environmental conditions. The growth data were used to model λ and μ_{max} as a function of temperature, pH, and NaCl. Rattanasomboon et al.[51] compared flow cytometry, turbidimetry, plate counts, and manual counts by microscopy for enumeration of *B. thermosphacta*. They found that turbidimetry overestimated the cell number as the *B. thermosphacta* cells changed morphology during growth, whereas flow cytometry gave a more accurate cell count than did plate counts when both were compared to manual counts.[51] This overestimation of cell number and hence μ could not be confirmed by Dalgaard and Koutsoumanis,[74] who found that turbidimetric measurements estimated μ_{max} and λ accurately.

Possible applications of flow cytometry include enumeration of microorganisms for both mono cultures and mixed cultures,[72,75] direct measurement of lag phase as described by Ueckert et al.,[76] separation of intermediate states between dead and culturable cells,[77] and detection of cell injury caused, e.g., by bacteriocins.[78]

However, flow cytometry also has the potential to be used for gaining more information on the microorganisms than just the number of cells. García-Ochoa et al.[79] recognized that in order to develop a structure kinetic model for the production of xanthan by *Xanthomonas capestris*, they needed quantitative data on intracellular compounds. They examined the DNA, RNA, and intracellular protein content by flow cytometry and traditional biochemical methods, enabling them to set up standard curves and thereby quantify these intracellular compounds by flow cytometry. It is also possible to determine other biochemical parameters such as intracellular esterase, protease, glycosidase, and phosphatase activities. One of the limitations in the use of flow cytometry is that it can be applied for liquid systems only. This problem was, however, partly overcome by de Alteriis et al.,[80] who studied the growth dynamics of *Saccharomyces cerevisiae* cells immobilized in a gelatin gel. When the cells were sampled for analysis, the gelatin was enzymatically liquefied with trypsin, thus enabling the cells to be analyzed by flow cytometry.

1.2.3.3 Microscopy and Colony Size

Microscopy is another method that is gaining interest as developments in image analysis programs and software tools for automation make the method more feasible. Microscopy enables direct studies of single cells, which give new opportunities for following the same cells for longer periods of time. One of the main advantages of microscopy and the measurement of colony size is the possibility of studying solid systems, which more closely resemble the situation in most food systems. It is, however, also possible to investigate growth in a liquid system. By the use of a microscope coverslip coated with, e.g., poly-L-lysine, it is possible to obtain immobilized cells in a liquid system, as has been demonstrated for both yeast[81] and bacteria.[82]

Reports on the use of microscopy for predictive modeling of single cells are still sparse. Wu et al.[83] recently compared the use of microscopy for determination of lag phase duration for individual cells of *L. monocytogenes* with the TTD method (described in Section 1.2.3.1). Microscopy has several advantages over the TTD method for the determination of λ of single cells. The method is a direct method allowing visual observation of the first cell division, whereas the TTD method depends on the time of detection, the growth rate, and extrapolation back to the single cell. Furthermore, any treatment that results in cells not dividing will not be detected by the TTD method.[83] A drawback when studying single cells by microscopy can be the difficulties in obtaining sufficient data for modeling purposes. Wright et al.[84] used a gel-cassette in which bacteria grow as colonies immobilized in gelatin gel, combined with a "laser gel-cassette scanner," to study the lag and doubling time of *Salmonella typhimurium* at different concentrations of NaCl and pH. The inoculated gel-cassette was continuously scanned, and the increase in fixed angle laser light scattering intensity was related to the increase in diameter of the individual nearly spherical bacterial colonies within the controlled environment of the gel-cassette. The system, however, needs extensive calibration; for example, it is necessary to recalibrate for each new experiment in order to relate laser scattering intensity to viable cell count.[84]

Radial growth of *L. monocytogenes* and *Yersinia enterocolitica* was studied on agar surfaces under different modified atmosphere conditions.[85] Growth of visible colonies was followed by image analysis and viable count per colony. A linear relationship was found between $\log_{10}$ viable cell number per colony and $\log_{10}$ colony radius and μ.[85] Dykes[86] used a similar method to investigate sublethal injury in *L. monocytogenes*. Cells subjected to either starvation or heat stress were plated onto Tryptic Soy Agar and incubated at 37°C for 48 h. The plates were photographed using a digital camera and the areas (mm^2) of individual colonies were determined using image analysis. The results were presented as histograms showing frequency distribution of colony area. The colony areas from nonstressed cells were normally distributed, whereas the colony areas from starved or heat-stressed cells had a skewed distribution due to an increased proportion of small colonies.[86] The growth of *Bacillus cereus* was also measured as radial growth at different concentrations of agar, NaCl, and potassium sorbate.[87] Agar plates were incubated at 30°C and photographs were taken at 30-min intervals. The colony diameters were measured on the slides, and the time to reach a diameter of at least 0.1 mm was called "time to visible growth." Growth was then evaluated as time to visible growth or radial growth rate.[87] Time to visible growth was also measured by Salvesen and Vadstein[88], although they defined a colony as visible when it reached a diameter of 2 mm. They studied seawater isolates and found an inverse relationship between the μ_{max} determined in liquid culture and the time necessary to form visible colonies on agar.[88]

In contrast to bacteria, the growth of molds is usually always measured as radial growth since molds are not unicellular. Gibson et al.[89] first modeled μ and the time to visible growth (diameter ≥ 3 mm) for fungi, where the growth of *Aspergillus flavus* was modeled at different water activities. Valík et al.[90] also modeled the effect of water activity but on *Penicillium roqueforti*. The diameter of the colonies was fitted to the model of Baranyi et al.[91] (see Chapter 2), and λ and μ modeled as a function of water activity. Later Valík and Piecková[92] used the same approach to model the effect of water activity on three different heat-resistant fungi, namely, *Byssochlamys fulva, Neosartorya fischeri,* and *Talaromyces avellaneus.* Recently, Rosso and Robinson[93] proposed a model to describe the effect of water activity on the radial growth of molds. The model is of the cardinal model family (see Chapter 3) and fitted successfully the radial μ of six different *Aspergillus* species as well as *Eurotium amstelodami, Eurotium chevalieri,* and *Xeromyces bisporus.*

1.2.3.4 Impedance

Microbiological impedance devices measure microbial metabolism in medium by tracking the movement of ions between two electrodes (conductance), or the storage of charge at the electrode–medium interface (capacitance). For bacterial growth, the conductivity of the growth medium increases with bacterial numbers because of the production of weakly charged organic molecules.[37] This production of charged molecules is due to, for example, the conversion of proteins to amino acids, carbo-hydrates to lactate, and lipids to acetate, all of which will increase the conductivity (G) of the growth medium.[94] When electrodes are immersed in a conductive medium, a dielectric field will build up at the electrode–solution interface. The medium will display a capacitance due to the polarization of the electrode–solution interface. An alternating sinusoidal potential applied to the system will therefore cause a resultant current depending on the impedance (Z) of the system, which is a function of its resistance (R, $G = 1/R$), its capacitance (C), and the applied frequency (f).[94]

$$Z = \sqrt{\left(\frac{1}{G}\right)^2 + \left(\frac{1}{2\pi fC}\right)^2}$$

Which signal should be measured (impedance, conductivity, or capacitance) depends on the instrument, and the microorganism and its metabolism. Generation times may be calculated based on TTD methods as described in Section 1.2.3.1, or from the time required for a doubling of the change in conductance.[37] Impedimetric instruments are often automated, allowing a large number of samples to run at the same time. Conductance has been used for modeling the growth of *Y. enterocolitica*[95] and impedance and conductance have been used for modeling the growth of *S. enteritidis*.[96,97] An indirect conductimetry method, in which CO_2 evolved during growth was trapped and measured, was proposed for the modeling of food spoilage by yeasts.[98]

1.3 CONCLUSION

It is important that a deliberate choice be made when choosing an experimental design or a method of data collection. The outcome of an experiment, and the ultimate value of the model, will be greatly influenced by the experimenter's choices. Selection of a data collection method involves some trade-off. The novel methods described above can roughly be divided into two groups, one that provides a possibility of automation and thereby allows a higher number of experiments to be analyzed, and another that gives additional information, e.g., on the physiological state of the microorganisms compared with viable counts. Turbidity and impedimetric methods are mainly in the first group, and flow cytometry and microscopy in the second. Although the viable count method probably remains the method of reference and of choice, it does not always give the correct answer, which was also pointed out in Section 1.2.3.2. It is expected that novel techniques for data collection will continue to increase in importance with the demand for more mechanistic models based on microbial physiology.

REFERENCES

1. Baranyi, J., Ross, T., McMeekin, T.A., and Roberts, T.A., Effects of parameterization on the performance of empirical models used in "predictive microbiology," *Food Microbiol.*, 13, 83, 1996.

2. Pin, C., Baranyi, J., and de Fernando, G., Predictive model for the growth of *Yersinia enterocolitica* under modified atmospheres, *J. Appl. Microbiol.*, 88, 521, 2000.

3. Masana, M.O. and Baranyi, J., Adding new factors to predictive models: the effect on the risk of extrapolation, *Food Microbiol.*, 17, 367, 2000.

4. Chhabra, A.T., Carter, W.H., Linton, R.H., and Cousin, M.A., A predictive model to determine the effects of pH, milk fat, and temperature on thermal inactivation of *Listeria monocytogenes, J. Food Prot.*, 62, 1143, 1999.

5. Uljas, H.E., Schaffner, D.W, Duffy, S., Zhao, L.H., and Ingham, S.C., Modeling of combined processing steps for reducing *Escherichia coli* O157:H7 populations in apple cider, *Appl. Environ. Microbiol.*, 67, 133, 2001.

6. Zhao, L., Montville, T.J., and Schaffner, D.W., Inoculum size of *Clostridium botulinum* 56A spores influences time-to-detection and percent growth-positive samples. *J. Food Sci.*, 65, 1369, 2000.

7. Buchanan, R.L. and Bagi, L.K., Microbial competition: effect of *Pseudomonas fluorescens* on the growth of *Listeria monocytogenes, Food Microbiol.*, 16, 523, 1999.

8. Rasch, M., The influence of temperature, salt and pH on the inhibitory effect of reuterin on *Escherichia coli, Int. J. Food Microbiol.*, 72, 225, 2002.

9. Membre, J.M., Ross, T., and McMeekin, T., Behaviour of *Listeria monocytogenes* under combined chilling processes, *Lett. Appl. Microbiol.*, 28, 216, 1999.

10. Farber, J.M., McKellar, R.C., and Ross, W.H., Modelling the effects of various parameters on the growth of *Listeria monocytogenes* on liver pate, *Food Microbiol.*, 12, 447, 1995.

11. Nerbrink, E., Borch, E., Blom, H., and Nesbakken, T., A model based on absorbance data on the growth rate of *Listeria monocytogenes* and including the effects of pH, NaCl, Na-lactate and Na-acetate, *Int. J. Food Microbiol.*, 47, 99, 1999.

12. Juneja, V.K. and Eblen, B.S., Predictive thermal inactivation model for *Listeria monocytogenes* with temperature, pH, NaCl, and sodium pyrophosphate as controlling factors, *J. Food Prot.*, 62, 986, 1999.

13. Juneja, V.K., Marmer, B.S., and Eblen, B.S., Predictive model for the combined effect of temperature, pH, sodium chloride, and sodium pyrophosphate on the heat resistance of *Escherichia coli* O157:H7, *J. Food Saf.*, 19, 147, 1999.

14. Khuri, A.I. and Cornell, J.A., *Response Surfaces. Design and Analyses,* Marcel Dekker, New York, 1987.

15. Battey, A.S. and Schaffner, D.W., Modelling bacterial spoilage in cold-filled ready to drink beverages by *Acinetobacter calcoaceticus* and *Gluconobacter oxydans, J. Appl. Microbiol.*, 91, 237, 2001.

16. Battey, A.S., Duffy, S., and Schaffner, D.W., Modelling mould spoilage in cold-filled ready-to-drink beverages by *Aspergillus niger* and *Penicillium spinulosum, Food Microbiol.*, 18, 521, 2001.

17. Battey, A.S., Duffy, S., and Schaffner, D.W., Modeling yeast spoilage in cold-filled ready-to-drink beverages with *Saccharomyces cerevisiae, Zygosaccharomyces bailii,* and *Candida lipolytica, Appl. Environ. Microbiol.*, 68, 1901, 2002.

18. Devlieghere, F., Debevere, J., and Van Impe, J., Concentration of carbon dioxide in the water-phase as a parameter to model the effect of a modified atmosphere on microorganisms, *Int. J. Food Microbiol.*, 43, 105, 1998.

19. Walls, I. and Scott, V.N., Validation of predictive mathematical models describing the growth of *Listeria monocytogenes, J. Food Prot.*, 60, 1142, 1997.

20. Guerzoni, M.E., Vannini, L., Lanciotti, R., and Gardini, F., Optimisation of the formulation and of the technological process of egg-based products for the prevention of *Salmonella enteritidis* survival and growth, *Int. J. Food Microbiol.*, 73, 367, 2002.

21. Lebert, I., Bégot, C., and Lebert, A., Development of two *Listeria monocytogenes* growth models in a meat broth and their application to beef meat, *Food Microbiol.*, 15, 499, 1998.

22. Lebert, I., Robles-Olvera, V., and Lebert, A., Application of polynomial models to predict growth of mixed cultures of *Pseudomonas* spp. and *Listeria* in meat, *Int. J. Food Microbiol.*, 61, 27, 2000.

23. Cheroutre-Vialette, M. and Lebert, A., Growth of *Listeria monocytogenes* as a function of dynamic environment at 10°C and accuracy of growth predictions with available models, *Food Microbiol.*, 17, 83, 2000.

24. Cheroutre-Vialette, M. and Lebert, A., Modelling the growth of *Listeria monocytogenes* in dynamic conditions, *Int. J. Food Microbiol.*, 55, 201, 2000.

25. Quignon, F., Huyard, A., Schwartzbrod, L., and Thomas, F., Use of Doehlert matrices for study of poliovirus-1 adsorption, *J. Virol. Methods*, 68, 33, 1997.

26. Terebiznik, M.R., Jagus, R.J., Cerrutti, P., de Huergo, M.S., and Pilosof, A.M.R., Combined effect of nisin and pulsed electric fields on the inactivation of *Escherichia coli*, *J. Food Prot.*, 63, 741, 2000.

27. Terebiznik, M., Jagus, R.J., Cerrutti, P., de Huergo, M.S., and Pilosof, A.M.R., Inactivation of *Escherichia coli* by a combination of nisin, pulsed electric fields, and water activity reduction by sodium chloride, *J. Food Prot.*, 65, 1253, 2002.

28. Bouttefroy, A., Linder, M., and Milliere, J.B., Predictive models of the combined effects of curvaticin 13, NaCl and pH on the behaviour of *Listeria monocytogenes* ATCC 15313 in broth, *J. Appl. Microbiol.*, 88, 919, 2000.

29. Bouttefroy, A., Mansour, M., Linder, M., and Milliere, J.B., Inhibitory combinations of nisin, sodium chloride, and pH on *Listeria monocytogenes* ATCC 15313 in broth by an experimental design approach, *Int. J. Food Microbiol.*, 54, 109, 2000.

30. Sautour, M., Rouget, A., Dantigny, P., Divies, G., and Bensoussan, M., Application of Doehlert design to determine the combined effects of temperature, water activity and pH on conidial germination of *Penicillium chrysogenum*, *J. Appl. Microbiol.*, 91, 900, 2001.

31. Versyck, K.J., Bernaerts, K., Geeraerd, A.H., and Van Impe, J.F., Introducing optimal experimental design in predictive modeling: a motivating example, *Int. J. Food Microbiol.*, 51, 39, 1999.

32. Bernaerts, K., Versyck, K.J., and Van Impe, J.F., On the design of optimal dynamic experiments for parameter estimation of a Ratkowsky-type growth kinetics at suboptimal temperatures, *Int. J. Food Microbiol.*, 54, 27, 2000.

33. Bernaerts, K., Servaes, R.D., Kooyman, S., Versyck, K.J., and Van Impe, J.F., Optimal temperature input design for estimation of the square root model parameters: parameter accuracy and model validity restrictions, *Int. J. Food Microbiol.*, 73, 145, 2002.

34. Ratkowsky, D.A., Olley, J., McMeekin, T.A., and Ball, A., Relationship between temperature and growth rate of bacterial cultures, *J. Bacteriol.*, 149, 1, 1982.

35. Baranyi, J. and Roberts, T.A., A dynamic approach to predicting bacterial growth in food, *Int. J. Food Microbiol.*, 23, 277, 1994.

36. Grijspeerdt, K. and Vanrolleghem, P., Estimating the parameters of the Baranyi model for bacterial growth, *Food Microbiol.*, 16, 593, 1999.

37. McMeekin, T.A., Olley, J.N., Ross, T., and Ratkowsky, D., *Predictive Microbiology: Theory and Application*, John Wiley & Sons, New York, 1993, 340 pp .

38. Neumeyer, K., Ross, T., and McMeekin, T.A., Development of a predictive model to describe the effects of temperature and water activity on the growth of spoilage pseudomonads, *Int. J. Food Microbiol.*, 38, 45, 1997.

39. Neumeyer, K., Ross, T., Thomson, G., and McMeekin, T.A., Validation of a model describing the effects of temperature and water activity on the growth of psychrotrophic pseudomonads, *Int. J. Food Microbiol.*, 38, 55, 1997.

40. McElroy, D.M., Jaykus, L.A., and Foegeding, P.M., Validation and analysis of modeled predictions of growth of *Bacillus cereus* spores in boiled rice, *J. Food Prot.*, 63, 268, 2000.

41. Miles, D.W., Ross, T., Olley, J., and McMeekin, T.A., Development and evaluation of a predictive model for the effect of temperature and water activity on the growth rate of *Vibrio parahaemolyticus, Int. J. Food Microbiol.*, 38, 133, 1997.

42. Benito, A., Ventoura, G., Casadei, M., Robinson, T., and Mackey, B., Variation in resistance of natural isolates of *Escherichia coli* O157 to high hydrostatic pressure, mild heat, and other stresses, *Appl. Environ. Microbiol.*, 65, 1564, 1999.

43. Oscar, T.P., Development and validation of a tertiary simulation model for predicting the potential growth of *Salmonella typhimurium* on cooked chicken, *Int. J. Food Microbiol.*, 76, 177, 2002.

44. Hudson, J.A., Comparison of response surface models for *Listeria monocytogenes* strains under aerobic conditions, *Food Res. Int.*, 27, 53, 1994.

45. Begot, C., Lebert, I., and Lebert, A., Variability of the response of 66 *Listeria monocytogenes* and *Listeria innocua* strains to different growth conditions, *Food Microbiol.*, 14, 403, 1997.

46. Whiting, R.C. and Golden, M.H., Variation among *Escherichia coli* O157:H7 strains relative to their growth, survival, thermal inactivation, and toxin production in broth, *Int. J. Food Microbiol.*, 75, 127, 2002.

47. Oscar, T.P., Variation of lag time and specific growth rate among 11 strains of *Salmonella* inoculated onto sterile ground chicken breast burgers and incubated at 25°C, *J. Food Saf.*, 20, 225, 2000.

48. Salter, M.A., Ross, T., and McMeekin, T.A., Applicability of a model for nonpathogenic *Escherichia coli* for predicting the growth of pathogenic *Escherichia coli*, *J. Appl. Microbiol.*, 85, 357, 1998.

49. McClure, P.J., Blackburn, C.W., Cole, M.B., Curtis, P.S., Jones, J.E., Legan, J.D., Ogden, I.D., Peck, M.W., Roberts, T.A., Sutherland, J.P., and Walker, S.J., Modelling the growth, survival and death of microorganisms in foods: the UK Food Micromodel approach, *Int. J. Food Microbiol.*, 23, 265, 1994.

50. Stewart, C.M., Cole, M.B., Legan, J.D., Slade, L., Vandeven, M.H., and Schaffner, D.W., Modeling the growth boundary of *Staphylococcus aureus* for risk assessment purposes, *J. Food Prot.*, 64, 51, 2001.

51. Rattanasomboon, N., Bellara, S.R., Harding, C.L., Fryer, P.J., Thomas, C.R., Al-Rubeai, M., and McFarlane, C.M., Growth and enumeration of the meat spoilage bacterium *Brochothrix thermosphacta, Int. J. Food Microbiol.*, 51, 145, 1999.

52. Gardini, F., Lanciotti, R., Sinigaglia, M., and Guerzoni, M.E., A head space gas chromatographic approach for the monitoring of the microbial cell activity and the cell viability evaluation, *J. Microbiol. Methods*, 29, 103, 1997.

53. Guerzoni, M.E., Lanciotti, R., Torriani, S., and Dellaglio, F., Growth modeling of *Listeria monocytogenes* and *Yersinia enterocolitica* in food model systems and dairy products, *Int. J. Food Microbiol.*, 24, 83, 1994.

54. Ellison, A., Anderson, W., Cole, M.B., and Stewart, G.S.A.B., Modeling the thermal inactivation of *Salmonella typhimurium* using bioluminescence data, *Int. J. Food Microbiol.*, 23, 467, 1994.

55. Farkas, J., Andrassy, E., Beczner, J., Vidacs, I., and Meszaros, L., Utilizing luminometry for monitoring growth of *Listeria monocytogenes* in its liquid or gelified monocultures and cocultures with "acid-only" *Lactococcus lactis, Int. J. Food Microbiol.,* 73, 159, 2002.

56. Dalgaard, P., Ross, T., Kamperman, L., Neumeyer, K., and McMeekin, T.A., Estimation of bacterial growth rates from turbidimetric and viable count data, *Int. J. Food Microbiol.,* 23, 391, 1994.

57. McClure, P.J., Cole, M.B., Davies, K.W., and Anderson, W.A., The use of turbidimetric data for the construction of kinetic models, *J. Ind. Microbiol.,* 12, 277, 1993.

58. Hudson, J.A. and Mott, S.J., Comparison of lag times obtained from optical-density and viable count data for a strain of *Pseudomonas fragi, J. Food Saf.,* 14, 329, 1994.

59. Huchet, V., Thuault, D., and Bourgeois, C.M., Development of a model predicting the effects of pH, lactic acid, glycerol and sodium chloride content on the growth of vegetative cells of *Clostridium tyrobutyricum* in a culture medium, *Lait,* 75, 585, 1995.

60. McKellar, R.C. and Knight, K., A combined discrete–continuous model describing the lag phase of *Listeria monocytogenes, Int. J. Food Microbiol.,* 54, 171, 2000.

61. Augustin, J.C., Rosso, L., and Carlier, V., Estimation of temperature dependent growth rate and lag time of *Listeria monocytogenes* by optical density measurements, *J. Microbiol. Methods,* 38, 137, 1999.

62. Chorin, E., Thuault, D., Cleret, J.J., and Bourgeois, C.M., Modelling *Bacillus cereus* growth, *Int. J. Food Microbiol.,* 38, 229, 1997.

63. Breand, S., Fardel, G., Flandrois, J.P., Rosso, L., and Tomassone, R., A model describing the relationship between lag time and mild temperature increase duration, *Int. J. Food Microbiol.,* 38, 157, 1997.

64. Cuppers, H.G.A.M. and Smelt, J.P.P.M., Time to turbidity measurement as a tool for modeling spoilage by *Lactobacillus, J. Ind. Microbiol.,* 12, 168, 1993.

65. Baranyi, J. and Pin, C., Estimating bacterial growth parameters by means of detection times, *Appl. Environ. Microbiol.,* 65, 732, 1999.

66. Jenkins, P., Poulos, P.G., Cole, M.B., Vandeven, M.H., and Legan, J.D., The boundary for growth of *Zygosaccharomyces bailii* in acidified products described by models for time to growth and probability of growth, *J. Food Prot.,* 63, 222, 2000.

67. Masana, M.O. and Baranyi, J., Growth/no growth interface of *Brochothrix thermosphacta* as a function of pH and water activity, *Food Microbiol.,* 17, 485, 2000.

68. McKellar, R.C., Lu, X., and Delaquis, P.J., A probability model describing the interface between survival and death of *Escherichia coli* O157:H7 in a mayonnaise model system, *Food Microbiol.,* 19, 235, 2003.

69. Vives-Rego, J., Lebaron, P., and Nebe-von Caron, G., Current and future applications of flow cytometry in aquatic microbiology, *FEMS Microbiol. Rev.,* 24, 429, 2000.

70. Walberg, M., Gaustad, P., and Steen, H.B., Rapid assessment of ceftazidime, ciprofloxacin, and gentamicin susceptibility in exponentially-growing *E. coli* cells by means of flow cytometry, *Cytometry,* 27, 169, 1997.

71. Smelt, J.P.P.M., Otten, G.D., and Bos, A.P., Modelling the effect of sublethal injury on the distribution of the lag times of individual cells of *Lactobacillus plantarum, Int. J. Food Microbiol.,* 73, 207, 2002.

72. Jacobsen, C.N., Rasmussen, J., and Jakobsen, M., Viability staining and flow cytometric detection of *Listeria monocytogenes, J. Microbiol. Methods,* 28, 35, 1997.

73. Sørensen, B.B. and Jakobsen, M., The combined effects of temperature, pH and NaCl on growth of *Debaryomyces hansenii* analyzed by flow cytometry and predictive microbiology, *Int. J. Food Microbiol.,* 34, 209, 1997.

74. Dalgaard, P. and Koutsoumanis, K., Comparison of maximum specific growth rates and lag times estimated from absorbance and viable count data by different mathematical models, *J. Microbiol. Methods*, 43, 183, 2001.

75. Budde, B.B., Siegumfeldt, H., and Rasch, M., The potential use of flow cytometry in predictive microbiology, in *COST 914: Predictive Modelling of Microbial Growth and Survival in Foods*, Roberts, T.A., Ed., Office for Official Publications of the European Communities, Luxembourg, 1999, p. 271.

76. Ueckert, J., Nebe-von Caron, G., Bos, A.P., and ter Steeg, P.F., Flow cytometric analysis of *Lactobacillus plantarum* to monitor lag times, cell division and injury, *Lett. Appl. Microbiol.*, 25, 295, 1997.

77. Nebe-von Caron, G., Stephens, P., and Badley, R.A., Assessment of bacterial viability status by flow cytometry and single cell sorting, *J. Appl. Microbiol.*, 84, 988, 1998.

78. Budde, B.B. and Rasch, M., A comparative study on the use of flow cytometry and colony forming units for assessment of the antibacterial effect of bacteriocins, *Int. J. Food Microbiol.*, 63, 65, 2001.

79. García-Ochoa, F., Santos, V.E., and Alcon, A., Intracellular compounds quantification by means of flow cytometry in bacteria: application to xanthan production by *Xanthomonas campestris, Biotech. Bioeng.*, 57, 87, 1998.

80. de Alteriis, E., Porro, D., Romano, V., and Parascandola, P., Relation between growth dynamics and diffusional limitations in *Saccharomyces cerevisiae* cells growing as entrapped in an insolubilised gelatin gel, *FEMS Microbiol. Lett.*, 195, 245, 2001.

81. Guldfeldt, L.U. and Arneborg, N., Measurement of the effects of acetic acid and extracellular pH on intracellular pH of nonfermenting, individual *Saccharomyces cerevisiae* cells by fluorescence microscopy, *Appl. Environ. Microbiol.*, 64, 530, 1998.

82. Shabala, L., Budde, B., Ross, T., Siegumfeldt, H., Jakobsen, M., and McMeekin, T., Responses of *Listeria monocytogenes* to acid stress and glucose availability revealed by a novel combination of fluorescence microscopy and microelectrode ion-selective techniques, *Appl. Environ. Microbiol.*, 68, 1794, 2002.

83. Wu, Y., Griffiths, M.W., and McKellar, R.C., A comparison of the Bioscreen method and microscopy for the determination of lag times of individual cells of *Listeria monocytogenes, Lett. Appl. Microbiol.*, 30, 468, 2000.

84. Wright, K.M., Coleman, H.P., Mackle, A.R., Parker, M.L., Brocklehurst, T.F., Wilson, D.R., and Hills, B.P., Determination of mean growth parameters of bacterial colonies immobilized in gelatin gel using a laser gel-cassette scanner, *Int. J. Food Microbiol.*, 57, 75, 2000.

85. Harrison, W.A., Peters, A.C., and Fielding, L.M., Growth of *Listeria monocytogenes* and *Yersinia enterocolitica* colonies under modified atmospheres at 4 and 8°C using a model food system, *J. Appl. Microbiol.*, 88, 38, 2000.

86. Dykes, G.A., Image analysis of colony size for investigating sublethal injury in *Listeria monocytogenes, J. Rapid Methods Autom. Microbiol.*, 7, 223, 1999.

87. Stecchini, M.L., Del Torre, M., Donda, S., and Maltini, E., Growth of *Bacillus cereus* on solid media as affected by agar, sodium chloride, and potassium sorbate, *J. Food Prot.*, 63, 926, 2000.

88. Salvesen, I. and Vadstein, O., Evaluation of plate count methods for determination of maximum specific growth rate in mixed microbial communities, and its possible application for diversity assessment, *J. Appl. Microbiol.*, 88, 442, 2000.

89. Gibson, A.M., Baranyi, J., Pitt, J.I., Eyles, M.J., and Roberts, T.A., Predicting fungal growth: the effect of water activity on *Aspergillus flavus* and related species, *Int. J. Food Microbiol.*, 23, 419, 1994.

90. Valík, L., Baranyi, J., and Gorner, F., Predicting fungal growth: the effect of water activity on *Penicillium roqueforti, Int. J. Food Microbiol.*, 47, 141, 1999.

91. Baranyi, J., Roberts, T.A., and McClure, P., A non-autonomous differential equation to model bacterial growth, *Food Microbiol.*, 10, 43, 1993.

92. Valík, L. and Piecková, E., Growth modelling of heat-resistant fungi: the effect of water activity, *Int. J. Food Microbiol.*, 63, 11, 2001.

93. Rosso, L. and Robinson, T.P., A cardinal model to describe the effect of water activity on the growth of moulds, *Int. J. Food Microbiol.*, 63, 265, 2001.

94. Firstenberg-Eden, R. and Eden, G., *Impedance Microbiology,* John Wiley & Sons, New York, 1984, 170 pp .

95. Lindberg, C.W. and Borch, E., Predicting the aerobic growth of *Yersinia enterocolitica* O-3 at different pH-values, temperatures and L-lactate concentrations using conductance measurements, *Int. J. Food Microbiol.*, 22, 141, 1994.

96. Fehlhaber, K. and Kruger, G., The study of *Salmonella enteritidis* growth kinetics using rapid automated bacterial impedance technique, *J. Appl. Microbiol.*, 84, 945, 1998.

97. Koutsoumanis, K., Tassou, C.C., Taoukis, P.S., and Nychas, G.J.E., Modelling the effectiveness of a natural antimicrobial on *Salmonella enteritidis* as a function of concentration, temperature and pH, using conductance measurements, *J. Appl. Microbiol.*, 84, 981, 1998.

98. Deak, T. and Beuchat, L.R., Use of indirect conductimetry for predicting growth of food spoilage yeasts under various environmental conditions, *J. Ind. Microbiol.*, 12, 301, 1993.

2 Primary Models

Robin C. McKellar and Xuewen Lu

CONTENTS

0-8493-1237-X/04/$0.00+$1.50
© 2004

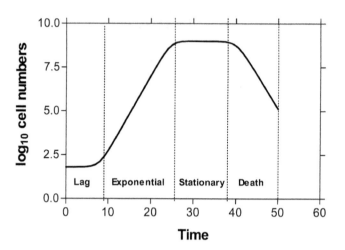

FIGURE 2.1 Stages of a bacterial growth curve.

2.1 GROWTH MODELS

2.1.1 INTRODUCTION

The concept of the primary model is fundamental to the field of predictive micro-biology (see the definition of a model in the Preface). A primary model for microbial growth aims to describe the kinetics of the process with as few parameters as possible, while still being able to accurately define the distinct stages of growth. A typical bacterial growth curve is shown in Figure 2.1. When the increase in population density (usually defined as the base 10 logarithm of cell numbers) is plotted against time, the resulting curve usually has four phases, referred to respectively as the lag, exponential, stationary, and death or decline phases.

In the only book published thus far that is devoted exclusively to the field of predictive microbiology, McMeekin et al.[1] provide an excellent review and discussion of the classical sigmoid growth functions, especially the modified logistic and Gompertz equations. As they point out, these are empirical applications of the original logistic and Gompertz functions. They lack mechanistic interpretability though the original logistic and Gompertz functions are considered mechanistic models. Over the last decade, a new generation of bacterial growth curve models have been developed that are purported to have a mechanistic basis: for example, the Baranyi model,[2,3] the Hills model,[4,5] the Buchanan model,[6] and the heterogeneous population model.[7] In addition to the book by McMeekin et al., other authors have provided reviews of microbial growth models.[3,8–11]

In this chapter, we will review the modified logistic and Gompertz equations as well as the new models that were not covered by McMeekin et al.[1] and discuss their applications. We will compare these models based on their performance in predictive microbiology applications.

2.1.2 THE LOGISTIC AND THE GOMPERTZ FUNCTIONS

Sigmoidal functions have been the most popular ones used to fit microbial growth data since these functions consist of four phases, similar to the microbial growth curve.[9] The most commonly used are the modified logistic (Equation 2.1) and the modified Gompertz (Equation 2.2) introduced by Gibson et al.[12]:

$$\log x(t) = A + \frac{C}{\left(1 + e^{(-B(t-M))}\right)} \tag{2.1}$$

$$\log x(t) = A + C \exp\{-\exp[-B(t-M)]\} \tag{2.2}$$

where $x(t)$ is the number of cells at time t, A the asymptotic count as t decreases to zero, C the difference in value of the upper and lower asymptote, B the relative growth rate at M, and M is the time at which the absolute growth rate is maximum.[1,9]

The above functions use $\log x(t)$ instead of $x(t)$ as the response variable. Thus, they are not simply reparameterizations of the original logistic[13,14] and Gompertz[15] functions, but are "modified" functions. The original logistic and Gompertz functions are considered to be mechanistic; however, the modified functions are empirical.

The parameters of the modified Gompertz equation can be used to characterize bacterial growth as follows[1]:

$$e = 2.718 \cdots$$

$$\text{lag time} = M - (1/B) + \frac{LogN(0) - A}{BC/e} \tag{2.3}$$

$$\exp\text{onential growth rate} = BC/e$$

$$\text{generation time} = \log(2)e/BC = 0.8183/BC$$

The expression in Equation 2.3 for lag time is different from the following Equation 2.4 proposed by Gibson et al.[12] and other workers[16,17]:

$$\text{lag time} = M - \frac{1}{B} \tag{2.4}$$

As explained by McMeekin et al.,[1] Equation 2.3 is a more general and correct expression for the lag time.

In order to simplify the fitting process, reparameterized versions of the Gompertz equation have been proposed[18,19]:

$$\log_{10} x = A + C \exp\left(-\exp\left[2.71\left(\frac{R_g}{C}\right)(\lambda - t) + 1\right]\right) \qquad (2.5)$$

where $A = \log_{10} x_0$ ($\log_{10}$ cfu $\times$ ml^{-1}), x_0 is the initial cell number, C the asymptotic increase in population density ($\log_{10}$ cfu $\times$ ml^{-1}), R_g the growth rate ($\log_{10}$ cfu h^{-1}), and λ is the lag-phase duration (h).

2.1.2.1 Applications of the Logistic Model

There have been limited examples of fitting of microbial growth data using the logistic function, since the Gompertz function, which is asymmetric about the point of inflection unlike the logistic function,[9,20,21] is generally preferred. Some recent examples include modeling of fish spoilage[22–24] and colony diameter of fungi.[25] A variation of the logistic model with a breakpoint at the transition between the lag phase and the exponential phase has also been used to model the lag phase of *Listeria monocytogenes*.[26]

2.1.2.2 Applications of the Gompertz Equation

The Gompertz equation has been used extensively by researchers to fit a wide variety of growth curves from different microorganisms. Some of the recent models developed with the Gompertz function include those for *Yersinia enterocolitica*,[27] *Staphylococcus aureus*,[28,29] *L. monocytogenes*,[30] *Vibrio parahaemolyticus*,[33] and *Bacillus cereus*.[32,33]

The Gompertz function has also been applied to growth curves based on turbidity data[34]; mixed cultures of *Pseudomonas* spp. and *Listeria* spp.[35,36]; *Lactobacillus curvatus*[37]; spoilage of vegetables,[38] beer,[39] and meat[40]; and germination and growth of *Clostridium botulinum*.[41]

There are, however, some limitations associated with the use of the Gompertz function. The Gompertz rate (μ_{max}) is always the maximum rate and occurs at an arbitrary point of inflection[42–44]; thus the generation time can be underestimated by as much as 13%.[31] In addition, since the slope of the function cannot be zero, the lower asymptote must be lower than the inoculum level, giving a negative λ for some data sets.[43] Another limitation is that, in order to get a good fit, experimental data are required over the whole growth range.[1,21]

2.1.3 BARANYI MODEL

In a series of papers,[2,3,10] Baranyi and coworkers introduced a mechanistic model for bacterial growth. Briefly, the lag phase is attributed to the need to synthesize an unknown substrate q that is critical for growth. Once cells have adjusted to the new environment, they grow exponentially until limited by restrictions dictated by the growth medium; thus:

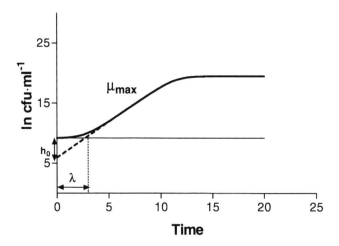

FIGURE 2.2 Example of a growth curve generated by the Baranyi and McKellar models. Parameters are defined in the text.

$$\frac{dx}{dt} = \frac{q(t)}{q(t)+1} \cdot \mu_{max} \cdot \left(1 - \left(\frac{x(t)}{x_{max}}\right)^m\right) x(t) \tag{2.6}$$

where x is the number of cells at time t, x_{max} the maximum cell density, and $q(t)$ is the concentration of limiting substrate, which changes with time:

$$\frac{dq}{dt} = \mu_{max} \cdot q(t) \tag{2.7}$$

The initial value of q (q_0) is a measure of the initial physiological state of the cells. A more stable transformation of q_0 may be defined as:

$$h_0 = \ln\left(1 + \frac{1}{q_0}\right) = \mu_{max}\lambda \tag{2.8}$$

The parameter m characterizes the curvature before the stationary phase. When $m = 1$ the function reduces to a logistic curve, a simplification of the model that is often assumed. Thus, the final model has four parameters: x_0, the initial cell number; h_0; x_{max}; and μ_{max}. The output of this model (and the relationship between h_0, λ, and μ_{max}) is shown in Figure 2.2.[45]

An explicit version of the Baranyi model has also been derived:

$$y(t) = y_0 + \mu_{max} A(t) - \frac{1}{m} \ln\left(1 + \frac{e^{m\mu_{max}A(t)} - 1}{e^{m(y_{max}-y_0)}}\right) \qquad (2.9)$$

$$A(t) = t + \frac{1}{v} \ln\left(\frac{e^{-vt} + q_0}{1 + q_0}\right) \qquad (2.10)$$

where $y(t) = \ln x(t)$, $y_0 = \ln x_0$, and v is the rate of increase of the limiting substrate, generally assumed to be equal to μ_{max}.

2.1.3.1 Applications of the Baranyi Model

Since its inception in the early 1990s, the Baranyi model has been used extensively to model microbial growth. The popularity of this model has been facilitated by the availability of two programs: DMFit, an Excel add-in; and MicroFit, a stand-alone fitting program, distributed by the Institute of Food Research in the U.K. (http://www.ifr.bbsrc.ac.uk/Safety/DMFit/default.html). The model was used for growth modeling of a wide variety of microorganisms, the results of which are included in the Food MicroModel software.[46] Some recent applications were related to *Listeria monocytogenes*,[47,48] *B. cereus*,[49] *Escherichia coli*,[50] *Y. enterocolitica*,[51] increasing colony diameter of heat-resistant fungi,[52] and spoilage in green asparagus and vegetable salad.[53,54]

One of the advantages of the Baranyi model is that it is readily available as a series of differential equations that allow modeling in a dynamic environment, generally resulting from nonisothermal temperature profiles. This form of the model was used to describe the behavior of *E. coli* at suboptimal temperatures,[55] and to develop and validate a dynamic growth model for *L. monocytogenes* in fluid whole milk.[56,57] It has also been used to study the influence of either slowly[58] or rapidly[59] changing temperature on the growth of *L. monocytogenes* and *Salmonella*.

2.1.4 HILLS MODEL

A general theory of spatially dependent bacterial growth in heterogeneous systems was developed by Hills and coworkers.[4,5] This was achieved by combining a structured-cell kinetic model with reaction-diffusion equations describing transport of nutrients.[4] The model was based in part on the concept of DNA synthesis and cell division being dependent on the excess cell biomass.

Assume M is the total biomass in the culture and N is the total number of cells in the culture. It can be shown that for inoculation with stationary-phase cells,

$$M(t) = M(0)\exp(At)$$

$$N(t) = N(0)[k_n \exp(At) + A\exp(-k_n t)] / (A + k_n) \qquad (2.11)$$

A and k_n are rate constants; in general, they depend on all the environment factors. The expression for $N(t)$ in Equation 2.11 has a much simpler form than the empirical

Gompertz function for fitting population growth, being a biexponential function where the second term, involving the rate of DNA synthesis, gives rise to the observed lag behavior. The lag time and the doubling time have the following relationships:

$$t_{LAG} = A^{-1} \log[1 + (A / k_n)]$$

$$t_{LAG} / t_D = (\ln 2)^{-1} \log[1 + (A / k_n)] \tag{2.12}$$

This shows that if the rate constants A and k_n have similar activation energies, the ratio of lag to doubling time should be nearly independent of temperature. This model takes no account of possible lag behavior in the total biomass (M).

The above model can also be generalized to spatially inhomogeneous systems such as food surfaces.[4] If more detailed kinetic information on cell composition is available, more complex multicompartment kinetic schemes can be incorporated. A two-compartment kinetic model of bacterial population dynamics has been developed that is capable of describing the phenomena of lethal and sublethal injury, resuscitation, and transient conditions. A more general three-compartment kinetic model has been developed to interpret lag behavior in total biomass. These models can be further generalized to describe growth in spatially heterogeneous systems.[5]

2.1.4.1 Applications of Hills Model

There have been few applications of the Hills model. The above two-compartment kinetic cell model was shown to fit batch-growth data for *L. monocytogenes*[4] and for *Salmonella typhimurium*.[5] More recently, the model was used for modeling viable counts of *S. typhimurium* in gel cassettes.[60]

2.1.5 BUCHANAN THREE-PHASE LINEAR MODEL

Buchanan et al.[6] proposed a three-phase linear model. It can be described by three phases: lag phase; exponential growth phase; and stationary phase:

Lag Phase:

$$\text{For } t \leq t_{LAG}, \quad N_t = N_0$$

Exponential Growth Phase:

$$\text{For } t_{LAG} < t < t_{MAX}, \quad N_t = N_0 + \mu(t - t_{LAG}) \tag{2.13}$$

Stationary Phase:

$$\text{For } t \geq t_{MAX}, \quad N_t = N_{MAX}$$

where N_t is the log of the population density at time t (log cfu ml^{-1}); N_0 the log of the initial population density (log cfu ml^{-1}); N_{MAX} the log of the maximum population

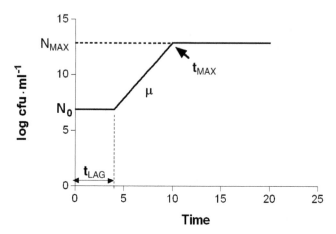

FIGURE 2.3 Example of a growth curve generated by the Buchanan model. Parameters are defined in the text.

density supported by the environment (log cfu ml^{-1}); t the elapsed time; t_{LAG} the time when the lag phase ends (h); t_{MAX} the time when the maximum population density is reached (h); and μ is the specific growth rate (log cfu ml^{-1} h^{-1}). The three-phase model is illustrated in Figure 2.3.

In this model, the growth rate was always at maximum between the end of the lag phase and the start of the stationary phase. The μ was set to zero during both the lag and stationary phases. The lag was divided into two periods: a period for adaptation to the new environment (t_a) and the time for generation of energy to produce biological components needed for cell replication (t_m). Thus, the lag phase is given by:

$$t_{LAG} = t_a + t_m \tag{2.14}$$

This implies that t_a and t_m can be estimated from data fitted with the linear model using the following relationships[6]:

$$t_m = \text{generation time}$$
$$t_a = t_{LAG} - \text{generation time} \tag{2.15}$$

2.1.5.1 Applications of the Buchanan Model

Surprisingly, this simple model has not been used extensively for fitting growth data. The original authors used the three-phase version of the model to fit experimental data for *E. coli* O157:H7.[6] As there is often little interest in modeling the stationary phase, a modified two-phase version has been proposed that fits only the lag and exponential phases. In a series of papers published in 1999, Oscar used the two-phase model to fit growth data for *S. typhimurium* in brain heart infusion broth,[61] and on cooked chicken[62] and ground chicken[63] breast meat. Fitting was accomplished using a useful nonlinear regression software package called Prism (GraphPad Software, San Diego, CA) in which an if–then statement defines the model:

$$N_t = N_0 + IF[t \leq t_{LAG}, 0, \mu \cdot (t - t_{LAG})] \qquad (2.16)$$

with symbols defined in Equation 2.13. A two-phase model was also used to model growth of *E. coli* O157:H7.[64]

2.1.6 MCKELLAR MODEL

One of the limitations of existing models is that they all assume a homogeneous population of cells. A heterogeneous population model was recently proposed in which growth was expressed as a function of two distinct cell populations.[7] Cells can exist in one of two "compartments" or states: growing or nongrowing. All growth was assumed to originate from a small fraction of the total population of cells that are present in the growing compartment at $t = 0$. Subsequent growth is based on the following logistic equation:

$$\frac{dG}{dt} = G \cdot \mu \cdot \left(1 - \frac{G}{N_{MAX}}\right) \qquad (2.17)$$

where G is the number of growing cells in the growing compartment. The majority of cells were considered not to contribute to growth, and remained in the nongrowing compartment, but were included in the total population. While this is an empirical model, it does account for the observation that growth in liquid culture is dominated by the first cells to begin growth, and that any cells that subsequently adapt to growth are of minimal importance.[7]

This model has an interesting relationship with the Baranyi model. It is derived from a different initial premise, that microbial populations are heterogeneous rather than homogeneous. It is based on two populations of cells that behave differently, rather than a single population. The sum of the two populations effectively describes the transition from lag to exponential phase, and defines a new parameter G_0, the initial population capable of growth. Reparameterization of the model led to the finding that a relationship existed between μ_{max} and λ, which is shown in Figure 2.2,[7] and which had been derived by Baranyi from a more mathematical argument.[3] Baranyi[65] later supported the geometric relationship in Figure 2.2, and stated that the initial physiological state of the whole population could reside in a small subpopulation. Thus, the McKellar model constitutes a simplified version of the Baranyi model, and has the same parameters.

The concept of heterogeneity in cell populations was extended further to the development of a combined discrete–continuous simulation model for microbial growth.[66] At the start of a growth simulation, all of the cells were assigned to the nongrowing compartment. A distribution of individual cell lag times was used to generate a series of discrete events in which each cell was transferred from the nongrowing to the growing compartment at a time corresponding to the lag time for that cell. Once in the growing compartment, cells start growing immediately according to Equation 2.17. The combination of the discrete step with the continuous growth function accurately described the transition from lag to exponential phase. This model was further modified to include a continuous adaptation phase prior to the discrete event.[67] A new physiological state parameter p_0 was proposed that represents the mean of the initial individual cell physiological states. This model is dynamic in both the lag and exponential phases, and thus is useful for simulating the behavior of individual cells in a changing environment.

2.1.6.1 Applications of the McKellar Model

This model has not been used extensively for modeling microbial growth partly because of its similarity to the Baranyi model. It is also a compartmental model, and as such cannot be fitted easily using conventional nonlinear regression programs. This model was fitted to data for growth of *L. monocytogenes* at 5 to 35°C, and compared to the Gompertz model.[7] Values for μ_{max} were slightly higher with this model, and λ were generally shorter than found with the Gompertz model. Goodness-of-fit analysis suggested that the McKellar model generally fit the data better than the Gompertz.

2.1.7 OTHER MODELS

There have been a large number of alternative models proposed for modeling microbial growth. Many of the earlier ones have been thoroughly discussed by McMeekin et al.,[1] and will not be discussed further here.

Whiting and Cygnarowicz-Provost[68] suggested a quantitative four-parameter model for the germination, growth, and decline of *C. botulinum*, and the growth *of L. monocytogenes*. Jones and Walker[69] developed an equation to predict growth, survival, and death of microorganisms based on data obtained using *Y. enterocolitica* in varying pH and sodium chloride concentrations at different temperatures. Van Impe et al.[70] proposed a dynamic first-order differential equation to predict both microbial growth and inactivation, with respect to both time and temperature. We are expecting more accurate and more mechanistic primary models when people gain more knowledge on the kinetics of individual cells and behavior of bacteria. Recently, a series of three models has been proposed in which μ can increase, remain constant, or decrease with time.[71] The latter two models bear some resemblance to those discussed earlier; however, the concept of μ increasing with time was designed to accommodate the observation that recombinant *E. coli* initially grew rapidly in a bioreactor because of high substrate concentrations.

2.1.8 EXAMPLES OF GROWTH MODEL FITTING

It seems appropriate at this point to provide an example of how some of the more popular and useful functions may be used to fit experimental growth data. The data selected (taken from an earlier study[7]) were for the growth of *L. monocytogenes* at 5°C (Table 2.1). The models used in this comparison were Gompertz using Equation

---

TABLE 2.1
Growth Data for *Listeria monocytogenes*
at 5°C

Time (d)	log cfu ml^{-1}
0	4.8
6	4.7
24	4.7
30	4.7
48	4.9
54	5.1
72	5.3
78	5.4
99	5.9
126	6.3
144	6.9
150	6.9
168	7.2
174	7.3
191	7.7
198	7.8
216	8.3
239	8.8
266	9.1
291	9.2
316	9.3
336	9.7
342	9.7
360	9.7
384	9.5

2.5, Baranyi using Equation 2.6 and Equation 2.7, McKellar using Equation 2.17, and Buchanan using Equation 2.13. Nonlinear regression analysis was done using the ModelMaker® software (Modelkinetix, Old Beaconsfield, Bucks, U.K., www.modelkinetix.com), which uses the Runge–Kutta method for solving differential equations. Initial parameter estimates were made using the simplex method, and regression was performed using the Marquardt algorithm. The Baranyi and McKellar models gave values for μ_{max} directly, since they were in the form of differential equations, and modeled the cell number rather than $\log_{10}$ cfu ml^{-1}. The Gompertz and Buchanan models were applied directly to $\log_{10}$ cfu ml^{-1} data, and thus the rate parameter (R_g) obtained from the fitting had to be converted to μ_{max} using the relationship $\mu_{max} = R_g \cdot \ln 10$. The λ parameter for the Gompertz and Buchanan models was obtained directly from the fitting, while the values for the Baranyi and McKellar models were derived from the h_0 parameter values using the following relationship: $h_0 = \mu_{max} \cdot \lambda$. The Baranyi model (Baranyi$_{MF}$) was also fitted using the MicroFit software, in which the model was reparameterized to fit λ directly. The

TABLE 2.2
Results of Model Fitting to Growth Data

Model[a]	μ (h^{-1})	λ (d)	Log x_0	Log x_{max}	DF	RMSE
Baranyi$_{MF}$	0.050	46.9	4.68	9.57	21	0.100
McKellar	0.049	44.9	4.63	9.57	21	0.112
Gompertz$_{PZ}$	0.054	54.7	4.68	10.0	21	0.119
Gompertz	0.058	68.4	4.76	9.87	21	0.139
Buchanan	0.048	53.8	4.84	9.49	20	0.157
Baranyi	0.056	61.4	4.72	9.32	21	0.179

Note: DF = residual degrees of freedom; RMSE = root mean square error.

[a] The McKellar, Gompertz, Buchanan, and Baranyi models were fit using the ModelMaker software. Baranyi$_{MF}$ is the Baranyi model fit using the MicroFit software. Gompertz$_{PZ}$ is the Gompertz model fit using the Prism software.

Gompertz model (Gompertz$_{PZ}$) was also fitted using Prism™ Version 3.03 (GraphPad Software, San Diego, CA, www.graphpad.com).

The results of the various fitting approaches are given in Table 2.2. The root mean square error (RMSE) was taken as the measure of goodness of fit, as suggested by Ratkowsky (Chapter 4). The models are placed in order of increasing RMSE.

The best model was Baranyi$_{MF}$, with the lowest RMSE. In contrast, the poorest fit was with the Baranyi model using ModelMaker, which also gave a higher μ_{max} and λ than did the Baranyi$_{MF}$ model. The next best model was the McKellar, with parameter values close to those for Baranyi$_{MF}$. The Gompertz model fitted using either Prism or ModelMaker gave larger μ_{max} and λ values than did the Baranyi$_{MF}$ and McKellar models, and the highest values of log x_{max} among all other models. The Buchanan model gave the lowest value of μ_{max} of all the models, and a shorter λ than all except the Baranyi$_{MF}$ and McKellar models.

The output of the four models fitted with ModelMaker is also shown in Figure 2.4. The steeper slope (μ_{max}) of the Gompertz and Baranyi models may be observed. The greatest difference between models occurred during the late-log early-stationary phase. The Gompertz model never reaches a plateau, which reflects its higher $\log_{10} x_{max}$ (Table 2.2). As expected, the Buchanan model has a sharp breakpoint, while the transition to the stationary phase appears smoother with both the McKellar and Baranyi models.

The results of the nonlinear regression fitting described above emphasize an important point: there is no single solution for nonlinear regression, in contrast to linear regression. The iterative approach used in nonlinear regression is dependent on the parameter starting values, and may find local, rather than global, optimum values. In addition, different software packages use different procedures for fitting, and thus the results obtained (such as those above) should be considered comparative rather than absolute. The fitting results do show that, while there are differences between the models and the software used, the parameter differences are often slight. It is worth noting that estimates of λ range from 44.9 to 68.4 days. Further discussion

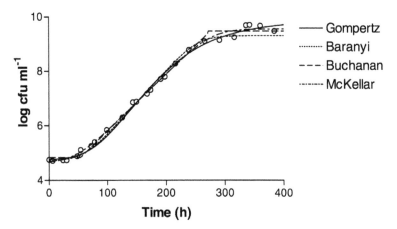

FIGURE 2.4 Comparison of growth models fitted to viable count data of *Listeria monocytogenes* grown at 5°C.

on the difficulties in modeling λ, and the role of the physiological state, can be found in Chapter 9, and a more complete discussion of model fitting can be found in Chapter 4.

2.1.9 COMPARISON OF EXISTING MODELS

Zwietering et al.[18] statistically compared several different modified sigmoidal functions (Logistic, Gompertz, Richards, Schnute, and Stannard) using the *t*-test and the *F*-test. In most of the cases, the modified Gompertz expression was regarded as the best model to describe the growth data both in terms of statistical accuracy and ease of use when compared to other sigmoidal functions.

Baranyi et al.[2] compared the output of their model with that of the Gompertz, and concluded that the goodness of fit was generally at least as good. They also showed that their model gave estimates for lag and growth rate that were slightly lower than in the Gompertz case. Baranyi et al. also compared their model to those of Hills[10] and Buchanan[72] and stated that these models are special cases of the Baranyi model. Baranyi argues that the Buchanan model has merit in its simplicity, but that the model lacks the capability of simulating dynamic behavior.[72] Buchanan et al.[6] asserted that their three-phase model is comparable to, and more robust than, either the Gompertz or the Baranyi models, especially when experimental data were minimal. The three-phase linear and Baranyi models predicted similar maximum population densities. These values were typically smaller than the values provided by the Gompertz model. Garthright[44] strongly supports the three-phase model, and points out its superiority in describing the lag and exponential phases as compared to the Gompertz. He concludes that the nonlinear approach does not achieve any advantage over the three-phase linear approach for environmental applications. This model appears particularly appropriate for modeling conditions where growth is poor, and an upper asymptote cannot be accurately fixed. The Baranyi model and the McKellar model can also be used when stationary-phase data are lacking.

Other comparisons between growth models have been made. A comparison of the logistic, Gompertz, and Baranyi models for fish spoilage showed that the logistic function was similar to the Baranyi but easier to fit.[73] A comparison between Gompertz and Baranyi models gave better fit with the Baranyi model, and a higher growth rate with Gompertz.[74] The Gompertz function was found to be more appropriate than the Baranyi model for monitoring CO_2 evolution as an indicator of bacterial growth.[75] Other workers have compared the Baranyi and Gompertz models, and have concluded that the Baranyi function gave better parameter estimates as compared to the Gompertz.[76]

At the present time it is not possible to select one growth model as the most appropriate representation of bacterial growth. If simple is better, then the three-phase model is probably sufficient to represent fundamental growth parameters accurately.[44,77] There does appear to be general agreement in relationship to underlying principles, and emphasis should be placed on the development and use of models and parameters that can be easily understood by food microbiologists.[77] However, in spite of Garthright's assertion that straight line simplicity is sufficient to model growth,[44] the development of more complex models (and subsequently more mechanistic models) will depend on an improved understanding of cell behavior at the physiological level.

2.2 SURVIVAL MODELS

2.2.1 INTRODUCTION

Our ability to understand and model the survival of pathogens in foods or during processing of food is critical to the safety of the food supply. Thus, models to describe microbial death due to heating have been used since the 1920s, and constitute one of the earliest forms of predictive microbiology. Much of the early work centered around the need to achieve destruction of *C. botulinum* spores in low acid canned foods,[17,78,79] and much effort has been put towards characterizing the kinetics of spore inactivation. In this section of the chapter we will focus on the evolution of survival modeling from the classical linear approach to the more complex models required to describe inactivation curves that deviate from linearity.

2.2.2 CLASSICAL LINEAR MODELS

It has always been assumed that spore inactivation follows simple first-order reaction kinetics under isothermal conditions:

$$\frac{dS_t}{dt} = -k'S_t \qquad (2.18)$$

where S_t is the survival ratio (N_t/N_0) and k' is the rate constant. Thus the number of surviving cells decreases exponentially:

$$S_t = e^{-k't} \tag{2.19}$$

and when expressed as $\log_{10}$, gives:

$$\log S_t = -kt \tag{2.20}$$

where $k = k'/\ln 10$. The well-known D-value (time required for a 1-log reduction) is thus equal to $1/k$, where k is the slope (Figure 2.5). The D-values can also be expressed as:

$$D\text{-value} = \frac{t}{\log N_0 - \log N_t} \tag{2.21}$$

When log D-values are plotted against the corresponding temperatures, the reciprocal of the slope is equal to the z-value, which is the increase in temperature required for a 1-log decrease in D-value (Figure 2.5; inset). The rate constant can also be related to the temperature by the Arrhenius equation:

$$k = N_0 e^{\left(-\frac{E_a}{RT}\right)} \tag{2.22}$$

where E_a is the activation energy, R the universal gas constant, and T is the temperature in Kelvin.

From the first-order reaction it is not possible to achieve complete destruction of all *C. botulinum* spores in a given volume of product; one spore will always be

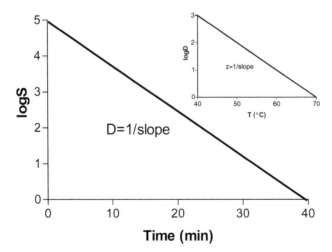

FIGURE 2.5 Geometric description of D- and z-values. (From McKellar, R.C., Modelling the effectiveness of pasteurization, in *Dairy Processing: Maximizing Quality*, Smit, G., Ed., CRC Press Inc./Woodhead Publishing, 2003 pp. 104–129. With permission.)

left in a can if a sufficient number of cans are examined. Thus it is generally assumed that a 12-log reduction (also known as 12D) is sufficient to achieve "commercial sterility," or an acceptable level of risk of survival of *C. botulinum*. Knowledge of the *D*-values of representative strains allows the determination of the F_0-value, which is the time required to achieve 12D, assuming a *z*-value of 10°C. At 121°C, F_0 is equal to 2.5 min for most strains of *C. botulinum*.[17]

Comparable standards for other food-borne pathogens do not exist; however, it is generally accepted that a 4- or 5-log reduction is considered adequate, depending on the product. An extensive amount of work has gone into the determination of *D*- and *z*-values for various pathogens. Thermal stability of pathogens such as *L. monocytogenes*,[80] salmonellae,[81] and *E. coli* O157:H7[82] has been well documented and summarized in recent reviews.

2.2.3 NONLINEAR MODELS

2.2.3.1 Nonlinearity Issues

The canning industry has enjoyed an enviable record of safety, and thus the concept of logarithmic death of microorganisms has persisted, and is now considered accepted dogma. In spite of this, nonlinear survival curves were reported for some bacteria almost 100 years ago.[83] In general there are two classes of nonlinear curves; those with a "shoulder" or lag prior to inactivation, and those that exhibit tailing. These two phenomena may be present together, or with other observed kinetics such as biphasic inactivation. A wide variety of complex inactivation kinetics have been reported, and several of these are shown in Figure 2.6. The theoretical basis for assuming logarithmic behavior for bacteria is based on the assumptions that bacterial populations are homogeneous with respect to thermal tolerance, and that inactivation is due to a single critical site per cell.[83] Both of these assumptions have been questioned, and thus concerns have been raised regarding the validity of extrapolation of linear inactivation curves.[84,85]

Stringer et al.[82] have summarized the possible explanations for nonlinear kinetics into two classes: those due to artifacts and limitations in experimental procedure and those due to normal features of the inactivation process. The first class encompasses such limitations as variability in heating procedure; use of mixed cultures or populations; clumping; protective effect of dead cells; method of enumeration; and poor statistical design. The second class includes such situations as possible multiple hit mechanisms; natural distribution of heat sensitivity; and heat adaptation. These two classes roughly parallel the two concepts reviewed by Cerf[85] to explain tailing in bacterial survival curves. The first of these (the "mechanistic" approach) also makes the assumption of homogeneity of cell resistance and proposes that thermal destruction follows a process analogous to a chemical reaction. In this approach, deviations from linearity are attributed mainly to artifacts; however, tailing is also related to the mechanism of inactivation or resistance. In the "vitalistic" approach, it is assumed that the cells possess a normal heterogeneity of heat resistance; thus survival curves should be sigmoidal or concave upward.[85]

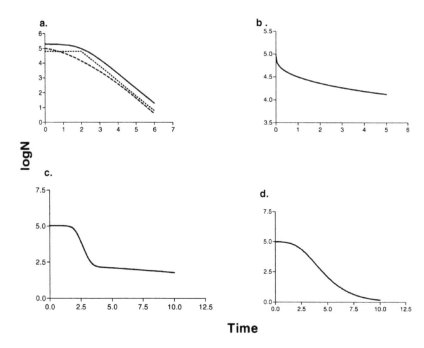

FIGURE 2.6 Examples of thermal death curves: (a) lag or shoulder, with either linear (dotted line), power law where $p > 1$ (broken line), or monophasic logistic (solid line) models; (b) concave with power law where $p < 1$; (c) biphasic logistic; and (d) sigmoidal.

There has been considerable controversy between the two schools of thought, and the literature is divided on the validity of nonlinear survival curves as representing the true state of the cell population. There is certainly evidence that inconsistencies in experimental protocols or the use of incorrect media can lead to artifacts; however, there is little convincing evidence that clumping of cells or the protective effect of dead cells is consistently responsible for nonlinear survivor curves. The current belief is, notwithstanding some contribution by artifacts, that cells do exhibit heterogeneity in thermal sensitivity, and the majority of modeling approaches now make this assumption. There is also inconsistency in actually defining what is meant by an artifact. If one assumes that an artifact in this context is anything that interferes with obtaining a linear death curve, then many of the situations currently classified as artifacts may be natural behavior of cell populations. This is particularly obvious in the study of spore inactivation where standardized suspensions are difficult to obtain, and much effort has been expended to remove artifacts such as genetic variants. The difficulty in obtaining linear kinetics may be a signal that, in most cases, nonlinearity is the norm.

The current theories of microbial inactivation must be revisited in light of recent improved understanding of the effect of heat on microorganisms. We now know that cells do not exist simply as alive or dead, but may also experience various degrees of injury or sublethal damage, which may give rise to apparent nonlinear survival curves.[82] The induction of heat resistance in food-borne pathogens due to expression of heat shock proteins has been extensively documented in recent years, and may contribute to apparent nonlinearity, particularly tailing.[82,86,87] Thus it appears important to model the actual conditions or situations experienced by bacteria in foods rather than relying on simplifications. Survival modeling should also include a more complete understanding of the molecular events underpinning microbial resistance to the environment.

It seems likely that heterogeneity within bacterial populations is responsible in most cases for nonlinear survival curves, and most recent attempts to model survival employ distributions. The use of distributions to account for nonlinearity is not new; log normal distributions had been suggested for this purpose as early as 1942.[83] Other distributions such as logistic, gamma, and Weibull have also been suggested; Weibull is the favored approach at the moment (see later). There is no complete agreement on the use of distributions,[83] and it is clear that this approach cannot adequately account for changes in heat resistance occurring during heating.

Our lack of understanding of the key physiological aspects of microbial inactivation and the complexities of nonlinear behavior suggest that a truly mechanistic model for thermal inactivation will not be developed in the near future. One approach to quantitating bacterial survival might be the thermal death point concept common to the canning industry. This approach allows one to define the conditions required to achieve a target log reduction, and makes no statement regarding the kinetics of that destruction. This approach has a number of attractive advantages; however, it would still be influenced by such artifacts as changes in heat resistance of a culture and cell injury.[83]

2.2.3.2 Shoulder/Tail Models

2.2.3.2.1 Linear Approach

Inactivation curves that deviate from simple exponential often have a lag or shoulder region prior to the exponential inactivation. This shape of inactivation curve is probably the most commonly experienced by researchers. A simple linear model to account for this behavior was developed by Whiting[88]:

$$\log N = \begin{cases} \log N_0 & \text{when } 0 < t < t_L \\ \log N_0 - \left(\dfrac{1}{D}\right)(t - t_L) & \text{when } t > t_L \end{cases} \tag{2.23}$$

where t_L is the lag prior to inactivation.

An example of the output of this model is the dotted line in Figure 2.6a. The advantage of this model is that linear regression can be used. This simple model has been used effectively to describe the nonthermal inactivation of *L. monocytogenes* as a function of organic acid and nitrite concentrations[89–92] and under reduced oxygen.[93] A similar two-phase linear model was described for thermal inactivation of *L. monocytogenes* by Bréand et al.[94]

It is quite common for the lag or shoulder region of the survival curve to be highly variable. This makes it difficult to develop secondary models to describe the influence of the environment on the lag. Thus, survival using this model is often described as the time required for a 4-log reduction (T_{4D})[89,92,95]:

$$t_{4D} = t_L + 4 \cdot D \tag{2.24}$$

2.2.3.2.2 Nonlinear Approach

Complex inactivation kinetics requires the use of nonlinear functions. It should be noted that nonlinearity as it relates to mathematical functions means that the parameters in the equation are nonlinear; the resulting curve may or may not appear linear. Linear regression can be easily performed by most spreadsheet programs; however, nonlinear regression is an iterative process that is supported by more specialized software. These software packages are readily available; thus considerable advances have been made in the development of nonlinear models.

Another of the more common shapes of survival curves is the concave curve, which has no lag, and a single, tailing population (Figure 2.6b). This function is best represented by the power law:

$$\log \frac{N}{N_0} = -\frac{t^p}{D} \tag{2.25}$$

where p is the power. A concave curve is produced when $p < 1$ (Figure 2.6b), and a convex (or shoulder) shape results from $p > 1$ (broken line in Figure 2.6a). A power law function has been used to model curvature in survival curves for *Enterococcus faecium*[96] and alkaline phosphatase[97] in milk. Other, seemingly novel, functions that have been derived to fit concave survival curves are really in fact power law functions.[98,99]

Tailing survival curves can also be represented by the exponentially damped polynomial model. In this model, deviation from simple linear kinetics, experienced while heating *Staphylococcus aureus* in skim milk, was fitted with the nonlinear function[100]:

$$\log \frac{N}{N_0} = -kte^{-\lambda t} \tag{2.26}$$

where k is the rate coefficient and λ is the damping coefficient.

As discussed earlier, a logistic equation may be used in growth modeling to modify the simple exponential growth to account for limiting the maximum population size as a result of nutrient limitation. In the same way, a logistic function can be used to account for death being limited by the amount of some stress factor or damage to the cell.[101] This "mirror image" of the logistic function is called the Fermi equation, and is used for sigmoidal decay curves, which are symmetric:

$$\log \frac{N}{N_0} = \log \left[\frac{1 + e^{-bt_L}}{1 + e^{b(t-t_L)}} \right] \tag{2.27}$$

where N is the population (cfu ml^{-1}) surviving at time t; N_0 is the population surviving at time 0; b is the maximum specific death rate; and t_L is the lag phase prior to inactivation. This equation has been modified to account for situations where one may find both a primary, heat-sensitive population and a secondary more heat-resistant population[88]:

$$\log \frac{N}{N_0} = \log\left[\frac{F(1+e^{-b_1 t_L})}{(1+e^{b_1(t-t_L)})} + \frac{(1-F)(1+e^{-b_2 t_L})}{(1+e^{b_2(t-t_L)})}\right] \tag{2.28}$$

where b_1 is the maximum specific death rate for the primary population and b_2 is the maximum specific death rate for the secondary population. Traditional D-values may be calculated as $2.3/b$ for each population. Lag phases are not always present, and this can be accounted for by setting the value of t_L to zero. An example of the output of this function is given in Figure 2.6c. The biphasic logistic model has been used to model inactivation of spores of *C. botulinum*,[102] and the nonthermal inactivation of *L. monocytogenes*[90,92,93] and *S. aureus*.[103] This model has also been applied to the thermal inactivation of bovine milk catalase[104] and *E. faecium*[96] during high-temperature short-time (HTST) pasteurization, and inactivation of *E. faecium* during bologna sausage cooking.[105] In situations where a single population exists, F can be set equal to 1 (solid line in Figure 2.6a).

Other variations of the logistic function have been suggested. A four-parameter logistic model was proposed by Cole et al.[106]:

$$y = \alpha + \frac{\omega - \alpha}{1 + e^{\frac{4\sigma(\tau-x)}{\omega-\alpha}}} \tag{2.29}$$

where $y = \log_{10}$cfu ml^{-1}; $x = \log_{10}$ time; α = upper asymptote; ω = lower asymptote; τ = position of maximum slope; and σ = maximum slope. This model was applied to the survival of *Y. enterocolitica* at suboptimal pH and temperature,[107] and the thermal inactivation of *Salmonella typhimurium*,[108] *C. botulinum*,[109] *Salmonella enteritidis*, and *E. coli*.[110]

As was shown earlier, the asymmetric Gompertz function has considerable advantages when fitting bacterial growth curves. In keeping with the trend to use mirror images of growth functions to describe inactivation, a reparameterized form of the Gompertz function was suggested by Linton et al.[111]:

$$\log \frac{N}{N_0} = C\exp(-\exp(A+Bt)) - C\exp(-\exp(A)) \tag{2.30}$$

This function has been used to fit nonlinear survival curves of *L. monocytogenes* in buffer[111] and infant formula.[112] An example of the Gompertz function is given in Figure 2.6d. Other applications for the Gompertz equation include the effect of combined high pressure and mild heat on the inactivation of *Escherichia coli* and *S. aureus* in milk and poultry,[113] and the inhibition of Enterobacteriaceae and clostridia during sausage curing.[114] In a similar fashion, the mirror image of the

TABLE 2.3
Survival Data from *Pediococcus* sp.
NRRL B2354 at 62°C

Time (min)	Log cfu ml^{-1}
0	8.4
5	8.3
10	8.1
15	8.1
20	7.7
25	7.3
30	6.9
35	6.4
40	6.2

TABLE 2.4
Parameter Estimates from Fitting the
Logistic and Linear Survival Models to
the Data in Table 2.3

Model	N_0	Lag	D-value	DF	RMSE
Logistic	8.11	12.3	12.0	6	0.089
Linear	8.27	12.8	12.7	6	0.102

Baranyi growth model (see earlier) has been used for fitting nonlinear survival curves for the thermal inactivation of *Brochothrix thermosphacta*[115] and *Salmonella enteritidis*.[116]

2.2.3.2.3 Examples of Model Fitting

It has often proven difficult to accurately fit survival data where a lag exists prior to inactivation. The models we have found most useful in this situation are the single-phase logistic (Equation 2.27) and the two-phase linear (Equation 2.23).

These models were fitted to unpublished data on survival of *Pediococcus* sp. NRRL B2354 heated at 62°C (Table 2.3), using Prism as described above. The results of the fitting are shown in Table 2.4, and in Figure 2.7. The logistic model was slightly better than the linear, with a smaller RMSE. Because of the sharp breakpoint between the shoulder and exponential decay, the D-value for the linear model was slightly larger, while the lag phase was only marginally greater than that in the logistic model. As was found with growth models, there is often little to choose between different models; thus personal preference and experience often dictate which model is generally used. A more complete discussion of model fitting may be found in Chapter 4.

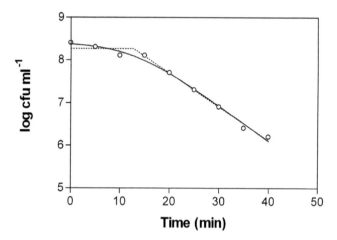

FIGURE 2.7 Example of fitting nonlinear survival data for *Pediococcus* NRRL B2354 using monophasic logistic (solid line) and two-phase linear (broken line) models. (From McKellar, R.C., Modelling the effectiveness of pasteurization, in *Dairy Processing: Maximizing Quality*, Smit, G., Ed., CRC Press Inc./Woodhead Publishing, 2003 pp. 104–129. With permission.)

2.2.4 DISTRIBUTIONS

One recent development in the modeling of bacterial survival is the use of distributions. This is based on the assumption that lethal events are probabilistic rather than deterministic. With a large initial population of cells, a continuous function can be used, much like with a chemical reaction (although a chemical reaction appears deterministic only because of the large number of molecules involved). The survival curve for a single cell is a step function, where a cell exists as either alive or dead[117]:

$$S_i(t) = \begin{cases} 1 & \text{(alive) for } t < t_c \\ 0 & \text{(dead) for } t \geq t_c \end{cases} \tag{2.31}$$

where t_c is the inactivation time. Since all cells would not be expected to die at the same time, values of t_c would follow some sort of distribution. The Weibull distribution is used in engineering to model time to failure, and so it seems appropriate for modeling bacterial inactivation. The distribution of t_c would then follow the probability density function (PDF) for the Weibull (solid line in Figure 2.8):

$$PDF = \frac{\beta}{\alpha}\left(\frac{t}{\alpha}\right)^{\beta-1} e^{\left(-\left(\frac{t}{\alpha}\right)^{\beta}\right)} \tag{2.32}$$

where α and β are parameters relating to the scale and shape of the distribution, respectively.[118] The survival curve is then the cumulative distribution function (CDF) (dotted line in Figure 2.8):

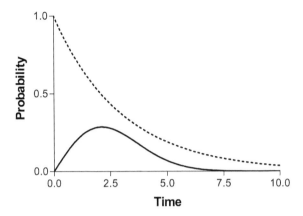

FIGURE 2.8 Probability density (solid line) and cumulative probability distribution (broken line) for the Weibull distribution.

$$CDF = e^{\left(-\left(\frac{t}{\alpha}\right)^{\beta}\right)}$$

(2.33)

It can be easily seen that the CDF of the Weibull distribution is essentially a reparameterization of the power law function (Equation 2.25). In the same fashion, the Fermi equation described earlier is the CDF of a log normal PDF.[21,86]

The Weibull parameter (β) has a very distinct influence on the shape of the survivor curve. When $\beta < 1$, a concave survival curve is obtained, and when $\beta > 1$, the curve is convex. Interestingly, the simple exponential model described earlier is a special case of the Weibull distribution when $\beta = 1$, providing further support for the use of the Weibull distribution as an effective modeling approach for microbial survival. Further, the value of β can have some implications for possible mechanisms of inactivation. When $\beta < 1$, there is an indication that the remaining cells are more resistant to the treatment, while when $\beta > 1$, an accumulation of the lethal effect is observed resulting in increasing rate of destruction with time. The classical D-value from linear survival curves can be related to the 90% percentile of the CDF, which is the time (t_d) required to reduce the number of microorganisms by a factor of 10[118]:

$$t_d = \alpha(2.303)^{\frac{1}{\beta}}$$

(2.34)

There have been a number of recent applications of the Weibull distribution to model survival curves for species of *Bacillus* and *Clostridium* spp.,[98] *Salmonella*,[119,120] and *E. coli*.[121] Van Boekel [118] has fitted the Weibull distribution to a large number of survival curves obtained from the literature. In almost all cases, the β values were different from 1, indicating that the classical linear model may not be generally applicable. Temperature had a significant effect on the α but not the β parameter. In order to determine if the Weibull distribution is appropriate for a

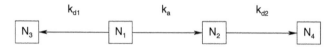

FIGURE 2.9 Model for spore activation and survival.

particular survival curve, a so-called hazard plot[118,121] of $\ln(-\ln S)$ vs. $\ln t$ should give a straight line. It should also be noted that, when survival curves are modeled using distributions, the presence of a "shoulder" can be attributed to the spread of a distribution being small relative to its mean or mode.[122]

2.2.5 SPORES

Modeling the inactivation of bacterial spores presents a unique problem. Spore-forming bacteria such as *Bacillus* and *Clostridium* spp. can exist in a dormant (spore) stage that is highly heat resistant. Germination of spores can be achieved by treatment with sublethal heat.[123] Because of the extreme heat resistance of some of these micro-organisms, activated spore preparations have traditionally been used to establish ster-ilization protocols in the canning and ultrahigh temperature industries.[124] As described earlier, the classical view of microbial thermal inactivation ascribes a first-order reaction to the process; however, it has been difficult to consistently achieve simple exponential inactivation with spore preparations. These variations manifest themselves as a shoulder on the decay curve, which has been attributed to activation of spores, and subsequent differences in the heat resistance of dormant and activated spores.[125] Consistent populations of activated spores are difficult to obtain; thus the shoulder is often ignored, and D-values are calculated from the linear portion of the decay curve.

More sophisticated models have been developed to account for the nonlinear aspects of survival curves. These include terms describing the germination of spores prior to inactivation (for descriptions of earlier models, see).[124–126] Figure 2.9 indi-cates the process of activation of dormant spores (N_1) into activated (N_2) spores with rate constant of k_a. The activated spores are then inactivated by heat treatment (N_3) at a rate equal to k_{d2}. The model also allows for possible inactivation of dormant spores (N_3) at a rate equal to k_{d1}. All reactions are considered to be independent first-order. The simplest form of this model was described by Shull et al.,[127] and assumes that only activated spores can be killed ($k_{d1} = 0$) and thus:

$$\frac{dN_1}{dt} = -k_a N_1 \qquad (2.35)$$

$$\frac{dN_2}{dt} = k_a N_1 - k_{d2} N_2 \qquad (2.36)$$

The model proposed by Rodriguez et al.[128,129] advances the Shull model by assuming that the dormant spores can also be inactivated:

$$\frac{dN_1}{dt} = -(k_{d2} + k_a)N_1 \tag{2.37}$$

and $k_{d2} = k_{d1}$. Sapru et al.[124] further extended the model to include a different rate of inactivation for dormant spores ($k_{d1} \neq k_{d2}$). The Sapru model is more general and includes the Shull and Rodriguez models as special cases. This model was proposed for use with *Bacillus stereothermophilus* at sterilization temperatures, and an explicit form has been presented[125]:

$$N_1(t) = N_1(0)e^{-(k_a+k_{d1})t} \tag{2.38}$$

$$N_2(t) = N_2(0)e^{-k_{d2}t} + B \cdot N_1(0)(1 - e^{-At})e^{-k_{d2}t} \tag{2.39}$$

with

$$B = \frac{k}{A} \tag{2.40}$$

$$A = k_a + k_{d1} - k_{d2} \tag{2.41}$$

and where $N_1(0)$ and $N_2(0)$ are the number of dormant and activated cells, respectively, at $t = 0$. An example output from the Sapru model is shown in Figure 2.10. With $N_1(0)$ at 1×10^8 and $N_2(0)$ at 1×10^5, the initial rapid increase in surviving cells is the result of spore activation. This is followed by an exponential decrease

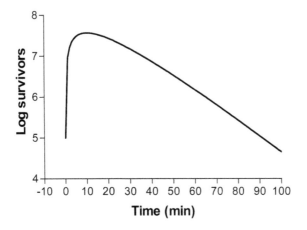

FIGURE 2.10 Output of model for spore activation and survival. (From McKellar, R.C., Modelling the effectiveness of pasteurization, in *Dairy Processing: Maximizing Quality*, Smit, G., Ed., CRC Press Inc./Woodhead Publishing, 2003 pp. 104–129. With permission.)

in surviving cells. This model has been expanded further to include subpopulations of spores having different heat resistances.[130]

2.2.6 Processing Models

2.2.6.1 Thermal

Thermal inactivation of microorganisms in static or batch systems is usually described by the D- and z-value concepts as discussed above, with temperature generally held constant. The situation in canning operations or continuous flow systems such as HTST pasteurization, sterilization, and ultrahigh temperature processes is somewhat more complex, due to nonisothermal conditions. In addition, the kinetics of inactivation in continuous systems differs from that in batch systems since in these systems there are additional factors such as pressure and shear forces that can influence microbial survival.[131] As most modern processes are continuous, it is necessary to have information on survival of microorganisms; however, few studies have been published in which laboratory or pilot plant continuous flow systems have been studied.[131]

In order to deal with nonisothermal conditions, Bigelow's[132] model has been the standard for the low-acid canned food industry for many decades. In this approach, the processing time F is determined by integrating the exposure time at various temperatures, $T[t]$, to time at a reference temperature, T_{Ref}[133]:

$$F = \int 10^{\frac{(T(t)-T_{\text{Ref}})}{z}}\, dt \qquad (2.42)$$

This model is considered to be an approximation of the Arrhenius model, which is valid over a wide range (4 to 160°C) of temperatures[133]:

$$PE = \frac{1}{t_0}\int_0^t e^{-\left(\frac{E_a}{R}\right)\left(\frac{1}{T}-\frac{1}{T_0}\right)}\, dt \qquad (2.43)$$

where PE = integrated lethal effect, or pasteurization effect; E_a = energy of activation, J mol^{-1}; $R = 8.314$ J mol^{-1} K^{-1}; T = temperature, K; T_0 = reference temperature, 345 K; t = time, s; t_0 = reference time, 15 s. The reference temperature (345 K or 72°C) and time (15 s) correspond to the International Dairy Federation standard for pasteurization.[134]

It is often necessary for food processors to demonstrate that the process they wish to use is effective in delivering the required lethal effect for the product and microorganism of concern. The integrated lethal effect is a useful concept, as it allows two or more processes that use different time/temperature combinations to be compared for efficacy against food-borne pathogens; however, there are few data available relating microbial survival to processing conditions. This is of particular concern in the case of pasteurization of milk, where the only accepted test for proper pasteurization is the alkaline phosphatase (AP) test. The relationship between AP inactivation and survival of food-borne pathogens is largely unknown, as is the response of AP to processing under alternative time/temperature combinations. Thus, modeling of HTST pasteurization of milk was studied extensively by McKellar and coworkers.

The residence times in each section of a pilot-scale HTST pasteurizer and in each of six holding tubes with nominal holding times of 3 to 60 s were calibrated using the standard salt test. Temperatures were taken at the beginning and end of each section using thermocouples. The PE could then be determined for each selected holding time/temperature combination using Equation 2.43. Raw milk at a constant flow rate was allowed to equilibrate at each time/temperature, and a sample was taken at the outflow for analysis. Residual enzyme activity or microbial survivors were matched with the corresponding PE for fitting.[97]

The fitting was accomplished using an iterative procedure in which the $\log_{10}\%$ initial activity or viable counts were regressed on PE, with the value of E_a/R varied to minimize the error sum of squares. Nonlinearity (generally concavity) in the data was modeled using a power transformation (Equation 2.25). The final model was of the form[97]:

$$\log_{10} \% \text{ initial activity} = a + b \cdot PE^c \tag{2.44}$$

where a = intercept, b = slope, and c = power. Generally, the parameter estimates for at least three trials were pooled, and the model for AP is shown in Table 2.5.

There is also a need to develop models for milk enzymes that might be used to confirm processing at temperatures above or below pasteurization. Higher temperatures ($\geq 75°C$) are appropriate for processing of more viscous products (such as ice-cream mix), while temperatures below pasteurization (63 to 65°C; termed subpasteurization or thermization temperatures) are used to extend the storage life of bulk

---

TABLE 2.5
Model Parameters for Inactivation of Various Milk Enzymes and Food-Borne Pathogens during High-Temperature Short-Time Pasteurization

Target	Trials	Intercept	Slope	Power	E_a/R (×1000)
Alkaline phosphatase	3	2.05	−4.05	0.50	66.5
γ-Glutamyl transpeptidase	3	2.00	−0.281	0.75	66.5
Lactoperoxidase	3	2.12	−0.10	0.75	59.0
Catalase	3	1.94	−2.65	0.50	82.0
α-L-Fucosidase	3	1.87	−17.6	1.00	39.8
Listeria innocua	5	1.86	−24.9	0.80	59.5
Listeria monocytogenes	3	1.68	−18.4	0.80	48.5
Enterobacter sakazakii	3	2.31	−24.4	0.65	59.5

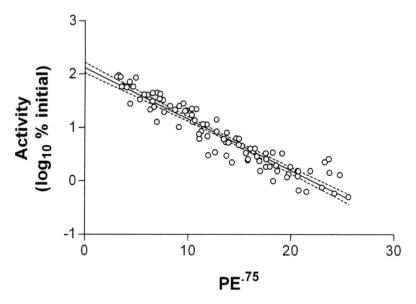

FIGURE 2.11 Linear model relating pasteurization effect (PE) and residual activity of lactoperoxidase during high-temperature short-time (HTST) pasteurization of bovine milk.

milk. Lactoperoxidase (LP) and γ-glutamyl transpeptidase (TP) are two naturally occurring milk enzymes that are inactivated at higher temperatures.[135] Model parameters for these two enzymes are given in Table 2.5. An example of an inactivation curve for LP is given in Figure 2.11, with the dotted lines representing the 95% confidence limits. There is close agreement among the three trials plotted, a characteristic common for all enzyme models. Models have also been developed for catalase[104] and α-L-fucosidase (FC),[136] which are appropriate for subpasteurization temperatures (Table 2.5).

Survival models for several food-borne pathogens have also been derived. *Listeria innocua*, a nonpathogen, is often used as a substitute for *L. monocytogenes* in situations (such as food processing environments) where it would be undesirable to introduce pathogens.[137] A model developed for *L. innocua* (Table 2.5) was shown to underpredict inactivation of *L. monocytogenes*; thus predictions are "fail-safe."[138] *Enterococcus faecium*, a nonpathogen, is also used as a model organism for pathogens, particularly in Europe.[139] The inactivation curve for this microorganism deviated strongly from linearity, and there were large intertrial variations. Thus, a random coefficient model using Equation 2.28 was used to fit the data.[96] The average ln D-values for the two populations were 0.825 and 2.856. Models were also generated for *Enterobacter sakazakii*, an "emerging" pathogen found contaminating infant formula.[140] Model parameters compared with those for *L. monocytogenes* (Table 2.5) showed that *E. sakazakii* was more sensitive to pasteurization.

Linear models for milk enzymes were characterized by limited intertrial variability (Figure 2.11). This allowed validation of models using data from other trials that were not used in the construction of the models. In contrast, considerable variation was noted in experiments with microorganisms; thus a different approach

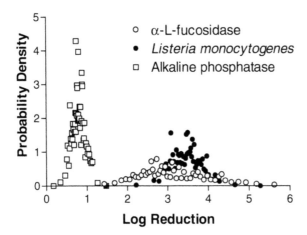

FIGURE 2.12 Probability densities for α-L-fucosidase, *Listeria monocytogenes,* and alkaline phosphatase generated from linear models for high-temperature short-time (HTST) pasteurization using Analytica®, with holding temperature and time of 66°C for 16 s.

was taken. Model parameters were incorporated into risk analysis software (@RISK, Palisade Corporation, Newfield, NY) as normal distributions, with means taken from Table 2.5 and standard deviations taken from the intertrial variations. When simulations were performed (1500 iterations), outcomes (log reduction in this case) were expressed as distributions.

Simulated log reductions were generated for AP, FC, and *L. monocytogenes* using a holding time of 65°C/15 s (thermization), and the probability density functions are shown in Figure 2.12. These conditions resulted in a narrow band of probabilities for AP, with greater predicted range for both FC and *L. monocytogenes.* AP is not completely inactivated, while FC (a potential indicator of thermization) experiences a >2 log reduction in most iterations. The mean log reduction of *L. monocytogenes* under these conditions is >3.

Models that can predict the probability of achieving a desired level of safety are an important addition to risk assessment models, which are still largely qualitative and based primarily on expert opinion (see Chapter 6 for a more complete discussion on expert systems). The pasteurization models described above have been incorporated into the risk analysis software Analytica® (Lumina Decision Systems, Los Gatos, CA), a commonly used software for building risk assessment models for the food industry. These models are now being incorporated into the USDA's Pathogen Modeling Program (available from http://www.arserrc.gov/mfs/pathogen.htm).

2.2.6.2 Alternative Technologies

Thermal treatment has been the traditional method for processing of many foods; however, with the increased consumer demand for fresh, less processed foods, new technologies have evolved. Some of these are based on temperature, such as microwave, radio frequency (RF), and ohmic heating, while others depend on other forms of microbial inactivation, such as high pressure (HP), pulse electric field (PEF),

pulsed or ultraviolet light, and ultrasound. In 1998, the U.S. Food and Drug Administration commissioned the Institute of Food Technologists to provide scientific
review and analysis of issues in food safety, food processing, and human health.
The first of these reports, entitled "Kinetics of Microbial Inactivation for Alternative
Food Processing Technologies," was released in 2000[141] and is available at the
following web site: http://vm.cfsan.fda.gov/~comm/ift-toc.html. Since this report
comprehensively reviews the scientific literature and makes recommendations for
future research, it is beyond the scope of this chapter to reproduce this body of work.
Instead, several key areas will be highlighted.

Many novel thermal technologies base their antimicrobial effect on temperature;
thus inactivation of microorganisms can be modeled using the traditional calculations
for D-value and z-value (see earlier). Processes that depend on other mechanisms
of inactivation such as HP and PEF require modified equations with different parameters. For example, HP effects on microbial population can be modeled using a
function similar to the traditional D-value[142]:

$$\log\left(\frac{D}{D_R}\right) = -\frac{(P-P_R)}{z_R} \tag{2.45}$$

where D_R = the decimal reduction time at a reference pressure P_R and z_R is the
pressure required for a 1-log reduction in D-value. An alternative model has been
proposed by Weemaes et al.[143]:

$$\ln(k) = \ln(k_R) - \left(\frac{V(P-P_R)}{RT_A}\right) \tag{2.46}$$

where k_R is the reaction rate constant, and P_R the reference pressure, V the activation
volume constant, P the pressure, and T_A the absolute temperature. With PEF processing, a model describing the influence of the electric field intensity on reduction
of microbial population can be described, which is similar to those used for thermal
and pressure processing:

$$\log\left(\frac{D}{D_R}\right) = -\frac{(E-E_R)}{z_E} \tag{2.47}$$

where D_R is the decimal reduction time at a reference field intensity E_R, and the
electric field coefficient z_E is the increase in the electric field intensity E required to
reduce the D-value by 1-log. An alternative model based on the Fermi equation was
proposed by Peleg[144]:

$$\frac{N}{N_0} = \frac{1}{1+e^{\frac{E-E_d}{K}}} \tag{2.48}$$

where E_d is the electric field intensity when the microbial population has been reduced to 50%, and K is a coefficient based on the slope of the survivor curve. A similar model was proposed by Hülsheger et al.[145]:

$$\frac{N}{N_0} = \left(\frac{t}{t_c}\right)^{-\frac{E-E_c}{K}} \tag{2.49}$$

where t is the treatment time, t_c the minimum treatment time for inactivation, E_c the minimum field strength for inactivation, and K is a specific rate constant. This function is similar to Equation 2.48 except that it also accounts for exposure time at a given electric field intensity.

The Institute of Food Technologies report has raised a number of relevant issues that would benefit from some discussion here. Kinetic parameters for microbial populations exposed to thermal treatment are well documented and provide a good basis upon which to develop models for alternative thermal processes. The nonthermal models described above assume that microbial inactivation is a first-order reaction; however, as mentioned earlier, there is little direct evidence supporting this view. It will be necessary to further evaluate the adequacy of linear survival models, and to hopefully develop a universal model applicable to both thermal and nonthermal processes. In addition, experimental protocols have been found to be inadequate to provide statistically reliable parameters for microbial reduction resulting from exposure to alternative technologies. This is particularly a problem with high pressure processing, where data are needed at different pressures with control of temperature and product. The inactivation mechanism for thermal destruction of microbes is generally well known, and evidence for additional independent mechanisms with processes such as ohmic heating is still lacking. Further work is needed to elucidate the mechanism of inhibition with alternative treatments such as PEF and HP, and to assess possible synergistic effects between alternative technologies and temperature.

2.2.7 INJURY/REPAIR MODELS

Almost without exception, available models for microbial growth and death have been developed using fully viable, unstressed cells; thus the resulting models represent the idealized scenario. It is well known that bacterial cells exposed to some form of sublethal stress require an adaptation or recovery period prior to growth; however, mathematical models do not incorporate the influence of stress. This was emphasized in a study designed to model the evolution of a log phase in *L. monocytogenes,* induced by acid, alkaline, and osmotic shocks.[146] When lag-phase cells (which are more sensitive to environmental stress than stationary-phase cells) were exposed to changes in pH or increased levels of NaCl, the subsequent generation times predicted by commercially available software were shorter than the observed experimental generation times.

The physiological events that account for microbial injury and repair are poorly understood; thus there have been very few attempts to apply mathematical models

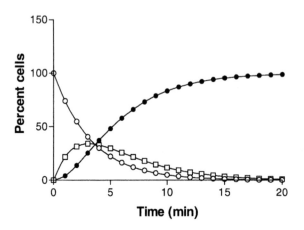

FIGURE 2.13 Change in populations of uninjured (○), injured (□), and dead (●) cells during sublethal heating according to Equation 2.50.

to the phenomena of bacterial cell injury and resuscitation. The models that do exist are of two general types: those that aim to quantitate the extent of injury with increased exposure to stress, and those that attempt to predict the time required for repair and recovery of viability.

Several attempts have been made to model the extension of the lag phase in response to stress. A model to describe the relationship between lag prior to growth and stress duration was proposed by Bréand et al.[147] This model was developed to reflect the observation that the lag increased with increasing stress duration, and then decreased to a minimum lag at longer stress times. The empirical model described the influence of stress on the lag with a linear function, followed by a logistic decrease. Cheroutre-Vialette and Lebert[148] proposed the use of a recurrent neural network to model the changes in lag phase and growth rate experienced by *L. monocytogenes* exposed to osmotic and pH shock. Lambert and van der Ouderaa[149] compared the relative ability of the Bioscreen (see Chapter 1) and viable counts to quantitate the inactivation of microorganisms by disinfection. They proposed a simple first-order inactivation reaction with accumulation of injured cells prior to complete loss of viability:

$$A_1 \xrightarrow{k_1} A_2 \xrightarrow{k_2} P \qquad\qquad (2.50)$$

where A_1 are the uninjured cells, A_2 the injured cells, and P are the dead cells. k_1 and k_2 are rate constants for injury and death, respectively. Populations of viable, injured, and dead cells were simulated based on the data of Lambert and van der Ouderaa,[149] and are shown in Figure 2.13. These responses were confirmed using image analysis of colony sizes on agar plates; viable and injured cells could be distinguished on the basis of size. Colony size was also used to quantitate cells of *L. monocytogenes* that had been injured by exposure to heat or starvation.[150] The colony size distribution was normal for uninjured cells, but demonstrated a right-hand skew with injured cells. Percent sublethal injury could be related to colony area using a linear function. More recent studies using the flow cytometer to measure the distribution of the lag times of individual cells of *Lactobacillus plantarum* also showed a deviation from normality with heat-treated cells. The extreme value distribution was found to be the best function for fitting both injured and uninjured cells:

$$F(x) = 1 - \exp\left(-\exp\left[\frac{x-a}{b}\right]\right) \tag{2.51}$$

where a and b are unknown parameters.

There are few studies that aim to model the recovery of cells from injury. Injured cells can be differentiated on the basis of increased sensitivity to selective media (e.g., 5% NaCl), and it is thus possible to develop models to predict the time required for cells to repair damage due to stress. This process is complicated by our lack of information on the true nature of injury in bacterial cells, and the mechanism by which cells recover. The two-compartment kinetic model developed by Hills and Mackey to describe bacterial growth[4] was revised and extended to account for cell injury and resuscitation.[5] In the revised model, there are rate constants for injury (R_i) and resuscitation (R_r), and parameters to describe the decrease (a) and increase (b) of the injury and resuscitation curves.[5] This model was used to fit data from the resuscitation of *L. monocytogenes* after exposure to sublethal heat.[151] It was shown that resuscitation could best be described with a reduced model with the parameter for increasing rate of recovery (b_r) eliminated. A quadratic regression model was subsequently derived that expressed the lag as a function of temperature and the initial number of injured cells.[151]

2.2.8 COMBINED GROWTH/DEATH MODELS

There have been a limited number of attempts to combine growth and death functions into single models. These are often simply combinations of functions such as the Gompertz or logistic with their mirror images. For example, a two-term model describing the behavior of *Lactobacillus* spp. during the ripening of fermented sausage incorporated a Gompertz function for both growth and death.[152] In a similar fashion, the logistic function and its mirror image, the Fermi equation, have been combined.[21,101] The latter model has been expanded to include a proposed distribution of cell resistances to stress, resulting in a death model that varies in shape.[21] The Baranyi model for growth was also combined with its mirror image to describe growth and death for *Brochothrix thermosphacta*.[115] In this model, a smoothing function was included to account for the transition between growth and death phases.

Other combined functions have used simple exponential growth and decline.[68,153,154] In one of these models,[68] the lag phase preceding growth was handled by a first-order step that represented spore germination, repair, or adaptation. Jones et al.[153] described the adaptation of cells to growth as a transition between cells in two states, immature and mature. This model reduced to a simple balance between growth and death, with variations in division and mortality rates being described by empirical functions.[153]

It is questionable if expressing death as a mirror image of growth is valid. There is little direct evidence that the lag phases preceding growth or death are due to similar physiological phenomena, although a convincing theoretical argument has been offered in support of this hypothesis.[65] It seems likely, however, that the stationary phase of growth and the "tailing" phase of inactivation are the result of different physiological processes. Models that address growth and death as different processes, and attempt to describe the response of microbes to their environment in terms of transitions between states, would seem to be the most useful for future development.[68,153]

REFERENCES

1. McMeekin, T.A., Olley, J.N., Ross, T., and Ratkowsky, D.A., *Predictive Microbiology: Theory and Application,* John Wiley & Sons, New York, 1993.
2. Baranyi, J., Roberts, T.A., and McClure, P., A non-autonomous differential equation to model bacterial growth, *Food Microbiol.,* 10, 43, 1993.
3. Baranyi, J. and Roberts, T.A., A dynamic approach to predicting bacterial growth in food, *Int. J. Food Microbiol.,* 23, 277, 1994.
4. Hills, B.P. and Wright, K.M., A new model for bacterial growth in heterogeneous systems, *J. Theor. Biol.,* 168, 31, 1994.
5. Hills, B.P. and Mackey, B.M., Multi-compartment kinetic models for injury, resuscitation, induced lag and growth in bacterial cell populations, *Food Microbiol.,* 12, 333, 1995.
6. Buchanan, R.L., Whiting, R.C., and Damert, W.C., When is simple good enough: a comparison of the Gompertz, Baranyi, and three-phase linear models for fitting bacterial growth curves, *Food Microbiol.,* 14, 313, 1997.
7. McKellar, R.C., A heterogeneous population model for the analysis of bacterial growth kinetics, *Int. J. Food Microbiol.,* 36, 179, 1997.
8. Shimoni, E. and Labuza, T.P., Modeling pathogen growth in meat products: future challenges, *Trends Food Sci. Technol.,* 11, 394, 2000.
9. Skinner, G.E., Larkin, J.W., and Rhodehamel, E.J., Mathematical modeling of microbial growth: a review, *J. Food Saf.,* 14, 175, 1994.
10. Baranyi, J. and Roberts, T.A., Mathematics of predictive food microbiology, *Int. J. Food Microbiol.,* 26, 199, 1995.
11. McDonald, K. and Sun, D.W., Predictive food microbiology for the meat industry: a review, *Int. J. Food Microbiol.,* 52, 1, 1999.
12. Gibson, A.M., Bratchell, N., and Roberts, T.A., The effect of sodium chloride and temperature on the rate and extent of growth of *Clostridium botulinum* type A in pasteurized pork slurry, *J. Appl. Bacteriol.,* 62, 479, 1987.
13. Causton, D.R., *A Biologist's Mathematics,* Edward Arnold, London, 1977.
14. Jason, A.C., A deterministic model for monophasic growth of batch cultures of bacteria, *Antonie Van Leeuwenhoek,* 49, 523, 1983.
15. Gompertz, B., On the nature of the function expressive of the law of human mortality, and on a new mode of determining the value of life contingencies, *Philos. Trans. R. Soc. Lond.,* 115, 513, 1825.
16. Buchanan, R.L., Stahl, H.G., and Whiting, R.C., Effects and interactions of temperature, pH, atmosphere, sodium chloride, and sodium nitrite on the growth of *Listeria monocytogenes, J. Food Prot.,* 52, 844, 1989.

17. Baker, D.A. and Genigeorgis, C. Predictive modeling, in *Clostridium botulinum: Ecology and Control in Foods,* Hauschild, A.H.W. and Dodds, K.L., Eds., Marcel Dekker, New York, 1993, pp. 343–412.

18. Zwietering, M.H., Jongenburger, I., Rombouts, F.M., and van't Riet, K., Modelling of the bacterial growth curve, *Appl. Environ. Microbiol.,* 56, 1875, 1990.

19. Willox, F., Mercier, M., Hendrickx, M., and Tobback, P., Modelling the influence of temperature and carbon dioxide upon the growth of *Pseudomonas fluorescens, Food Microbiol.,* 10, 159, 1993.

20. Kochevar, S.L., Sofos, J.N., Bolin, R.R., Reagan, J.O., and Smith, G.C., Steam vacuuming as a pre-evisceration intervention to decontaminate beef carcasses, *J. Food Prot.,* 60, 107, 1997.

21. Peleg, M., Modeling microbial populations with the original and modified versions of the continuous and discrete logistic equations, *CRC Crit. Rev. Food Sci. Nutr.,* 37, 471, 1997.

22. Dalgaard, P., Mejlholm, O., and Huss, H.H., Application of an iterative approach for development of a microbial model predicting the shelf-life of packed fish, *Int. J. Food Microbiol.,* 38, 169, 1997.

23. Koutsoumanis, K.P., Taoukis, P.S., Drosinos, E.H., and Nychas, G.J.E., Applicability of an Arrhenius model for the combined effect of temperature and CO_2 packaging on the spoilage microflora of fish, *Appl. Environ. Microbiol.,* 66, 3528, 2000.

24. Koutsoumanis, K. and Nychas, G.J.E., Application of a systematic experimental procedure to develop a microbial model for rapid fish shelf life predictions, *Int. J. Food Microbiol.,* 60, 171, 2000.

25. Membre, J.M. and Kubaczka, M., Predictive modelling approach applied to spoilage fungi: growth of *Penicillium brevicompactum* on solid media, *Lett. Appl. Microbiol.,* 31, 247, 2000.

26. Augustin, J.C., Brouillaud-Delattre, A., Rosso, L., and Carlier, V., Significance of inoculum size in the lag time of *Listeria monocytogenes, Appl. Environ. Microbiol.,* 66, 1706, 2000.

27. Sutherland, J.P. and Bayliss, A.J., Predictive modelling of growth of *Yersinia enterocolitica:* the effects of temperature, pH and sodium chloride, *Int. J. Food Microbiol.,* 21, 197, 1994.

28. Sutherland, J.P., Bayliss, A.J., and Roberts, T.A., Predictive modelling of growth of *Staphylococcus aureus:* the effects of temperature, pH and sodium chloride, *Int. J. Food Microbiol.,* 21, 217, 1994.

29. Eifert, J.D., Hackney, C.R., Pierson, M.D., Duncan, S.E., and Eigel, W.N., Acetic, lactic, and hydrochloric acid effects on *Staphylococcus aureus* 196E growth based on a predictive model, *J. Food Sci.,* 62, 174, 1997.

30. Murphy, P.M., Rea, M.C., and Harrington, D., Development of a predictive model for growth of *Listeria monocytogenes* in a skim milk medium and validation studies in a range of dairy products, *J. Appl. Bacteriol.,* 80, 557, 1996.

31. Miles, D.W., Ross, T., Olley, J., and McMeekin, T.A., Development and evaluation of a predictive model for the effect of temperature and water activity on the growth rate of *Vibrio parahaemolyticus, Int. J. Food Microbiol.,* 38, 133, 1997.

32. Chorin, E., Thuault, D., Cleret, J.J., and Bourgeois, C.M., Modelling *Bacillus cereus* growth, *Int. J. Food Microbiol.,* 38, 229, 1997.

33. McElroy, D.M., Jaykus, L.A., and Foegeding, P.M., Validation and analysis of modeled predictions of growth of *Bacillus cereus* spores in boiled rice, *J. Food Prot.,* 63, 268, 2000.

34. Zhao, L., Montville, T.J., and Schaffner, D.W. Inoculum size of *Clostridium botulinum* 56A spores influences time-to-detection and percent growth-positive samples, *J. Food Sci.*, 65, 1369, 2000.

35. Buchanan, R.L. and Bagi, L.K., Microbial competition: effect of *Pseudomonas fluorescens* on the growth of *Listeria monocytogenes, Food Microbiol.*, 16, 523, 1999.

36. Lebert, I., Robles-Olvera, V., and Lebert, A., Application of polynomial models to predict growth of mixed cultures of *Pseudomonas* spp. and *Listeria* in meat, *Int. J. Food Microbiol.*, 61, 27, 2000.

37. Wijtzes, T., Rombouts, F.M., Kant-Muermans, M.L.T., van't Riet, K., and Zwietering, M.H., Development and validation of a combined temperature, water activity, pH model for bacterial growth rate of *Lactobacillus curvatus, Int. J. Food Microbiol.*, 63, 57, 2001.

38. Riva, M., Franzetti, L., and Galli, A., Microbiological quality and shelf life modeling of ready-to-eat cicorino, *J. Food Prot.*, 64, 228, 2001.

39. Membre, J.M. and Tholozan, J.L., Modeling growth and off-flavours production of spoiled beer bacteria, *Pectinatus frisingensis, J. Appl. Microbiol.*, 77, 456. 1994.

40. Devlieghere, F., Van Belle, B., and Debevere, J., Shelf life of modified atmosphere packed cooked meat products: a predictive model, *Int. J. Food Microbiol.*, 46, 57, 1999.

41. Juneja, V.K. and Marks, H.M., Proteolytic *Clostridium botulinum* growth at 12–48°C simulating the cooling of cooked meat: development of a predictive model, *Food Microbiol.*, 16, 583, 1999.

42. Garthright, W.E., Refinements in the prediction of microbial growth curves, *Food Microbiol.*, 8, 239, 1991.

43. Baranyi, J., McClure, P.J., Sutherland, J.P., and Roberts, T.A., Modeling bacterial growth responses, *J. Ind. Microbiol.*, 12, 190, 1993.

44. Garthright, W.E., The three-phase linear model of bacterial growth: a response, *Food Microbiol.*, 14, 193, 1997.

45. Baranyi, J., Stochastic modelling of bacterial lag phase, *Int. J. Food Microbiol.*, 73, 203, 2002.

46. McClure, P.J., Blackburn, C.D., Cole, M.B., Curtis, P.S., Jones, J.E., Legan, J.D., Ogden, I.D., Peck, M.W., Roberts, T.A., Sutherland, J.P., Walker, S.J., and Blackburn, C.D.W., Modelling the growth, survival and death of microorganisms in foods: the UK food micromodel approach, *Int. J. Food Microbiol.*, 23, 265, 1994.

47. McClure, P.J., Beaumont, A.L., Sutherland, J.P., and Roberts, T.A., Predictive modelling of growth of *Listeria monocytogenes:* the effects on growth of NaCl, pH, storage temperature and NaNO$_2$, *Int. J. Food Microbiol.*, 34, 221, 1997.

48. Fernandez, P.S., George, S.M., Sills, C.C., and Peck, M.W., Predictive model of the effect of CO$_2$, pH, temperature and NaCl on the growth of *Listeria monocytogenes, Int. J. Food Microbiol.*, 37, 37, 1997.

49. Sutherland, J.P., Aherne, A., and Beaumont, A.L., Preparation and validation of a growth model for *Bacillus cereus*: the effects of temperature, pH, sodium chloride, and carbon dioxide, *Int. J. Food Microbiol.*, 30, 359, 1996.

50. Sutherland, J.P., Bayliss, A.J., Braxton, D.S., and Beaumont, A.L., Predictive modelling of *Escherichia coli* O157:H7: inclusion of carbon dioxide as a fourth factor in a pre-existing model, *Int. J. Food Microbiol.*, 37, 113, 1997.

51. Pin, C., Baranyi, J., and deFernando, G., Predictive model for the growth of *Yersinia enterocolitica* under modified atmospheres, *J. Appl. Microbiol.*, 88, 521, 2000.

52. Valik, L. and Pieckova, E., Growth modelling of heat-resistant fungi: the effect of water activity, *Int. J. Food Microbiol.*, 63, 11, 2001.

53. Garcia-Gimeno, R.M., Castillejo-Rodriguez, A.M., Barco-Alcala, E., and Zurera-Cosano, G., Determination of packaged green asparagus shelf-life, *Food Microbiol.*, 15, 191, 1998.

54. Garcia-Gimeno, R.M. and Zurera-Cosano, G., Determination of ready-to-eat vegetable salad shelf-life, *Int. J. Food Microbiol.*, 36, 31, 1997.

55. Bernaerts, K., Versyck, K.J., and Van Impe, J.F. On the design of optimal dynamic experiments for parameter estimation of a Ratkowsky-type growth kinetics at suboptimal temperatures, *Int. J. Food Microbiol.*, 54, 27, 2000.

56. Alavi, S.H., Puri, V.M., Knabel, S.J., Mohtar, R.H., and Whiting, R.C., Development and validation of a dynamic growth model for *Listeria monocytogenes* in fluid whole milk, *J. Food Prot.*, 62, 170, 1999.

57. Alavi, S.H., Puri, V.M., and Mohtar, R.H., A model for predicting the growth of *Listeria monocytogenes* in packaged whole milk, *J. Food Process Eng.*, 24, 231, 2001.

58. Bovill, R., Bew, J., Cook, N., D'Agostino, M., Wilkinson, N., and Baranyi, J., Predictions of growth for *Listeria monocytogenes* and *Salmonella* during fluctuating temperature, *Int. J. Food Microbiol.*, 59, 157, 2000.

59. Bovill, R.A., Bew, J., and Baranyi, J., Measurements and predictions of growth for *Listeria monocytogenes* and *Salmonella* during fluctuating temperature. II. Rapidly changing temperatures, *Int. J. Food Microbiol.*, 67, 131, 2001.

60. Wright, K.M., Coleman, H.P., Mackle, A.R., Parker, M.L., Brocklehurst, T.F., Wilson, D.R., and Hills, B.P., Determination of mean growth parameters of bacterial colonies immobilized in gelatin gel using a laser gel-cassette scanner, *Int. J. Food Microbiol.*, 57, 75, 2000.

61. Oscar, T.P., Response surface models for effects of temperature, pH, and previous growth pH on growth kinetics of *Salmonella typhimurium* in brain–heart infusion broth, *J. Food Prot.*, 62, 106, 1999.

62. Oscar, T.P., Response surface models for effects of temperature and previous growth sodium chloride on growth kinetics of *Salmonella typhimurium* on cooked chicken breast, *J. Food Prot.*, 62, 1470, 1999.

63. Oscar, T.P., Response surface models for effects of temperature and previous temperature on lag time and specific growth rate of *Salmonella typhimurium* on cooked ground chicken breast, *J. Food Prot.*, 62, 1111, 1999.

64. Cornu, M., Delignettemuller, M.L., and Flandrois, J.P., Characterization of unexpected growth of *Escherichia coli* O157:H7 by modeling, *Appl. Environ. Microbiol.*, 65, 5322, 1999.

65. Baranyi, J. and Pin, C., A parallel study on bacterial growth and inactivation, *J. Theor. Biol.*, 210, 327, 2001.

66. McKellar, R.C. and Knight, K.P., A combined discrete–continuous model describing the lag phase of *Listeria monocytogenes, Int. J. Food Microbiol.*, 54, 171, 2000.

67. McKellar, R.C., Development of a dynamic continuous–discrete–continuous model describing the lag phase of individual bacterial cells. *J. Appl. Microbiol.*, 90, 407, 2001.

68. Whiting, R.C. and Cygnarowicz-Provost, M., A quantitative model for bacterial growth and decline, *Food Microbiol.*, 9, 269, 1992.

69. Jones, J.E. and Walker, S.J., Advances in modeling microbial growth, *J. Ind. Microbiol.*, 12, 200, 1993.

70. Van Impe, J.F., Nicolai, B.M., Martens, T., Baerdemaeker, J., and Vandewalle, J., Dynamic mathematical model to predict microbial growth and inactivation during food processing, *Appl. Environ. Microbiol.*, 58, 2901, 1992.

71. Diaz, C., Lelong, P., Dieu, P., Feuillerat, C., and Salome, M., On-line analysis and modeling of microbial growth using a hybrid system approach, *Process. Biochem.*, 34, 39, 1999.

72. Baranyi, J., Simple is good as long as it is enough, *Food Microbiol.*, 14, 189, 1997.

73. Dalgaard, P., Modelling of microbial activity and prediction of shelf life for packed fresh fish, *Int. J. Food Microbiol.*, 26, 305, 1995.

74. Graham, A.F., Mason, D.R., and Peck, M.W., A predictive model of the effect of temperature, pH and sodium chloride on growth from spores of non-proteolytic *Clostridium botulinum*, *Int. J. Food Microbiol.*, 31, 69, 1996.

75. Gardini, F., Lanciotti, R., Sinigaglia, M., and Guerzoni, M.E., A head space gas chromatographic approach for the monitoring of the microbial cell activity and the cell viability evaluation, *J. Microbiol. Methods*, 29, 103, 1997.

76. Membre, J.M., Ross, T., and McMeekin, T., Behaviour of *Listeria monocytogenes* under combined chilling processes, *Lett. Appl. Microbiol.*, 28, 216, 1999.

77. Buchanan, R., The three-phase linear model of bacterial growth: response, *Food Microbiol.*, 14, 399, 1997.

78. Hersom, A.C. and Hulland, E.D., Principles of thermal processing, in *Canned Foods. Thermal Processing and Microbiology*, Hersom, A.C. and Hulland, E.D., Eds., Churchill Livingstone, New York, 1980, pp. 177–207.

79. Jay, J.M., *Modern Food Microbiology*, 4th ed., Van Nostrand Reinhold, New York, 1992.

80. Doyle, M.E., Mazzotta, A.S., Wang, T., Wiseman, D.W., and Scott, V.N., Heat resistance of *Listeria monocytogenes*, *J. Food Prot.*, 64, 410, 2001.

81. Doyle, M.E. and Mazzotta, A.S., Review of studies on the thermal resistance of salmonellae, *J. Food Prot.*, 63, 779, 2000.

82. Stringer, S.C., George, S.M., and Peck, M.W., Thermal inactivation of *Escherichia coli* O157:H7, *J. Appl. Microbiol.*, 88, 79S, 2000.

83. Moats, W.A., Dabbah, R., and Edwards, V.M., Interpretation of nonlogarithmic survivor curves of heated bacteria, *J. Food Sci.*, 36, 523, 1971.

84. Campanella, O.H. and Peleg, M., Theoretical comparison of a new and the traditional method to calculate *Clostridium botulinum* survival during thermal inactivation, *J. Sci. Food Agric.*, 81, 1069, 2001.

85. Cerf, O., Tailing of survival curves of bacterial spores, *J. Appl. Bacteriol.*, 42, 1, 1977.

86. Augustin, J.C., Carlier, V., and Rozier, J., Mathematical modelling of the heat resistance of *Listeria monocytogenes*, *J. Appl. Microbiol.*, 84, 185, 1998.

87. Shadbolt, C.T., Ross, T., and McMeekin, T.A., Nonthermal death of *Escherichia coli*, *Int. J. Food Microbiol.*, 49, 129, 1999.

88. Whiting, R.C., Modeling bacterial survival in unfavorable environments, *J. Ind. Microbiol.*, 12, 240, 1993.

89. Buchanan, R.L. and Golden, M.H., Interaction of citric acid concentration and pH on the kinetics of *Listeria monocytogenes* inactivation, *J. Food Prot.*, 57, 567, 1994.

90. Buchanan, R.L., Golden, M.H., Whiting, R.C., Phillips, J.G., and Smith, J.L., Nonthermal inactivation models for *Listeria monocytogenes*, *J. Food Sci.*, 59, 179, 1994.

91. Buchanan, R.L., Golden, M.H., and Phillips, J.G., Expanded models for the nonthermal inactivation of *Listeria monocytogenes*, *J. Appl. Microbiol.*, 82, 567, 1997.

92. Buchanan, R.L. and Golden, M.H. Interactions between pH and malic acid concentration on the inactivation of *Listeria monocytogenes*, *J. Food Saf.*, 18, 37, 1998.

93. Buchanan, R.L. and Golden, M.H., Model for the non-thermal inactivation of *Listeria monocytogenes* in a reduced oxygen environment, *Food Microbiol.*, 12, 203, 1995.

94. Breand, S., Fardel, G., Flandrois, J.P., Rosso, L., and Tomassone, R., Model of the influence of time and mild temperature on *Listeria monocytogenes* nonlinear survival curves, *Int. J. Food Microbiol.,* 40, 185, 1998.
95. Buchanan, R.L., Golden, M.H., and Whiting, R.C., Differentiation of the effects of pH and lactic or acetic acid concentration on the kinetics of *Listeria monocytogenes* inactivation, *J. Food Prot.,* 56, 474, 1993.
96. Ross, W.H., Couture, H., Hughes, A., Gleeson, T., and McKellar, R.C., A non-linear mixed effects model for the destruction of *Enterococcus faecium* in a high-temperature short-time pasteurizer, *Food Microbiol.,* 15, 567, 1998.
97. McKellar, R.C., Modler, H.W., Couture, H., Hughes, A., Mayers, P., Gleeson, T., and Ross, W.H., Predictive modeling of alkaline phosphatase inactivation in a high-temperature short-time pasteurizer, *J. Food Prot.,* 57, 424, 1994.
98. Mafart, P., Couvert, O., Gaillard, S., and Leguerinel, I., On calculating sterility in thermal preservation methods: application of the Weibull frequency distribution model, *Int. J. Food Microbiol.,* 72, 107, 2002.
99. Huang, L.H. and Juneja, V.K., A new kinetic model for thermal inactivation of microorganisms: development and validation using *Escherichia coli* O157:H7 as a test organism, *J. Food Prot.,* 64, 2078, 2001.
100. Daughtry, B.J., Davey, K.R., Thomas, C.J., and Verbyla, A.P., Food processing: a new model for the thermal destruction of contaminating bacteria, in *Engineering and Food at ICEF 7,* Academic Press, Sheffield, U.K., 1997, pp. A113–A116.
101. Pruitt, K.M. and Kamau, D.N., Mathematical models of bacterial growth, inhibition and death under combined stress conditions, *J. Ind. Microbiol.,* 12, 221, 1993.
102. Juneja, V.K., Marmer, B.S., Phillips, J.G., and Miller, A.J., Influence of the intrinsic properties of food on thermal inactivation of spores of nonproteolytic *Clostridium botulinum*: development of a predictive model, *J. Food Saf.,* 15, 349, 1995.
103. Whiting, R.C., Sackitey, S., Calderone, S., Merely, K., and Phillips, J.G., Model for the survival of *Staphylococcus aureus* in nongrowth environments, *Int. J. Food Microbiol.,* 31, 231, 1996.
104. Hirvi, Y., Griffiths, M.W., McKellar, R.C., and Modler, H.W., Linear-transform and non-linear modelling of bovine catalase inactivation in a high-temperature short-time pasteurizer, *Food Res. Int.,* 29, 89, 1996.
105. Zanoni, B., Peri, C., Garzaroli, C., and Pierucci, S., A dynamic mathematical model of the thermal inactivation of *Enterococcus faecium* during bologna sausage cooking, *Food Sci. Technol. Lebensm. Wiss.,* 30, 727, 1997.
106. Cole, M.B., Davies, K.W., Munro, G., Holyoak, C.D., and Kilsby, D.C., A vitalistic model to describe the thermal inactivation of *Listeria monocytogenes, J. Ind. Microbiol.,* 12, 232, 1993.
107. Little, C.L., Adams, M.R., Anderson, W.A., and Cole, M.B., Application of a log-logistic model to describe the survival of *Yersinia enterocolitica* at sub-optimal pH and temperature, *Int. J. Food Microbiol.,* 22, 63, 1994.
108. Duffy, G., Ellison, A., Anderson, W., Cole, M.B., and Stewart, G.S.A.B., Use of bioluminescence to model the thermal inactivation of *Salmonella typhimurium* in the presence of a competitive microflora, *Appl. Environ. Microbiol.,* 61, 3463, 1995.
109. Anderson, W.A., McClure, P.J., Baird-Parker, A.C., and Cole, M.B., The application of a log-logistic model to describe the thermal inactivation of *Clostridium botulinum* 213B at temperatures below 121.1°C, *J. Appl. Bacteriol.,* 80, 283, 1996.

110. Blackburn, C.D., Curtis, L.M., Humpheson, L., Billon, C., and McClure, P.J., Development of thermal inactivation models for *Salmonella enteritidis* and *Escherichia coli* O157:H7 with temperature, pH and NaCl as controlling factors, *Int. J. Food Microbiol.*, 38, 31, 1997.

111. Linton, R.H., Carter, W.H., Pierson, M.D., and Hackney, C.R., Use of a modified Gompertz equation to model nonlinear survival curves for *Listeria monocytogenes* Scott A, *J. Food Prot.*, 58, 946, 1995.

112. Linton, R.H., Carter, W.H., Pierson, M.D., Hackney, C.R., and Eifert, J.D., Use of a modified Gompertz equation to predict the effects of temperature, pH, and NaCl on the inactivation of *Listeria monocytogenes* Scott A heated in infant formula, *J. Food Prot.*, 59, 16, 1996.

113. Patterson, M.F. and Kilpatrick, D.J., The combined effect of high hydrostatic pressure and mild heat on inactivation of pathogens in milk and poultry, *J. Food Prot.*, 61, 432, 1998.

114. Bello, J. and Sanchezfuertes, M.A., Application of a mathematical model for the inhibition of Enterobacteriaceae and clostridia during a sausage curing process, *J. Food Prot.*, 58, 1345, 1995.

115. Baranyi, J., Jones, A., Walker, C., Kaloti, A., Robinson, T.P., and Mackey, B.M., A combined model for growth and subsequent thermal inactivation of *Brochothrix thermosphacta*, *Appl. Environ. Microbiol.*, 62, 1029, 1996.

116. Koutsoumanis, K., Lambropoulou, K., and Nychas, G.J.E., A predictive model for the non-thermal inactivation of *Salmonella enteritidis* in a food model system supplemented with a natural antimicrobial, *Int. J. Food Microbiol.*, 49, 63, 1999.

117. Peleg, M., On calculating sterility in thermal and non-thermal preservation methods, *Food Res. Int.*, 32, 271, 1999.

118. Van Boekel, M.A.J.S., On the use of the Weibull model to describe thermal inactivation of microbial vegetative cells, *Int. J. Food Microbiol.*, 74, 139, 2002.

119. Mattick, K.L., Legan, J.D., Humphrey, T.J., and Peleg, M., Calculating *Salmonella* inactivation in nonisothermal heat treatments from isothermal nonlinear survival curves, *J. Food Prot.*, 64, 606, 2001.

120. Periago, P.M., Palop, A., Martinez, A., and Fernandez, P.S., Exploring new mathematical approaches to microbiological food safety evaluation: an approach to more efficient risk assessment, *Dairy Food Environ. Sanit.*, 22, 18, 2002.

121. Hutchinson, T.P., Graphing the death of *Escherichia coli*, *Int. J. Food Microbiol.*, 62, 77, 2000.

122. Peleg, M., Microbial survival curves: the reality of flat "shoulders" and absolute thermal death times, *Food Res. Int.*, 33, 531, 2000.

123. Stumbo, C.R., *Thermobacteriology in Food Processing*, Academic Press, New York, 1965.

124. Sapru, V., Teixeira, A.A., Smerage, G.H., and Lindsay, J.A., Predicting thermophilic spore population dynamics for UHT sterilization processes, *J. Food Sci.*, 57, 1248, 1257, 1992.

125. Sapru, V., Smerage, G.H., Teixeira, A.A., and Lindsay, J.A., Comparison of predictive models for bacterial spore population resources to sterilization temperatures, *J. Food Sci.*, 58, 223, 1993.

126. Teixeira, A.A. and Rodriguez, A.C., Microbial population dynamics in bioprocess sterilization, *Enzyme Microb.Technol.*, 12, 469, 1990.

127. Shull, J.J., Cargo, G.T., and Ernst, R.R., Kinetics of heat activation and thermal death of bacterial spores, *Appl. Microbiol.*, 11, 485, 1963.

128. Rodriguez, A.C., Smerage, G.H., Teixeira, A.A., Lindsay, J.A., and Busta, F.F., Population model of bacterial spores for validation of dynamic thermal processes, *J. Food Process. Eng.*, 15, 1, 1992.

129. Rodriguez, A.C., Smerage, G.H., Teixeira, A.A., and Busta, F.F., Kinetic effects of lethal temperatures on population dynamics of bacterial spores, *Trans. Am. Soc. Agric. Eng.*, 31, 1594, 1988.

130. Smerage, G.H. and Teixeira, A.A., Dynamics of heat destruction of spores: a new view, *J. Ind. Microbiol.*, 12, 211, 1993.

131. Fairchild, T.M., Swartzel, K.R., and Foegeding, P.M., Inactivation kinetics of *Listeria innocua* in skim milk in a continuous flow processing system, *J. Food Sci.*, 59, 960, 1994.

132. Bigelow, W.D., Logarithmic nature of thermal death time curves, *J. Infect. Dis.*, 29, 538, 1921.

133. Nunes, R.V., Swartzel, K.R., and Ollis, D.F., Thermal evaluation of food processes: the role of a reference temperature, *J. Food Eng.*, 20, 1, 1993.

134. Kessler, H.G., Considerations in relation to some technological and engineering aspects, in Monograph on Pasteurized Milk, IDF Bulletin 200, International Dairy Federation, Brussels, 1986, pp. 80–86.

135. McKellar, R.C., Liou, S., and Modler, H.W., Predictive modelling of lactoperoxidase and gamma-glutamyl transpeptidase inactivation in a high-temperature short-time pasteurizer, *Int. Dairy J.*, 6, 295, 1996.

136. McKellar, R.C. and Piyasena, P., Predictive modelling of the inactivation of bovine milk α-L-fucosidase in a high-temperature short-time pasteurizer, *Int. Dairy J.*, 10, 1, 2000.

137. Fairchild, T.M. and Foegeding, P.M., A proposed nonpathogenic biological indicator for thermal inactivation of *Listeria monocytogenes*, *Appl. Environ. Microbiol.*, 59, 1247, 1993.

138. Piyasena, P., Liou, S., and McKellar, R.C., Predictive modelling of inactivation of *Listeria* spp. in bovine milk during HTST pasteurization, *Int. J. Food Microbiol.*, 39, 167, 1998.

139. Gagnon, B., Canadian code of recommended manufacturing practices for pasteurized/modified atmosphere packed/refrigerated foods, in *Guidelines for the "Code of Practice on Processed Refrigerated Foods,"* Agri-Food Safety Division, Agriculture Canada, 1989.

140. Nazarowec-White, M., McKellar, R.C., and Piyasena, P., Predictive modelling of *Enterobacter sakazakii* inactivation in bovine milk during high-temperature short-time pasteurization, *Food Res. Int.*, 32, 375, 1999.

141. Institute of Food Technologists, Kinetics of microbial inactivation for alternative food processing technologies. A report of the Institute of Food Technologists for the Food and Drug Administration, *J. Food Sci.*, 65, S4, 2000.

142. Zook, C.D., Parish, M.E., Braddock, R.J., and Balaban, M.O., High pressure inactivation kinetics of *Saccharomyces cerevisiae* ascospores in orange and apple juice, *J. Food Sci.*, 64, 533, 1999.

143. Weemaes, C., Ooms, V., Indrawati, L., Ludikhuyze, I., Ven den Broeck, A., Van Loey, A., and Hendrickx, M., Pressure–temperature degradation of green color in broccoli juice, *J. Food Sci.*, 64, 504, 1999.

144. Peleg, M., A model of microbial survival after exposure to pulsed electric fields, *J. Sci. Food Agric.*, 67, 93, 1995.

145. Hülsheger, H., Pottel, J., and Niemann, E.G., Killing of bacteria with electric pulses of high field strength, *Radiat. Environ. Biophys.*, 20, 53, 1981.

146. Cheroutre-Vialette, M. and Lebert, A., Growth of *Listeria monocytogenes* as a function of dynamic environment at 10°C and accuracy of growth predictions with available models, *Food Microbiol.,* 17, 83, 2000.

147. Bréand, S., Fardel, G., Flandrois, J.P., Rosso, L., and Tomassone, R., A model describing the relationship between lag time and mild temperature increase duration, *Int. J. Food Microbiol.,* 38, 157, 1997.

148. Cheroutre-Vialette, M. and Lebert, A., Modelling the growth of *Listeria monocytogenes* in dynamic conditions, *Int. J. Food Microbiol.,* 55, 201, 2000.

149. Lambert, R.J.W. and van der Ouderaa, M.L.H., An investigation into the differences between the Bioscreen and the traditional plate count disinfectant test methods, *J. Appl. Microbiol.,* 86, 689, 1999.

150. Dykes, G.A., Image analysis of colony size for investigating sublethal injury in *Listeria monocytogenes, J. Rapid Methods Autom. Microbiol.,* 7, 223, 1999.

151. McKellar, R.C., Butler, G., and Stanich, K., Modelling the influence of temperature on the recovery of *Listeria monocytogenes* from heat injury, *Food Microbiol.,* 14, 617, 1997.

152. Bello, J. and Sanchezfuertes, M.A., Application of a mathematical model to describe the behaviour of the *Lactobacillus* spp. during the ripening of a Spanish dry fermented sausage (chorizo), *Int. J. Food Microbiol.,* 27, 215, 1995.

153. Jones, J.E., Walker, S.J., Sutherland, J.P., Peck, M.W., and Little, C.L., Mathematical modelling of the growth, survival and death of *Yersinia enterocolitica, Int. J. Food Microbiol.,* 23, 433, 1994.

154. Membre, J.M., Thurette, J., and Catteau, M., Modelling the growth, survival and death of *Listeria monocytogenes, J. Appl. Microbiol.,* 82, 345, 1997.

3 Secondary Models

Thomas Ross and Paw Dalgaard

CONTENTS

3.1 INTRODUCTION

Changes in populations of microorganisms in foods over time (i.e., "microbial kinetics") are governed by storage conditions ("extrinsic" factors) and product characteristics ("intrinsic" factors). Collectively these have been termed "environmental parameters." They may represent simple situations, e.g., where the storage temperature is the only important factor influencing microbial kinetics, but in many foods the environmental parameters that influence microbial kinetics are complex and dynamic and include the combined effects of extrinsic factors such as temperature and storage atmosphere; intrinsic factors such as water activity, pH, naturally occurring organic acids, and added preservatives; and interactions between groups of microorganisms.

Consistent with the widely accepted terminology introduced by Whiting and Buchanan (1993), we term those models that describe the response of microorganisms to a single set of conditions *over time* as "primary" models (see Chapter 2). Models that describe the effect of environmental conditions, e.g., physical, chemical, and biotic features, on the values of the parameters of a primary model are termed "secondary" models.

Knowledge of the environmental parameters that most influence growth of microorganisms in foods is essential for the development, as well as for the practical use, of predictive microbiology models. Secondary models that do not include all the environmental parameters important in a food are said to be "incomplete" (Ross, Baranyi, McMeekin, 2000) and require expansion (or simple "calibration" if those factors are constant) to accommodate their effect on microbial kinetics. The environmental parameters that are important for particular foods, however, are not always known. In those situations the systematic approach of predictive microbiology can help to elucidate the microbial ecology of the product.

In this chapter we consider a range of types of secondary models including those that model the probability that a predicted kinetic response will occur. The chapter includes descriptions and comparison of models, as considerations for development of robust, secondary models. Appendix A3.1 details methods to measure environmental factors of importance — an essential element of the application of predictive microbiology.

3.1.1 Philosophy, Terminology, and Methodology

The history of predictive microbiology, including the philosophical motivations of Roberts and Jarvis (1983), who first proposed the concept, was traced by Ross and McMeekin (1994). From a purely pragmatic perspective, predictive microbiology aims to collect and make accessible computerized data on the behavior of microbial populations in response to defined environmental conditions, but mathematical modeling also provides a useful and rigorous framework for the hypothetico-deductive scientific process. To develop a consistent framework that enables us to understand and predict the microbial ecology of foods it is desirable to integrate the patterns of microbial behavior revealed in predictive modeling studies with knowledge of the physiology of microorganisms and physical and chemical processes and phenomena that occur in foods and food processes (Ross, Baranyi, McMeekin, 2000).

Various types and categorizations of models are recognized. Empirical models are, essentially, pragmatic and simply describe a set of data in a convenient mathematical relationship with no consideration of underlying phenomena. Mechanistic models are built up from theoretical bases and, if they are correctly formulated, can allow the response to be interpreted in terms of known physical, chemical, and biological phenomena. An advantage of mechanistic approaches is that they tend to provide a better foundation for subsequent development and expansion of models; i.e., taken to their logical extreme, models for specific situations would simply be special, or reduced, cases of a much larger and holistic model that describes, quantitatively, the microbial ecology of foods. The process of developing models that are able to be integrated readily with other models so as to describe more complex phenomena has been termed "nesting" or "embedding." A fuller explanation of the benefits of that approach was provided by Baranyi and Roberts (1995).

In one sense, a model is the mathematical expression of a hypothesis. If this approach is adopted, it follows that the parameters in such models might be readily interpretable properties of the system under study, and that the mathematical form of the model would enable interpretation of the interactions between those factors. Interpretability of model parameters is a feature highly valued by many authors in the predictive microbiology literature (e.g., Augustin and Carlier, 2000a,b; Rosso et al., 1993; Wijtzes et al., 1995). Although the development of predictive microbiology has seen the embedding of more and more mechanistic elements, or at least models whose structure and parameterization reflects known or hypothesized underlying phenomena, in practice many models currently available in predictive microbiology are not purely empirical, and none are purely mechanistic (Ross, Baranyi, McMeekin, 2000).

Another, often cited, advantage of mechanistic models is that if they are built on sound theory they are more likely to facilitate prediction by extrapolation. Conversely, as none of the models in use in predictive microbiology can be considered to be mechanistic, they can only be used to make predictions by interpolation. (Determination of the interpolation region encompassed by a model is discussed in Sections 3.2.5 and 3.4.3.4.) It is perhaps ironic, then, that 20 years of experience in predictive microbiology has not demonstrated the practical usefulness of mechanistic models that have been proposed to date (see Section 3.2.4). In general, even with good quality data the mechanistic models do not provide better fit and are usually harder to work with than quasi-mechanistic or empirical models currently used.

Predictive microbiology is a specific application of the field of mathematical modeling and, as such, the same rules of modeling as are applied in those other disciplines are relevant to the development of predictive food microbiology models. These have been discussed by various authors (Draper and Smith, 1981; McMeekin et al., 1993; Ratkowsky, 1993), and an overview is presented in Table 3.1.

Experimental methods and design considerations relevant to kinetic models were discussed in detail in McMeekin et al. (1993; Chapter 2), Davies (1993), and Legan et al. (2002) and are also discussed in Chapter 1. Two points that we feel are necessary to reiterate are the limitations of the central composite design in predictive microbiology studies, and consideration of spoilage domains when growth of spoilage microorganisms is studied. Legan et al. (2002) accentuated the importance of experimental design in growth modeling studies stating:

> in other disciplines, such as engineering, central composite designs are commonly used for developing response surface models. For microbiological modeling, however, these designs have serious limitations and should be avoided. Central composite designs concentrate treatments in the centre of the design space and have fewer treatments in the extreme regions where biological systems tend to exhibit much greater variability.

Microbial food spoilage is dynamic and in some cases relatively small changes in environmental parameters cause a complete shift in the microflora responsible for product spoilage. Thus, to avoid modeling growth of spoilage microorganisms under conditions where they have no influence on quality, a product-oriented approach that includes determination of the spoilage domain of specific micoorganisms is often required (Dalgaard, 2002).

We will not comment further on methodology appropriate to development of kinetic models, other than to say that to develop reliable secondary models an understanding of microbial physiology and its interaction with food environments and storage and processing conditions must be borne in mind in the design of experiments including the preparation of cultures and interpretation of measurements of population changes. This issue is particularly explored and exemplified in Section 3.4.4.4 concerning experimental considerations relevant to the development of growth limits models.

3.2 SECONDARY MODELS FOR GROWTH RATE AND LAG TIME

Implicit in the appropriate development and use of secondary models in predictive food microbiology is the ability to characterize foods, and the environment they

TABLE 3.1
Some Considerations in the Selection of Models

Subject	Reasons
Parameter estimation properties	Relates to the procedure and reliability of estimating the model parameters. In general, models should have parameters that are independent, identically distributed, normal or "iidn" (see, e.g., Ratkowsky, 1993)
Stochastic assumption	The form of the model, and choice of response variables, should be such that the difference between prediction and observations (or some mathematical transformation of them) is normally distributed, and that the magnitude of the error is independent of the magnitude of the response modeled. If not, the fitting can be dominated by some data, at the expense of other data
Parameter interpretability	As noted in the text, it is useful if the parameters have biological/physical/chemical interpretations that can be readily related to the independent and dependent variables. This can simplify the process of model creation and also aid in understanding of the model (This may be less important than the behavior and performance of the model.)
Parsimony	Models should have no more parameters than are required to describe the underlying behavior studied. Too many parameters can lead to a model that fits the *error* in the data, i.e., generates a model that is specific to a particular set of observations. Nonparsimonious models have poor predictive ability
Interpolation region	No models in predictive microbiology can be considered to be mechanistic and predictions can be made by interpolation only. Thus, the interpolation region defines the useful range of applicability of the model. The interpolation region is affected by not only the range of individual variables, but also the experimental design (see Section 3.2.5)
Correct qualitative features	In mathematical terms, these are the analytical properties of the model function. They include convexity, monotonity, locations of extreme, and zero values. If biological considerations prescribe any of these, the model should reflect those properties accurately
"Extendibility" (embedding, nesting)	When a model is developed further (such as to include more or dynamically changing environmental factors) the new, more complex model should embody the old, simpler model as a special case

Source: Modified from Ross, T., Baranyi, J., and McMeekin, T.A. In *Encyclopaedia of Food Microbiology,* Robinson, R., Batt, C.A., and Patel, P. (Eds.), Academic Press, London, 2000, pp. 1699–1710.

present to contaminating microorganisms, in terms of those biotic and abiotic elements that affect the dynamics of the microbial population of interest. Methods to characterize the physicochemical environment, including temperature, gaseous atmosphere, salt and/or water activity, pH and organic acids, spices, smoke, and other components, are discussed in detail in Appendix A3.1. These topics are also considered in Chapter 5, including discussion of the influence of other organisms and heterogeneity in the environment. Another important element is the ability to characterize temporal changes in the environment. Techniques for modeling microbial population dynamics under time-varying conditions are considered in Chapter 7.

Within predictive microbiology the development and application of secondary models for growth rates and lag times have been extensively reviewed (Buchanan, 1993b; Davey, 1999; Farber, 1986 ; ICMSF, 1996a,b; McDonald and Sun, 1999; McMeekin et al., 1993; Ross, 1999a,b; Ross and McMeekin, 1994; Skinner et al., 1994; Whiting, 1995). This section describes types of secondary growth rate and lag time models that are currently available, but with particular focus on more recent developments, and also includes a detailed tabulation of models available for specific microorganisms.

3.2.1 SQUARE-ROOT-TYPE MODELS

3.2.1.1 Temperature

As discussed later (Section 3.2.4), in many cases the classical Arrhenius equation is inappropriate to describe the effect of suboptimal temperature on growth rates of microorganisms because the (apparent) activation energy (E_a) itself is temperature dependent. To overcome this problem Ratkowsky et al. (1982) suggested a simple empirical model (Equation 3.1). When this model was fitted to experimental growth rates the data were square-root transformed to stabilize their variance and this simple model and its numerous expansions are named square-root-type, Ratkowsky-type, or Bêlerádek-type models (McMeekin et al., 1993). These models, and the closely related cardinal parameter models (see Section 3.2.3), are probably the most important group of the secondary models within predictive microbiology.

$$\sqrt{\mu_{max}} = b \cdot (T - T_{min}) \tag{3.1}$$

where b is a constant and T is the temperature. The parameter T_{min}, a theoretical minimum temperature for growth, is the intercept between the model and the temperature axis (Figure 3.1). T_{min} is a model parameter and its value can be 5 to 10°C lower than the lowest temperature at which growth is actually observed. This interpretation differs from that embodied in the cardinal parameter models, as discussed in Section 3.2.3 and Chapter 4).

From growth rates measured at several different constant temperatures the values of b and T_{min} in Equation 3.1 can be determined by classical model fitting techniques (see Chapter 4). Recently it was suggested that b and T_{min} could be estimated from a single, optimally designed, experiment where growth resulting from a temperature profile is recorded (Bernaerts et al., 2000). These authors concluded that such an optimal, dynamic, one-step experiment would reduce the experimental work required to develop a model significantly and would have substantial potential within predictive microbiology. So far this technique has not found wider use within predictive microbiology and its ability to estimate model parameters accurately remains to be confirmed for different microorganisms and environmental parameters.

Ratkowsky et al. (1983) expanded Equation 3.1 to include the entire biokinetic range of growth temperatures (Equation 3.2, Figure 3.1). From this model the optimal

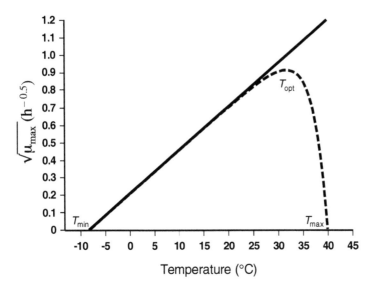

FIGURE 3.1 Simulation of Equation 3.1 (solid line) and Equation 3.2 (dashed line). $b = 0.025$ $h^{0.5}$/°C, $T_{min} = -8$°C, $c = 0.30$°C^{-1}, and $T_{max} = 40$°C.

growth temperature can be determined by solving the following equation: $c \times (T_{opt} - T_{min}) = \exp[c \times (T_{opt} - T_{min})] - 1$ (McMeekin et al., 1993).

$$\sqrt{\mu_{max}} = b \cdot (T - T_{min}) \cdot (1 - \exp(c(T - T_{max}))) \tag{3.2}$$

where b and c are constants, T is the temperature, T_{min} the theoretical minimum temperature below which no growth is possible, and T_{max} is the theoretical maximum temperature beyond which growth is not possible.

While Ratkowsky et al. (1982, 1983) settled for an exponent of 2, the original Bêlerádek models had a variable exponent value. Dantigny (1998) and Dantigny and Molin (2000) used the concepts of dimensionless growth rate variables (effectively the same as the gamma factor concept; see Section 3.2.3) to explore the most appropriate value of the exponent for bacterial growth rate data using Bêlerádek-type models. They reported a correlation between the estimate of T_{min} and the exponent value used and found that when T_{min} and the exponent were simultaneously fitted by nonlinear regression, thermophiles had lower fitted exponent values than did mesophiles or psychrotrophic organisms. They reported that the use of the square-root model leads to an underestimation of the minimum temperature for growth when the exponent value is significantly less than 2.

3.2.1.2 Water Activity

McMeekin et al. (1987) found that growth responses of *Staphylococcus xylosus* followed Equation 3.1 at different values of water activity. T_{min} was constant and thus independent of water activity and Equation 3.3 was suggested to describe the combined effect of temperature and water activity (McMeekin et al., 1987).

$$\sqrt{\mu_{max}} = b \cdot (T - T_{min}) \cdot \sqrt{a_w - a_{w\,min}} \qquad (3.3)$$

where b and T_{min} are as previously defined, a_w is the water activity, and $a_{w\,min}$ is the theoretical minimum water activity below which growth is not possible.

Later, Miles et al. (1997) suggested that Equation 3.4 be used to take into account the effect of the entire biokinetic ranges of both temperature and water activity.

$$\sqrt{\mu_{max}} = b \cdot (T - T_{min}) \cdot (1 - \exp(c(T - T_{max}))) \cdot \sqrt{(a_w - a_{w\,min})(1 - \exp(d(a_w - a_{w\,max})))}$$
$$(3.4)$$

where b, c, T, T_{min}, T_{max}, a_w, and $a_{w\,min}$ are as previously defined, d is a fitted constant, and $a_{w\,max}$ is a theoretical maximum water activity beyond which growth is not possible.

Most food-related microorganisms grow at water activities very close to 1.000 and in those cases the expanded water activity term (i.e., containing $a_{w\,max}$) in Equation 3.4 is not needed to predict growth in foods. However, some microorganisms, e.g., several marine bacteria, have a substantial requirement for minerals. To model growth responses of these microorganisms, the inhibitory effect of high water activities, i.e., low salt concentrations, must be taken into account. For the human pathogen *Vibrio parahaemolyticus*, $a_{w\,max}$ has been determined to be 0.998. Some seafood spoilage bacteria are more inhibited by high water activity; e.g., growth of *Halobacterium salinarium* was only observed at a_w values below 0.9 (Chandler and McMeekin, 1989; Doe and Heruwati, 1988; Miles et al., 1997).

3.2.1.3 pH

For *Yersinia enterocolitica*, Adams et al. (1991) found that growth responses followed Equation 3.1 at different values of pH. Again, T_{min} was constant and Equation 3.5 was suggested.

$$\sqrt{\mu_{max}} = b \cdot (T - T_{min}) \cdot \sqrt{pH - pH_{min}} \qquad (3.5)$$

where pH_{min} is the theoretical minimum pH below which growth is not possible and other parameters are as previously defined.

On the basis of the observation of Cole et al. (1990) that growth rate was proportional to hydrogen ion concentration, Presser et al. (1997) introduced the following quasi-mechanistic term to describe the effect of pH on bacterial growth:

$$\mu_{max} = \mu_{opt} \times (1 - 10^{pH_{min} - pH}) \quad (3.6a)$$

By analogy, another term was introduced for superoptimal (i.e., alkaline) pH conditions, leading to the following model for the entire biokinetic pH range:

$$\mu_{max} = \mu_{opt} \times (1 - 10^{pH_{min} - pH}) \times (1 - 10^{pH - pH_{max}}) \quad (3.6b)$$

The validity of that term was evaluated against an extensive data set for *Escherichia coli* growth, including variables of temperature, water activity, and lactic acid concentration for a range of acid and alkaline environmental pH levels (see Equation 3.10).

Wijtzes et al. (1995, 2001) continued the development of square-root-type models and suggested Equation 3.7 for growth responses of *Lactobacillus curvatus* at different temperatures, a_w values, and pH

$$\mu = b \cdot (a_w - a_{w\,min}) \cdot (pH - pH_{min}) \cdot (pH - pH_{max}) \cdot (T - T_{min})^2 \quad (3.7)$$

3.2.1.4 Other Factors

Equation 3.8 was suggested to model the effect of carbon dioxide-enriched (%CO_2) atmospheres on growth of the specific spoilage organism *Photobacterium phosphoreum* on fish (Dalgaard, 1995; Dalgaard et al., 1997). Later, similar but square-root-transformed terms were used to model the effect of CO_2 and sodium lactate (NaL) on growth of *Lactobacillus sake* and *Listeria monocytogenes* at a constant pH (Equation 3.9; Devlieghere et al., 1998, 2000a,b, 2001).

$$\sqrt{\mu_{max}} = b(T - T_{min}) \times \frac{(\%CO_{2\,max} - \%CO_2)}{\%CO_{2\,max}} \quad (3.8)$$

$$\sqrt{\mu_{max}} = b
\begin{aligned}
&\cdot (T - T_{min}) \\
&\cdot \sqrt{a_w - a_{w\,min}} \\
&\cdot \sqrt{CO_{2\,max} - CO_2} \\
&\cdot \sqrt{NaL_{max} - NaL}
\end{aligned} \quad (3.9)$$

As noted above, a more comprehensive square-root-type model that includes the effects of temperature, pH, water activity, and lactic acid has been suggested and developed in a series of publications (Presser et al., 1997; Ross, 1993a,b; Salter et al., 1998; Tienungoon, 1988) and has been applied to *Listeria monocytogenes* and *Escherichia coli* growth rates. It was presented in its most complete form in Ross et al. (2003):

$$\sqrt{\mu_{max}} = c$$
$$\times (T - T_{min}) \times (1 - \exp(d \times (T - T_{max})))$$
$$\times \sqrt{a_w - a_{w\,min}} \times (1 - \exp(g \times (a_w - a_{w\,max})))$$
$$\times \sqrt{1 - 10^{pH_{min} - pH}}$$
$$\times \sqrt{1 - 10^{pH - pH_{max}}} \qquad (3.10)$$
$$\times \sqrt{1 - \frac{LAC}{U_{min} \times (1 + 10^{pH - pk_a})}}$$
$$\times \sqrt{1 - \frac{LAC}{D_{min} \times (1 + 10^{pk_a - pH})}}$$

where c, d, and g are fitted parameters, LAC is the lactic acid concentration (mM), U_{min} the minimum concentration (mM) of undissociated lactic acid that prevents growth when all other factors are optimal, D_{min} the minimum concentration (mM) of dissociated lactic acid that prevents growth when all other factors are optimal, pK_a is the pH for which concentrations of undissociated and dissociated lactic acid are equal, reported to be 3.86 (Budavari, 1989), and all other terms are as previously defined.

One of the advantages of the square-root-type models, and the cardinal parameters models, is that their form enables them to be readily simplified into models for special cases; e.g., in Equation 3.10, if one factor is held constant then the terms involving that factor simply reduce to constants.

An example is a model developed for *Listeria monocytogenes* (Ross et al., in press; WHO/FAO, in press), in which the superoptimal water activity term is not relevant, and in which a term for the effect of nitrite on *L. monocytogenes* growth rate was also included. That novel term was based on analysis of the predictions of the Pathogen Modeling Program (Buchanan, 1993a; www.arserrc.gov/mfs/pathogen.htm). The fitted model is shown in Equation 3.11.

$$\sqrt{\mu_{max}} = 0.1626$$
$$\times (T - 0.60) \times (1 - \exp(0.129 \times (T - 51.0)))$$
$$\times \sqrt{(a_w - 0.925)}$$
$$\times \sqrt{1 - 10^{4.93 - pH}}$$
$$\times \sqrt{1 - \frac{LAC}{4.55 \times (1 + 10^{pH - 3.86})}} \qquad (3.11)$$
$$\times \sqrt{\left(\left(493 - NIT \times \left(1 + \frac{(6.5 - pH)}{2}\right)\right) \Big/ 493\right)}$$

where NIT is the concentration of nitrite and all other terms are as previously defined.

As shown in Figure 3.2, Equation 3.10 and Equation 3.11 represent a new generation of square-root-type models where the level of lactic acid influences the range of pH values for which growth is theoretically observed, reflecting the known interaction between pH and undissociated lactic acid, and also the individual growth rate suppressing effects of hydrogen ion concentration and undissociated lactic acid concentration. This was not the case for the environmental parameters included in Equation 3.1 to Equation 3.9. In those models, each term expressed how an environmental factor reduced the growth rate of a microorganism. However, for those models the expected multidimensional growth space was not influenced by levels of the different environmental parameters. This limitation of predictive models for growth rate has been recognized and has led, in part, to the development of growth/no growth models (discussed in Section 3.2.3 and Section 3.4). To make accurate predictions, a model can include terms to force the predicted growth rate to zero (Augustin and Carlier, 2000b; Le Marc et al., 2002). Alternatively, the probability of growth under the test conditions can first be assessed using a growth boundary model. If growth is possible, a growth rate model in combination with a lag time model can be used to estimate the extent of growth (Ross et al., in press; WHO/FAO, in press).

It is also notable that the pH and lactic acid terms in Equation 3.10 are effectively gamma-model type terms (see Section 3.2.2), in which the effect of the level of growth rate inhibitor is scaled between 0 and 1, where 1 represents no inhibition, i.e., the optimal level of that environmental factor. In the case of lactic acid, the optimal level would be 0, while for pH the optimum is ~7. This illustrates the close relationship between square-root-type models, and those that embody the gamma concept, such as the cardinal parameter models.

3.2.2 THE GAMMA CONCEPT

The concept of dimensionless growth factors, now known as the gamma (γ) concept, was introduced in predictive microbiology by Zwietering et al. (1992). Later, minor changes and new developments were added (Wijtzes et al., 1998, 2001; Zwietering, 1999; Zwietering et al., 1996).

The gamma (γ) concept relies on:

1. The observation (e.g., Adams et al., 1991; McMeekin et al., 1987) that many factors that affect microbial growth rate act independently, and that the effect of each measurable factor on growth rate can be represented by a discrete term that is multiplied by terms for the effect of all other growth rate affecting factors, i.e.:

$$\mu = f(\text{temperature}) \times f(a_w) \times f(\text{pH}) \times f(\text{organic acid})$$
$$\times f(\text{other}_1) \times f(\text{other}_2) \times \ldots.f(\text{other}_n)$$

2. That the effect on growth rate of any factor can be expressed as a fraction of the maximum growth rate (i.e., the rate when that environmental factor is at the optimum level)

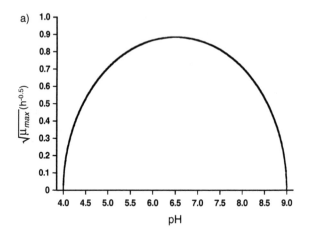

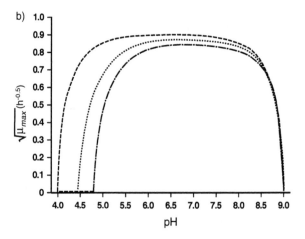

FIGURE 3.2 Simulation of Equation 3.7 (a) and Equation 3.10 (b) at a fixed temperature and water activity. pH_{min} is 4.0 and pH_{max} is 9.0. For Equation 3.10, $U_{min} = 10$ mM and $D_{max} = 1000$ mM. The concentrations of lactic acid (LAC) depicted are 0 mM (dashed line), 50 mM (dotted line), and 100 mM (dash-dotted line).

Under completely optimal conditions each microorganism has a reproducible maximum growth rate, notwithstanding the potential effect of strain variability. As any environmental factor becomes suboptimal the growth rate declines in a predictable manner, and the extent of that inhibition can be related to the optimum growth rate by calculating the relative rate at the test condition compared to that at the optimum. Thus, under the gamma concept approach, the cumulative effect of many factors poised at suboptimal levels can be estimated from the product of the relative inhibition of growth rate due to each factor, as indicated by Equation 3.12. The relative inhibitory effect of a specific environmental variable is described by a growth factor "gamma" (γ), a dimensionless measure that has a value between 0 and 1 (e.g., Equation 3.13 to Equation 3.15).

The relative inhibitory effect can be determined from the "distance" between the optimal level of the factor and the minimum (or maximum) level that completely inhibits growth by recourse to a predictive model. In the gamma model approach, the reference growth rate is μ_{max}, so that reference levels of temperature, water activity, etc. are those that are the optimum for growth rate, usually represented as T_{opt}, $a_{w\ opt}$, pH_{opt}, etc. The combined effect of several environmental factors is then determined by multiplication of their respective γ factors (Equation 3.16).

$$\gamma = \frac{\text{Growth rate at actual environmental conditions}}{\text{Growth rate at optimal environmental conditions}}$$

$$= \frac{\mu_{max}(T, a_w, pH, \text{etc.})}{\mu_{max\ opt}} \tag{3.12}$$

$$\gamma(T) = \left(\frac{T - T_{min}}{T_{opt} - T_{min}}\right)^2 \tag{3.13}$$

$$\gamma(a_w) = \frac{a_w - a_{w\ min}}{1 - a_{w\ min}} \tag{3.14}$$

$$\gamma(pH) = \frac{(pH - pH_{min}) \cdot ((pH_{max} - pH))}{(pH_{opt} - pH_{min}) \cdot (pH_{max} - pH_{opt})} \tag{3.15}$$

$$\mu_{max} = \mu_{max\ opt} \cdot \gamma(T) \cdot \gamma(a_w) \cdot \gamma(pH) \tag{3.16}$$

The effect of environmental parameters like carbon dioxide, sodium lactate, and nitrite has also been included in square-root-type models (see, e.g., Equation 3.8 to Equation 3.11). The absence of these inhibitory substances is optimal for growth and therefore the calculation of γ factors requires information only about the lowest concentration of each substance that prevents growth (or, similarly, the maximum level that can be tolerated before growth ceases) analogous to minimum inhibitory concentrations (MICs).

$$\gamma(CO_2) = \left(\frac{\%CO_{2\ max} - \%CO_2}{\%CO_{2\ max} - \%CO_{2\ opt}}\right)^2 = \left(\frac{\%CO_{2\ max} - \%CO_2}{\%CO_{2\ max}}\right)^2 \tag{3.17}$$

3.2.2.1 Expanding Existing Models

Given that there is a finite number of models (see Table 3.5 and Table 3.6), and that few models include factors of relevance to all foods, some workers have attempted to integrate terms for specific variables from one model into another to suit a specific food and the conditions of interest. Because of the assumption of independent action of growth rate inhibitors, the dimensionless γ factors can, in principle, be readily exchanged between existing models and, at the time of writing, this is increasingly being done. Values of parameters like μ_{max}, μ_{opt}, T_{min}, T_{opt}, $a_{w\,min}$, pH_{min}, pH_{opt}, pH_{max}, and $\%CO_{2\,max}$ from which gamma factors can be derived are known for a considerable number of food-related pathogenic microorganisms. The approach was possibly taken to its logical conclusion by Augustin and Carlier (2000a,b) who collated, and integrated into a single model, literature data and observations for more than 15 factors in foods that affect the growth rate of *L. monocytogenes*.

For spoilage bacteria from chilled foods, growth kinetics at low temperatures are often well characterized but values of μ_{max}, μ_{opt}, T_{opt}, pH_{opt}, and pH_{max} are frequently unknown or have not been determined accurately. This is the case, for example, for the specific spoilage organisms *Photobacterium phosphoreum*, *Shewanella putrefaciens*, and *Brochothrix thermosphacta*. In this situation the classical gamma concept cannot be used to develop a secondary model. However, when a simple square-root-type model including the effect of temperature and, e.g., CO_2, has been developed for chilled product stored at a known pH (pH_{ref}) and water activity ($a_{w\,ref}$) then these models can be expanded at suboptimal growth conditions by addition of γ-like factors, as shown in Equation 3.18 (Dalgaard et al., 2003).

$$
\begin{aligned}
\sqrt{\mu_{max}} \quad = \quad & b \\
\cdot \quad & (T - T_{min}) \\
\cdot \quad & (\%CO_{2\,max} - \%CO_2)\,/\,\%CO_{2\,max} \\
\cdot \quad & \sqrt{(a_w - a_{w\,min})\,/\,(a_{w\,ref} - a_{w\,min})} \\
\cdot \quad & \sqrt{(pH - pH_{min})\,/\,(pH_{ref} - pH_{min})}
\end{aligned}
\tag{3.18}
$$

Clearly, this approach should be used with some caution because the assumption of independent action has not been tested for all environmental factor combinations. Thus, the range of applicability of the expanded model should be evaluated, e.g., by comparison with data from challenge tests or naturally contaminated products (Giménez and Dalgaard, in press). (Section 3.2.5 discusses the expansion of existing polynomial models.)

3.2.3 CARDINAL PARAMETER MODELS

Cardinal parameter models (CPMs) were introduced to predictive microbiology in 1993 and have become an important group of empirical secondary models (Augustin and Carlier, 2000a,b; Le Marc et al., 2002; Messens et al., 2002; Pouillot et al., 2003; Rosso, 1995, 1999; Rosso et al., 1993, 1995; Rosso and Robinson, 2001). The basic idea behind CPMs is to use model parameters that have a biological or graphical interpretation. When models are fitted to experimental data by nonlinear regression (see Chapter 4), this has the obvious advantage that appropriate starting values are easy to determine. General CPMs rely on the assumption that the inhibitory effect of environmental factors is multiplicative, an assumption that was formalized in the gamma (γ) concept discussed above (Section 3.2.2). Thus, general CPMs consist of a discrete term for each environmental factor, with each term expressed as the growth rate *relative to that when that factor is optimal*; i.e., each term has a numerical value between 0 and 1. At optimal growth conditions all terms have a value of 1 and thus μ_{max} is equal to μ_{opt} (Equation 3.19).

Equation 3.19 to Equation 3.21 show a CPM that includes the effect of temperature (T), water activity (a_w), pH, inhibitory substances (c_i) and qualitative factors (k_j) on μ_{max} (Augustin and Carlier, 2000a). This extensive CPM was developed from available literature data from many studies for growth of *Listeria monocytogenes*. The inhibitory substances included (1) undissociated acetic acid, lactic acid, and citric acid, (2) Na-benzoate, K-sorbate, and the undissociated form of sodium nitrite, and (3) glycerol monolaurin, butylated hydroxyanisole, butylated hydroxytoluene, *tert*-butylhydroquinone, CO_2, caffeine, and phenol. In addition, the effect of competitive growth of microorganisms and the inhibitory effect due to specific types of foods were included in the model as qualitative factors.

$$\mu_{max} = \mu_{opt} \cdot CM_2(T) \cdot CM_2(a_w) \cdot CM_1(pH) \cdot \prod_{i=1}^{n} \gamma(c_i) \cdot \prod_{j=1}^{p} k_j \qquad (3.19)$$

$$CM_n = \qquad (3.20)$$

$$\begin{cases} 0, & X \le X_{min} \\[2mm] \dfrac{(X - X_{max}) \cdot (X - X_{min})^n}{(X_{opt} - X_{min})^{n-1} \cdot [(X_{opt} - X_{min}) \cdot (X - X_{opt}) - (X_{opt} - X_{max}) \cdot \\ ((n-1) \cdot X_{opt} + X_{min} - n \cdot X)]}, & X_{min} < X < X_{max} \\[2mm] 0, & X \ge X_{max} \end{cases}$$

$$\gamma(c_i) = \begin{cases} (1 - c_i / MIC_i)^2, & c_i < MIC_i \\ 0, & c_i \ge MIC_i \end{cases} \qquad (3.21)$$

where X is temperature, water activity, or pH. X_{min} and X_{max} are, respectively, the values of X_i below and above which no growth occurs, X_{opt} is the value at which μ_{max} is equal to its optimal value μ_{opt}. MIC_i is the minimal inhibitory concentration of specific compounds above which no growth occurs.

Within predictive microbiology various CPMs were developed during the 1990s and in the same period different cardinal parameter temperature models were independently developed in other fields, e.g., to predict the effect of temperature on growth rates (r) of crops (Equation 3.22; Yan and Hunt, 1999; Yin and Wallace, 1995).

$$r = r_{max} \left(\frac{T - T_{min}}{T_{opt} - T_{min}} \right) \left(\frac{T_{max} - T}{T_{max} - T_{opt}} \right)^{\frac{T_{max} - T_{opt}}{T_{opt} - T_{min}}} \qquad (3.22)$$

TABLE 3.2
Parameter Values in Square-Root Type (Sqrt) and Cardinal Parameter
Models (CPM)

Organism	T_{min} Sqrt	CPM	T_{opt} Sqrt	CPM	T_{max} Sqrt	CPM	μ_{opt} Sqrt	CPM	Reference
Escherichia coli	2.9	4.9	41.0	41.3	49.2	47.5	2.3	2.3	Rosso et al. (1993)
Salmonella Typhimurium	3.8	5.7	39.8	40.0	51.1	49.3	1.7	1.7	Oscar (2002)

	pH_{min} Sqrt	CPM	pH_{opt} Sqrt	CPM	pH_{max} Sqrt	CPM	μ_{opt} Sqrt	CPM	
Listeria monocytogenes	4.2	4.6	7.0	7.1	9.8	9.4	1.0	0.95	Rosso et al. (1995)

In several ways CPMs resemble square-root models and responses of the two types of models can be practically identical, e.g., for the effect of temperature, water activity, and pH (Oscar, 2002; Rosso et al., 1993, 1995). Parameters in the two types of models are typically named T_{min}, T_{max}, $a_{w\ min}$, $a_{w\ max}$, pH_{min}, and pH_{max}. However, these model parameters are not defined in entirely the same way for CPMs and square-root-type models. In fact, when identical data are fitted to the two types of models square-root-type models estimate lower T_{min}, $a_{w\ min}$, and pH_{min} values and higher T_{max}, $a_{w\ max}$, and pH_{max} values (Table 3.2; see also Chapter 4).

T_{min} values estimated by CPMs and square-root-type models often differ by ~2°C as shown in Table 3.2. Table 3.3 shows that a 2°C difference of a T_{min} value has a pronounced effect on μ_{max} values predicted by both a square-root-type model and a CPM. Thus, parameter values estimated by using one of these types of models

TABLE 3.3
Effect of T_{min} Values (−1°C and +1°C) on μ_{max} Values Predicted by a
Square-Root and a Cardinal Parameter Model at 4, 8, and 12°C

Temperature (°C)	Square-Root Model[a] μ_{max} (h⁻¹) $T_{min} = -1°C$	$T_{min} = +1°C$	% Difference	Cardinal Parameter Model[b] μ_{max} (h⁻¹) $T_{min} = -1°C$	$T_{min} = +1°C$	% Difference
4	0.0216	0.0078	64	0.0216	0.0087	60
8	0.0700	0.0424	40	0.0718	0.0485	33
12	0.1461	0.1046	28	0.1549	0.1237	20

[a] The model of Ratkowsky et al. (1983) used with values of the model parameters b and c selected to obtain a T_{opt} value ~37°C and a μ_{opt} value of ~ 1.0 h⁻¹. T_{max} was 45.0°C.
[b] The model of Rosso et al. (1993) used with T_{opt} of 37°C, μ_{opt} 1.0 h⁻¹, and T_{max} 45.0°C.

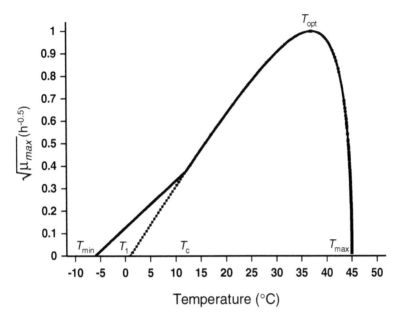

FIGURE 3.3 Simulation of the model $\mu_{max} = \mu_{opt} \times CM_2(T)$, with $CM_2(T)$ given by Equation 3.23 and with T_{min} of $-6°C$, T_1 of $1°C$, T_c of $12°C$, T_{opt} of $37°C$, and T_{max} of $45°C$. μ_{opt} was $1.0\ h^{-1}$.

cannot be used with the other type of model. This situation is similar to the estimation of μ_{max} values by some primary growth models. Modified Gompertz models (Gibson et al., 1987; Zwietering et al., 1990), e.g., overestimate μ_{max} by ~15% (Dalgaard et al., 1994; Membré et al., 1999; Whiting and Cygnarowicz-Provost, 1992) and their growth rate values should not be used together with the exponential, the logistic, or other Richards family of growth models that rely on accurate μ_{max} values.

Classical CPMs (Equation 3.19 and Equation 3.20) as well as square-root-type models describe a straight line relation between suboptimal temperatures and $\sqrt{\mu_{max}}$ (Figure 3.1 and Figure 4.4 [Chapter 4]). It has been reported by Bajard et al. (1996) that a different, biphasic, relationship can be observed for some strains of *Listeria monocytogenes*. More recently, Le Marc et al. (2002) observed a biphasic relationship for a strain of *Listeria innocua*. Le Marc et al. (2002) suggested an expanded CPM (Equation 3.23) to simulate this type of growth response (Figure 3.3). In Equation 3.23, T_c is the change temperature and T_1 corresponds to the T_{min} value in a classical cardinal temperature model (Rosso et al., 1993). McMeekin et al. (1993), however, cautioned against the interpretation of apparently continuously curved relationships as the combination of two linear responses, and provided a simple illustration of the effect. It should also be noted that other workers (e.g., Nichols et al., 2002a,b; see also Chapter 4) have not observed the "curvature" in the low temperature region of *L. monocytogenes* growth.

$CM_2(T) =$

$$
\begin{cases}
\dfrac{(T - T_1)^2 \cdot (T - T_{max})}{(T_{opt} - T_1) \cdot [(T_{opt} - T_1) \cdot (T - T_{opt}) - (T_{opt} - T_{max}) \cdot (T_{opt} + T_1 - 2 \cdot T)]} & , T_c < T < T_{max} \\[3mm]
\dfrac{(T_c - T_1)^2 \cdot (T - T_{max})}{(T_{opt} - T_1) \cdot [(T_{opt} - T_1) \cdot (T_c - T_{opt}) - (T_{opt} - T_{max}) \cdot (T_{opt} + T_1 - 2 \cdot T_c)]} \cdot \left(\dfrac{T - T_{min}}{T_c - T_{min}} \right)^2 & , T_{min} < T < T_c
\end{cases}
$$

$$(3.23)$$

As stated above, general CPMs rely on the assumption that different environmental parameters have independent and thereby multiplicative effects on μ_{max} (Equation 3.19). The successful use of many general CPMs and square-root-type models has shown this assumption to be reasonable for wide ranges of environmental conditions. However, numerous studies have shown that the growth range of a microorganism to one environmental condition is affected by other environmental factors (see Section 3.4). This suggests that the predictive accuracy of general CPMs can be improved by taking into account interactions between environmental parameters, particularly where one factor is sufficiently stringent that it reduces the growth range of the organism in other environmental "dimensions."

Various approaches have been suggested to describe growth limits under the influence of multiple variables (see Section 3.4.4). Two such approaches have been suggested for direct incorporation in CPMs and are discussed briefly here. Augustin and Carlier (2000b) developed a global secondary model for *L. monocytogenes,* including terms for interactions that prevented growth. Absolute minimal cardinal values X^0_{min} were estimated by assuming that all inhibitory substances were absent. Similarly, absolute minimal inhibitory concentrations MIC_i^0 were estimated for optimal concentrations of other environmental parameters ($X = X_{opt}$). Then, interaction between environmental parameters was taken into account by modifying each of the X^0_{min} values (Equation 3.24) and the MIC_i^0 values, depending on levels of other environmental parameters. After calculation of appropriate T_{min}, $a_{w\,min}$, pH_{min}, and MIC_i values, growth rates were then predicted by using Equation 3.19 to Equation 3.21.

$$ X_{min} = X_{opt} - (X_{opt} - X^0_{min}) \cdot $$

$$ \left(\prod_{i=1}^{n} \left(1 - \dfrac{c_i}{MIC_i^0} \right) - \left(\dfrac{Y_{opt} - Y}{Y_{opt} - Y^0_{min}} \right)^3 - \left(\dfrac{Z_{opt} - Z}{Z_{opt} - Z^0_{min}} \right)^3 \right)^{1/3} \qquad (3.24) $$

with X, Y, and Z being temperature, pH, or water activity.

A different approach was used by Le Marc et al. (2002) to model the interactive effects of temperature, pH, and concentration of undissociated organic acids (HA) on growth of *Listeria innocua.* Cardinal parameter values were kept constant and the space of environmental factors was divided into (1) the independent effect space ($\xi = 1$), (2) the interaction space ($0 < \xi < 1$), and (3) the no growth space ($\xi = 0$) (Equation 3.25 to Equation 3.27).

$$\mu_{max} = \mu_{opt} \cdot CM_2(T) \cdot CM_1(pH) \cdot \tau([HA]) \cdot \xi(T, pH, [HA]) \qquad (3.25)$$

$$\xi(\varphi(T, pH, ULA)) = \begin{cases} 1 & \psi \leq \theta \\ 2(1-\psi) & \theta < \psi < 1 \\ 0 & \psi \geq 1 \end{cases} \qquad (3.26)$$

$$\psi = \sum_i \frac{\varphi_{e_i}}{2 \prod_{j \neq i} (1 - \varphi_{e_j})} \qquad (3.27)$$

with $\varphi_T = (1 - \sqrt{CM_2(T)})^2$, $\varphi_{pH} = (1 - CM_1(pH))^2$, and $\varphi_{Undissociated\ Lactic\ Acid\ (ULA)}$ = $1 - (ULA/MIC_{ULA})$) and where e_i are the environmental factors. For calculation of $CM_2(T)$ and $CM_1(pH)$, see Equation 3.20. Le Marc et al. (2002) selected a value of 0.5, which was used for θ.

The performance of the two approaches to model interaction between environmental parameters is considered in greater detail in Section 3.4.4. As shown above, CPMs that take into account the effect of interaction between environmental parameters are relatively complicated models. Thus, these models are not fully in agreement with the originally cardinal parameter modeling approach, i.e., that CPM uses only simple biological meaningful parameters that microbiologists are familiar with and that are easy to use by biologists (Rosso et al., 1993), and raises questions about whether those models are the most parsimonious forms available.

The model suggested by Augustin and Carlier (2000b) predicts the effect of interaction between temperature, pH, and lactic acid concentration on growth of *Listeria monocytogenes* to be more pronounced than the effect predicted for *Listeria innocua* by the model of Le Marc et al. (2002). For example, the Augustin and Carlier (2000b) model predicts no growth of *Listeria monocytogenes* at 8°C, pH 6.0, and with 200 m*M* of lactic acid, whereas at this condition the model of Le Marc et al. (2002) predicts growth and also that there is no interactive effect of the environmental factors ($\xi = 1$). Recently, Giménez and Dalgaard (in press) found the model of Augustin and Carlier (2000b) to substantially underestimate growth of *Listeria monocytogenes* in cold-smoked salmon. This could indicate that the model is in fact overestimating the importance of the interaction between at least some sets of environmental factors.

In a similar vein Ratkowsky and Ross (1995), recognizing the relationship between absolute limits for each environmental factor and their relationship to the parameters of square-root-type models and CPMs, experimented with the use of a kinetic model as the basis of a growth boundary model using linear logistic regression. This approach is discussed later (see Section 3.4.3.2).

The classical CPMs, in particular those including the effect of temperature, water activity, or pH, are now popular and used for many purposes within predictive microbiology (see Table 3.5 and Table 3.6). As one example a cardinal temperature and pH model has been combined with classical models of microbial kinetics, i.e., models that rely on yield factors and maintenance constants. In this way, production of curvacin A by *Lactobacillus curvatus* LTH 1174 growing in MRS broth was successfully modeled between 20 and 38°C and at pH values from 4.8 to 7.0 (Messens et al., 2002). Other examples include the use for CPMs to predict the radial growth rate of molds on solidified laboratory media (Panagou et al., 2003; Rosso and Robinson, 2001; Sautour et al., 2001). The ability of these models to predict growth in foods deserves further study.

For practical use of secondary predictive models it is important to know the precision of the predicted responses. With CPMs it has been suggested to determine cardinal parameters values for a number of different strains within each of the microbial species of interest (Membré et al., 2002). In this way a measure of intra-species variability can be obtained. As an example, variability in the pH_{min} value for 10 strains of *E. coli* was ±0.20 corresponding to approximately four times the experimental error (Membré et al., 2002). More recently Pouillot et al. (2003) suggested the use of a CPM together with a Bayesian procedure for parameter estimation. This approach includes the use of hyperparameters and allows uncertainty (due to imperfect knowledge or data) and true variability (e.g., due to difference between strains) to be determined separately (see also Chapter 4). The approach seems most interesting and definitely deserves to be studied further for different secondary predictive models.

3.2.3.1 Secondary Lag Time Models and the Concept of Relative Lag Time

When exponentially growing microorganisms are transferred from one environment into another, similar environment, growth usually continues without delay, i.e., a lag time is rarely observed. However, when the two environments differ, a lag time is often observed. Similarly, when microorganisms in the lag or stationary phases are transferred into identical or new environmental conditions a lag time may continue or result, respectively. Depending on the physiological state of the microorganisms, the magnitude of the shift in the environmental conditions, and the new environmental conditions themselves, the duration of the lag time may range from 0 to infinity.

Development of secondary lag time models is complicated by the fact that lag time is influenced not only by the actual environmental conditions but also by previous environmental conditions and the physiological status of the cell, i.e., the growth phase of microorganisms at the time of transfer between environments and their "enzymatic readiness" to exploit the specific carbon and energy resources within the new environment. Within predictive microbiology, two main approaches have been used for development of secondary lag time models: (1) models where lag time and growth rate are modeled independently and (2) models where lag time is assumed proportional to the generation time. The latter group of models typically rely on the assumption that microorganisms need to perform a given amount of work to adapt to a new environment and that the rate at which this work can be done depends on the growth rate potential of the organism in the new environment (Robinson et al., 1998).

In the former approach, lag times or lag rates (i.e., reciprocal of lag time) are typically log-transformed to stabilize the variance of these data. Frequently, polynomial models (see Section 3.2.5) or artificial neural networks (see Section 3.2.6) have been used to develop independent secondary lag time models (Table 3.5). To model the effect of temperature downshifts, temperature upshifts, and physiological status of cells (e.g., exponential phase, stationary phase, starved, frozen, dried), separate polynomial models have been used for the different physiological conditions (Whiting and Bagi, 2002). When square-root-type and Arrhenius-type models are used for lag time modeling, lag *rates* are modeled or reciprocal forms of the growth rate models are used (see Section 3.2.1 and Section 3.2.4; Table 3.5 and Table 3.6).

Zwietering et al. (1994), e.g., used a square-root model (Equation 3.2) with identical values of the parameters T_{min}, c, and T_{max} to model lag time and growth rate — only the value of b differed between the two models. Specific secondary lag time models for particular environmental parameters have also been suggested, e.g., a hyperbola model for the effect of temperature (Equation 3.28; Oscar, 2002; Zwietering et al., 1994):

$$\lambda = \left[\frac{p}{T-q} \right]^{m} \tag{3.28}$$

where λ is the lag time, T the temperature, p the rate of change of lag time as a function of temperature, q the temperature at which lag time is infinite, and m is an exponent to be estimated.

Baranyi and Roberts (1994), Smith (1985), and McMeekin et al. (1993) have observed that lag times for identical inocula introduced to (at least some) environmental conditions are inversely proportional to growth rates and thus proportional to generation times (T_{gen}). This generalization has limits, however, as discussed further below and probably is most relevant to changes in environmental *temperature*. For example, Zwietering et al. (1994) showed that for the effect of temperature on *Lactobacillus plantarum* the product of μ_{max} and lag time (λ) was constant and had an average value close to 2. In these situations secondary lag time models can be derived directly from a growth rate model by using the simple concept of relative lag time (RLT; Equation 3.29) in common use but first defined by Mellefont and Ross (2003). Clearly, RLT reflects the physiological status of microorganisms introduced into a new environment as well as the difference between their actual and their previous environments, and can be interpreted as the amount of work the cell has to do to change its physiology (e.g., enzymes, membrane composition, number of ribosomes) to be able to grow at μ_{max} in that new environment.

Baranyi and Roberts (1994) suggested a primary model to estimate lag times from microbial growth curves and this model allowed determination of the parameters h_0, q_0, and α_0 all of which reflect the physiological state of microorganisms and, thereby, their readiness to grow in a given environment (Equation 3.29; Chapter 2). It can be seen that the parameter RLT is directly proportional to h_0.

$$\frac{\lambda}{T_{gen}} = RLT, \qquad \lambda = \frac{RLT \cdot \ln(2)}{\mu_{max}},$$

$$\lambda \cdot \mu_{max} = RLT \cdot \ln(2) = h_o = \ln\left(1 + \frac{1}{q_o}\right) = -\ln(\alpha_o) \tag{3.29}$$

where all parameters have meanings as indicated earlier.

Experimental methods to determine the physiological status of low levels of microorganisms in foods remain to be developed. Thus, for the time being these parameters have mainly theoretical importance.

The RLT concept is practically very useful for development of secondary lag time models, but it should be used with caution. Delignette-Muller (1998) evaluated data from nine studies where the effect of temperature, pH, NaCl, and NaNO$_2$ on lag time and generation time on different food-borne microorganisms had been modeled independently. In four of the nine studies, RLT was constant and an independent lag time model was not needed. However, primarily pH and NaCl influenced RLT in the remaining studies. On the basis of large amounts of experimental data, Ross (1999a) showed the distribution of RLT of *B. stearothermophilus, Clostridium perfringens, E. coli, L. monocytogenes, Salmonella,* and *S. aureus* included peaks in the range 3 to 6 under a very wide range of experimental conditions. These distributions were similar to those presented by Augustin and Carlier (2000a), who observed a median RLT of 3.09 for *L. monocytogenes* (*n* = 1176). Using extreme environmental shifts, and severely growth-limiting outgrowth conditions, the hypothesis that RLT values have an upper limit was tested (Mellefont et al., 2003, in press). It was found that most RLTs were in the range 4 to 6, and that RLTs greater than 8 could not be induced within the experimental system employed. These observations suggest that while lag time is apparently highly variable, RLT is more uniform and reproducible. Distributions of RLT can be used in stochastic modeling studies, for example, microbial food safety risk assessments, where they could be used as plausible default assumptions if specific lag time information was not available. This approach can also simplify the growth modeling process because use of the RLT as a variable enables the effects of growth rate and lag to be predicted by a single growth rate model, as explained above.

The RLT concept implies that λ is at a minimum value (λ_{min}) when the growth rate is optimal (μ_{opt}). This relation has been used together with CPMs to obtain simple secondary lag time models (Equation 3.30 and Equation 3.31; Augustin and Carlier, 2000a; Le Marc et al., 2002; Pouillot et al., 2003; Rosso, 1995, 1999a,b).

$$\lambda = \frac{\lambda_{min} \cdot \mu_{opt}}{\mu_{max}} \tag{3.30}$$

$$\lambda = \frac{\lambda_{min}}{CM_2(T) \cdot CM_2(a_w) \cdot CM_1(\text{pH})} \qquad (3.31)$$

For RLT models to be used in practice it must be known if, and to what extent, abrupt or smooth shifts in environmental parameters like temperature, pH, and water activity influence RLT.

Data presented by Rosso (1999a,b) suggested that the effect of shifts in temperature and pH on growth of *E. coli* during fermentation of yoghurt was appropriately predicted by a CPM that relied on assumption of a constant RLT. Augustin et al. (2000) suggested a model to take into account the effect of growth phase and temperature history of *L. monocytogenes* on its RLT. For temperature downshifts the RLT increased from ~0 for a temperature shift of 0–5°C to ~2 for a downshift of 30–35°C. To model the effect of temperature downshifts and upshifts on RLT of *L. monocytogenes,* Delignette-Muller et al. (2003) recently used the data of Whiting and Bagi (2002) and suggested simple biphasic linear models. Separate models were used for inoculum with different physiological states. For *E. coli,* Mellefont and Ross (2003) found a similar effect of temperature downshifts whereas temperature upshifts had no systematic effect on RLT. For abrupt downshifts and upshifts in water activity the data of Mellefont et al. (2003) suggest that simple biphasic linear models, with different slopes for down- and upshifts, may be appropriate to predict RLT of both Gram-negative and Gram-positive food-borne bacteria. The universality of these responses remains unclear. For example, RLTs of *S. aureus* and *L. monocytogenes* were largely unaffected by abrupt osmotic shifts over a wide range of salt concentrations, whereas RLT of Gram-negative cells was strongly affected. More research is required before models that are as reliable as existing growth rate models can be developed for lag time, or RLT.

3.2.4 SECONDARY MODELS BASED ON THE ARRHENIUS EQUATION

3.2.4.1 The Arrhenius Equation

The empirical Arrhenius–van't Hoff relationship:

$$\text{rate} = A \exp(\Delta E_a / RT) \qquad (3.32)$$

or its mechanistic interpretation and modification due to Eyring (1935) based in absolute reaction-rate theory:

$$rate = KT \exp(\Delta H^{\ddagger} / RT) \qquad (3.33)$$

where the parameters may be interpreted as follows: A is a constant related to the number of collisions between reactants per unit time, E_a the activation energy, R the gas constant (8.314 J/K/mol), T the temperature in Kelvin, K is similar to A but includes steric and entropic effects, and $\Delta H^{\ddagger}$ is the enthalpy difference between the transition state complex and the reactants, are well established in chemistry to describe the effects of temperature on the rate of chemical reactions. Taking the logarithm of both sides of Equation 3.32:

$$\ln\ (rate) = \ln(A) \times \Delta E/RT$$

and reparameterizing the equation becomes:

$$\ln\ (rate) = A' + \left(\frac{\Delta E}{R}\right) \times \left(\frac{1}{T}\right)$$

Thus, if ln(rate) is plotted against $\left(\dfrac{1}{T}\right)$, the resulting plot is a straight line over temperature ranges relevant to microbial growth and allows estimation of the "activation energy" of the reaction, as shown in Figure 3.4. The activation energy can be used to characterize the reaction.

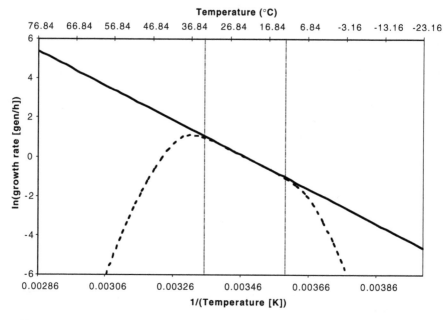

FIGURE 3.4 Diagram showing the effect of temperature on reaction rate predicted using the Arrhenius model (Equation 3.33; solid line) and the effect of temperature on microbial growth rate (dashed line) for a representative mesophilic organism. The "activation energy" is estimated from the slope of the solid line, multiplied by the universal gas constant. Over a narrow range of temperatures, the microbial growth rate follows the Arrhenius model prediction (Equation 3.29). This range has been termed the "normal physiological range" (NPR). At temperatures above or below the NPR, microbial growth rate deviates markedly from that predicted by the Arrhenius model.

It has been argued that because all life processes are the result of chemical reactions, the growth rate of organisms that cannot achieve thermal homeostasis should also be described by Arrhenius kinetics. Within a narrow range of temperature this is true. In practice, however, when microbial growth rate data for the full biokinetic temperature range are presented as an Arrhenius plot, the data are far from linear as shown in Figure 3.4, and confirmed by numerous studies (Heitzer et al., 1991; McMeekin et al., 1993; Schoolfield et al., 1981).

A range of secondary models, based on adherence to the reaction kinetics described by the Arrhenius model, but including terms to account for the observed deviations, have been proposed. These models fall into two groups:

1. Those based on putative mechanistic modifications of the Arrhenius models
2. Those based on empirical modifications

3.2.4.2 Mechanistic Modifications of the Arrhenius Model

Models in this category include those of Johnson and Lewin (1946) to describe the high-temperature growth of bacteria, Hultin (1955) to describe rates of enzymatic catalysis in the low temperature region, Sharpe and DeMichele (1977) who synthesized these two equations to produce a model for the temperature dependence of bacterial growth rate in the entire biokinetic region, the model of Schoolfield et al. (1981), which is a reparameterization of the Sharpe and DeMichele model to overcome difficulties in fitting by nonlinear regression, and the models of McMeekin et al. (1993) and Ross (1993a, 1999b). The latter models incorporate contemporary knowledge of the thermodynamics of protein folding to overcome failures in the Schoolfield et al. model related to unrealistic parameter estimate (Ratkowsky et al., 1991).

The above models were originally developed to provide an interpretation of microbial growth rates or enzyme-catalyzed reaction rates, in response to temperature but their mechanistic basis makes them attractive for use as secondary models.

This class of secondary models have previously been reviewed (McMeekin et al., 1993; Ratkowsky et al., 1991; Ross, 1999b; Ross and McMeekin, 1994). In summary, all of the models are based on the assumption that there is a single, enzyme-catalyzed, rate-limiting reaction in any microorganism. This reaction is characterized by an activation energy, which governs the rate of reaction in response to temperature, according to Arrhenius kinetics. Enzymes are proteins, however, and are themselves subject to the effects of temperature. The functional activity of enzymes is dependent upon their shape, or conformation, but they are flexible — the flexibility being required to achieve their catalytic function. Because temperature affects the bonds in the molecule, if the temperature changes too much, the conformation becomes so disrupted that denaturation takes place, both at high and low temperatures. These denaturation events are reversible, but at high temperatures if the temperature increases sufficiently, irreversible denaturation takes place (Ross, 1999b). Thus, these models include terms to model the probability, as a function of temperature, that the enzyme is in its metabolically active conformation and use this estimate to modify the predictions of the Arrhenius model. Equation 3.34 to Equation 3.36 are examples of this form of model.

Model of Hinshelwood (1946):

$$\text{rate} = A_1 \exp(-E_a/RT) - A_2 \exp(-E_{a\,high}/RT) \tag{3.34}$$

where R, T, A, and E_a have the same meaning as above. $E_{a\,high}$ is the activation energy of the high-temperature denaturation of the rate-limiting enzyme.

Model of Schoolfield et al. (1981):

$$\frac{1}{K} = \frac{\rho_{(25)}\dfrac{T}{298}\exp\left\{\dfrac{H_A}{R}\left(\dfrac{1}{298}-\dfrac{1}{T}\right)\right\}}{1+\exp\left\{\dfrac{H_L}{R}\left(\dfrac{1}{T_{1/2_L}}-\dfrac{1}{T}\right)\right\}+\exp\left\{\dfrac{H_H}{R}\left(\dfrac{1}{T_{1/2_H}}-\dfrac{1}{T}\right)\right\}} \tag{3.35}$$

where T is the absolute temperature, R is the universal gas constant, and, for modeling bacterial growth, the other parameters have been interpreted as follows: K is the response (e.g., generation) time, $\rho_{(25)}$ a scaling factor equal to the response rate ($1/K$) at 25°C, H_A the activation energy of the rate-controlling reaction, H_L the activation energy of denaturation of the growth-rate-controlling enzyme at low temperatures, H_H the activation energy of denaturation of the growth-rate-controlling enzyme at high temperatures, $T_{1/2_L}$ the lower temperature at which half of the growth-rate-controlling enzyme is denatured, and $T_{1/2_H}$ is the higher temperature at which half of the growth-rate-controlling enzyme is denatured.

Model of Ross (1999b):

$$\textit{rate} =$$

$$\frac{CT\exp(\Delta H^{\ddagger}/RT)}{1+\exp(-n(\Delta H^*-T\Delta S^*+\Delta C_p[(T-T^*_H)-T\ln(T/T^*_S)])/RT)} \tag{3.36}$$

where C is a parameter whose value must be estimated, $\Delta H^{\ddagger}$ the activation enthalpy of the reaction catalyzed by the enzyme controlling the overall reaction rate, ΔC_p the difference in heat capacity (per mole amino acid residue) between the native (catalytically active) and denatured state of the enzyme, T_H^* the temperature (K) at which the ΔC_p contribution to enthalpy is 0, T_S^* the temperature (K) at which the ΔC_p contribution to entropy is 0, ΔH^* the value of enthalpy at T_H^* per mole amino acid residue, ΔS^* the value of entropy at T_S^* per mole amino acid residue, T the temperature (K), R the gas constant (8.314 J/K/mol), and n is the number of amino acid residues in the enzyme.

Equation 3.34 to Equation 3.36 include the simple Arrhenius model in the numerator of the equation. The denominators in Equation 3.35 and Equation 3.36, however, model the probability that the enzyme is in its active conformation. When

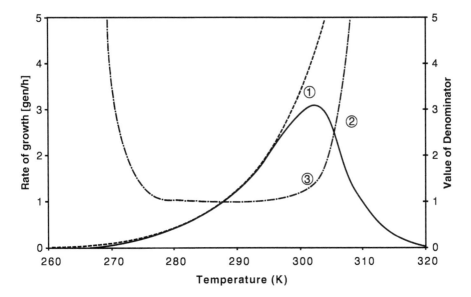

FIGURE 3.5 Diagram showing the interaction of terms in mechanistic models for microbial growth rate response to temperature. Curve 1 (dashed line) is the predicted growth rate in the absence of master enzyme denaturation, i.e., Arrhenius kinetics as modeled by the numerator of Equation 3.36. Curve 2 (dot-dash line) is the inverse of the probability of the "master enzyme" being in the active conformation, i.e., the denominator of Equation 3.36. Curve 3 (solid line) is the overall predicted rate from the model, i.e., the quotient of values in Curve 1 divided by values in Curve 2.

that probability is high, the denominator takes values close to 1, so that the overall rate is close to that predicted by the Arrhenius equation in the numerator. When the probability is lower, the value of the denominator increases, so that the observed rate is lower than that predicted by the numerator alone. These relationships are shown in Figure 3.5 for Equation 3.36, presented as rate vs. temperature for clarity of interpretation.

In practice, few of these types of models have been routinely applied in predictive microbiology, possibly because the models are highly nonlinear, and initial parameter estimates are difficult to determine. Furthermore, it is currently not possible to independently measure the values of the parameters of the model because the putative master reaction has not been identified, and the concept that a single reaction is rate limiting under all environmental conditions seems improbable (Daughtry et al., 1997; Ross, 1999b). Finally, several workers (Heitzer et al., 1991; Ratkowsky, D.A., personal communication, 2003; Ross, 1993b) demonstrated that even with good quality data, square-root-type models provide an equally good fit as those "mechanistic" models, and are usually easier to work with. Examples of their use include Broughall et al. (1983) and Broughall and Brown (1984) who used the Schoolfield model, but also extended it to model the effect of water activity and pH, by replacing some terms in the model with polynomial expressions in a_w and pH. Adair et al. (1989) used a reparameterized form of the Schoolfield et al. model, which was essentially another form of the Johnson and Lewin (1946) model. Recent studies (Ratkowsky et al., unpublished) have confirmed that Equation 3.36 does describe bacterial temperature–growth rate curves well for a wide range of species, and, in contrast with earlier models, that the estimated parameter values are realistic and consistent with the theoretical bases of the model.

3.2.4.3 Empirical Modifications of the Arrhenius Model

A second class of Arrhenius-based models for growth rate and reciprocal of lag time have been presented by Davey and coworkers. Davey (1989) introduced an Arrhenius-type model for the effects of temperature and water activity, which is linear and thus allows for explicit solution of the optimum parameter values. This model has the form:

$$\ln(\text{rate}) = C_0 + \frac{C_1}{T} + \frac{C_2}{T^2} + C_3 a_w + C_4 a_w^2 \tag{3.37a}$$

where T is temperature (K), a_w has its usual meaning, and C_0, C_1, C_2, C_3, C_4 are coefficients to be determined.

Davey (1989) reported that the model described well seven data sets from the literature and subsequently demonstrated the ability of the model to also describe the reciprocal of lag phase duration (Davey, 1991). Davey (1994) fitted a variation of the model to the data of Adams et al. (1991) for *Yersinia enterocolitica* growth. The model included terms for temperature and pH, and is analogous to Equation 3.37a:

$$\ln(\text{rate}) = C_0 + \frac{C_1}{T} + \frac{C_2}{T^2} + C_3 pH + C_4 pH^2 \tag{3.37b}$$

where T is temperature (K), pH has its usual meaning, and C_0, C_1, C_2, C_3, C_4 are coefficients to be determined.

On the basis of these observations, Davey (1994) extended his earlier proposed general model structure for linear Arrhenius models (Davey, 1989) to account for the effect of multiple environmental factors affecting growth rate to the following form:

$$\ln(\text{rate}) = C_0 + \sum_{i=1}^{j} (C_{2i-1} V_i + C_{2i}^2 V_i) \tag{3.37c}$$

where j environmental factors, V, act in combination to affect the growth of the modeled organism, and C_0, C_1, C_2, ..., C_j are coefficients to be determined.

This general form was applied by Davey and Daughtry (1995) to data of Gibson et al. (1988) for *Salmonella* growth in response to temperature, NaCl, and pH. Thus, their equation had the form:

$$\ln(rate) = C_0 + \frac{C_1}{T} + \frac{C_2}{T^2} + C_3 S + C_4 S^2 + C_5 pH + C_6 pH^2 \qquad (3.37d)$$

where S is salt concentration (% w/v).

While the above model forms are empirical, they also recognize implicitly the temperature dependence of microbial growth rates. Daughtry et al. (1997) invoked chemical reaction rate theory to develop an alternative mechanistic model based on the Arrhenius equation. Those workers cited Levenspiel (1972) as stating that curvature in Arrhenius plots can arise if there are two, or more, reactions that "compete" to limit the reaction rate and dominate under different conditions so that the overall effect of temperature is the synthesis of the individual activation energies for the rate-limiting reactions at different temperatures. Daughtry et al. (1997) considered that bacterial growth was likely to be such a system.

By assuming that the "heat of reaction" (equivalent to the activation energy or activation enthalpy in the above discussion) is a polynomial function of temperature, the following modified Arrhenius model was developed:

$$\ln(rate) = C_0 + \frac{C_1}{T} + C_2 \ln T \qquad (3.38)$$

This model fitted experimental data as well as the temperature-only form of Equation 3.37a.

The "linear Arrhenius" or "Davey" models have been used to model growth of molds on solid microbiological media (Molina and Giannuzzi, 1999; Panagou et al., 2003). Panagou et al. (2003) preferred cardinal parameter and gamma-concept-type models (see Sections 3.2.2 and 3.2.3) over the Davey model because of their interpretable parameter values. Davey models have also been applied to UV and thermal inactivation and data describing the combined effects of pH and water activity on thermal inactivation, including vitamin denaturation (see Section 3.3), but they have not been widely adopted by other workers. McMeekin et al. (1993) and Davey (2001) identified a close correlation between estimates of coefficients C_1 and C_2, and C_3 and C_4, of Equation 3.37a, suggesting that the model was overparameterized.

3.2.4.4 Application of the Simple Arrhenius Model

For the entire biokinetic temperature range, growth rates of microorganisms are described less appropriately by the Arrhenius-type equations (Equation 3.34 to Equation 3.36) than by square-root-type and cardinal parameter models (see Section 3.2.4.2; Rosso et al., 1993; Zwietering et al., 1991). However, Arrhenius-type models remain useful as secondary kinetic models for less extensive ranges of storage temperatures (Table 3.5 and Table 3.6). Koutsoumanis and Nychas (2000) used Equation 3.32 to model the effect of temperatures between 0 and 15°C on μ_{max} and reciprocal lag time of naturally occurring pseudomonads growing aerobically on a type of Mediterranean fish. Koutsoumanis et al. (2000) also expanded the classical

Arrhenius model to take into account the combined effect of temperature and CO_2 on growth rates of spoilage bacteria in modified atmosphere packed fresh fish (Equation 3.39).

$$\ln(\mu_{max}) = \frac{E_a}{R} \times \left(\frac{1}{T_{ref}} - \frac{1}{T} \right) + \ln(\mu_{ref} - d_{CO_2} \times \% \, CO_2) \qquad (3.39)$$

where T, E_a, and R have their usual meaning, $\%CO_2$ is the equilibrium concentration of CO_2 in the headspace gas, d_{CO_2} is a constant expressing the effect of CO_2 on μ_{max} and T_{ref} and μ_{ref} are temperature and maximum specific growth rate, respectively, at 273 K and 0 $\%CO_2$. The term including d_{CO_2} in Equation 3.39, describing CO_2 inhibition of growth rate, was previously suggested by Kalina (1993).

The simple Arrhenius model has also been used to calculate relative rates of spoilage (RRS) (Equation 3.37). RRS for a food product is defined as the shelf life (determined by sensory evaluation) at a reference temperature (T_{ref}) divided by the shelf life observed at the actual storage temperature (Equation 3.40).

$$RRS = \frac{\text{Shelf life at } T_{ref}}{\text{Shelf life at } T} = \exp\left[\frac{E_A}{R} \times \left(\frac{1}{T} - \frac{1}{T_{ref}} \right) \right] \qquad (3.40)$$

where T, E_a, and R have their usual meaning and T_{ref} is a reference temperature at which the shelf life is known.

RRS models are interesting because they enable shelf life to be predicted at different temperatures and for products where the specific spoilage organisms or the type of reaction responsible for spoilage are not known. For an unusually temperature-sensitive modified atmosphere packed shrimp product ($E_a > 100$ kJ/mol), Equation 3.40 described the effect of temperature (0 to 25°C) on shelf life more appropriately than a similarly formulated RRS model relying on the square-root model (Equation 3.1). However, a simple exponential RRS model was as useful as Equation 3.40. That the Arrhenius and exponential RRS models performed better than the square-root model was due to the fact that different groups of microorganisms were responsible for spoilage at low and high storage temperatures, respectively (Dalgaard and Jørgensen, 2000). This situation is common and a reason why entirely empirical RRS models can be more appropriate for shelf-life prediction than kinetic models relying on growth of known spoilage microorganisms. In fact, kinetic models for growth of spoilage bacteria are generally useful only for shelf-life prediction within the spoilage domain of a specific microorganism (Dalgaard, 2002).

3.2.5 POLYNOMIAL AND CONSTRAINED LINEAR POLYNOMIAL MODELS

Of the types of secondary models applied within predictive microbiology polynomial models are probably the most common. As shown in Table 3.5 and Table 3.6, the effect of many different environmental parameters (e.g., temperature, NaCl/a_w, pH, nitrite, CO_2, organic acids, and natural antimicrobials) has been described by these linear models. Polynomial models were extensively used during the 1990s and they remain widely applied although square-root-type and CPMs are becoming increasingly popular (Table 3.5). Polynomial models are attractive, first, because they are relatively easy to fit to experimental data by multiple linear regression, which is available in most statistical packages. Second, polynomial models allow virtually any of the environmental parameters and their interactions to be taken into account. Thus, application of polynomial models is a simple way to summarize information from a data set. Once the coefficients in a polynomial model have been estimated, the information is easy to use particularly if the model is included in application software. In fact, the application software packages Pathogen Modeling Program and Food MicroModel rely primarily on the use of polynomial models (www.arserrc.gov/mfs/pathogen.htm; Buchanan, 1993a; McClure et al., 1994a).

To illustrate the use of polynomial models a quadratic equation used by McClure et al. (1993) is shown below (Equation 3.41):

$$\ln y = p_1 + p_2 x_1 + p_3 x_2 + p_4 x_3 + p_4 x_1 x_2 + p_6 x_1 x_3$$
$$+ p_7 x_2 x_3 + p_8 x_1^2 + p_9 x_2^2 + p_{10} x_3^2 + e$$
(3.41)

where $\ln y$ denotes the natural logarithm of the modeled growth responses ($y = \mu_{max}$, lag time or maximum population density [MPD], or the modified Gompertz model parameters B or M); p_i ($i = 1, \ldots, 10$) are the coefficients to be estimated; x_1 is the temperature (°C); x_2 is the pH; x_3 is NaCl (% w/v); e is a random error supposed to have zero mean and constant variance.

As shown by Equation 3.41 the same polynomial equation can be used to model different microbial growth responses. Actually, many studies have modeled the effect of environmental conditions on specific parameters in primary growth models, particularly B, M, and C in the modified Gompertz model. Measures of lag time, growth rate, or time for, e.g., a 1000-fold increase in the cell concentration are then calculated at specific environmental conditions from the predicted value of B, M, and C (Buchanan and Phillips, 2000; Eifert et al., 1997; Skinner et al., 1994; Zaika et al., 1998). Growth responses to be modeled are typically ln- or $\log_{10}$-transformed (Equation 3.41) and it is common practice to transform the growth response without transforming the model.

However, polynomial models have properties that limit their usefulness as secondary predictive models. Polynomials include many coefficients that have no biological interpretation. As an example, Equation 3.41 uses 10 coefficients to model the effect of three environmental parameters. With four environmental parameters, polynomials with 15 coefficients are frequently used. The high number of coefficients and their lack of biological interpretation make it difficult to compare polynomial models with other secondary predictive models. The important information included in, e.g., the T_{min} parameter of a square-root-type model, is not provided by a polynomial model.

Higher order polynomial models, e.g., cubic or quadratic models have been criticized for being too flexible and for attempting to model, rather than eliminate, experimental error (Chapter 4; Baranyi et al., 1996; Sutherland et al., 1996). Because of the very flexible nature of higher order polynomial models they should not be used as secondary models within predictive microbiology unless very high quality experimental data are available and support the application of these models. Furthermore, because quadratic polynomial models are highly flexible they should only be used to provide predictions by interpolation. Baranyi et al. (1996) pointed out that the interpolation region of a polynomial model is the minimum convex polyhedron (MCP) defined by the ranges of the environmental parameters used to develop the model, i.e., the experimental design. These authors also stressed that the interpolation region (Figure 3.10) can be substantially smaller than the rectangular parallelepiped whose sides are given by the endpoints of the ranges of environmental parameters, termed the "nominal variable space" (Baranyi et al., 1996).

Determination of the interpolation region of a polynomial model is not self-evident and requires information about ranges of the environmental parameters used to develop the model. Pin et al. (2000) suggested a method to determine if a specific environmental condition is inside or outside the interpolation region of a particular polynomial model. This method relies on the iterative algorithm used by the Solver add-in of Microsoft Excel and thus is readily accessible to many users. However, we believe for it to become widely used the calculation of interpolation regions should be included in dedicated predictive modeling application software.

To overcome the problem that quadratic polynomial models can be too flexible, and therefore in some situations provide predictions that are not logical, the application of constrained polynomial models was recently suggested (Geeraerd et al., in press). With this approach, the basic idea is to combine *a priori* information about the effect of environmental parameters on growth responses with classical polynomial models. For example, at suboptimal conditions it was assumed that the growth rate should always increase for increasing temperature and a_w values and decrease for increasing CO_2 levels. Thus, the partial derivative of the model with respect to temperature and a_w should always be positive whereas the partial derivative of the model with respect to CO_2 should always be negative. Coefficients of the polynomial model were then fitted with the constraints obeyed at all edges of the experimental design. The constrained polynomial model was fitted by the usual process of minimizing the sum of squared errors and the fitting was carried out using the Optimization Toolbox within the MatLab software (Geeraerd et al., in press). As compared to classical polynomial models, constrained polynomial models have the clear advantage of being more robust but the clear disadvantage of being substantially more difficult to fit. Simplification of the fitting process seems necessary before constrained polynomial models find wide application in predictive microbiology.

Masana and Baranyi (2000a) described methods for integration of new data into existing polynomial models, pointing out that the interpolation region of the newly developed model can be unexpectedly small and also presenting methods for quantifying the increased risk of inadvertent extrapolation (Baranyi et al., 1996). Polynomial models feature many cross-product terms, making the addition of new terms much more complex than with models embodying the gamma concept (Section

3.2.2). Nonetheless, when expanding a model by the addition of data for a new variable, Masana and Baranyi (2000b) demonstrated that the original model can be retained as a special case of the expanded model, by holding the terms of the original model, i.e., those that do not contain the new variable, as constants during the model fitting process for the expanded model.

3.2.6 ARTIFICIAL NEURAL NETWORKS

Artificial neural networks (ANNs) are algorithms that can be used to perform complex statistical modeling between a set of predictor variables and response variables. Their particular advantage is that they have the potential to approximate underlying relationships of any complexity between those variables. They have been used to generate secondary models for microbial growth rates and lag times (Garcia-Gimeno et al., 2002, 2003; Geeraerd et al., 1998a; Jeyamkondan et al., 2001; Lou and Nakai, 2001; Najjar et al., 1997), growth under fluctuating environmental conditions (Cheroutre-Vialette and Lebert, 2000; Geeraerd et al., 1998a), microbial inactivation (Geeraerd et al., 1998b), and have been proposed as an alternative to logistic regression modeling techniques (Tu, 1996). Their potential to replace logistic regression for growth limits modeling (see Section 3.4) has also been described (Hajmeer and Basheer, 2002, 2003a,b) in which context they have been termed "probabilistic neural networks" (PNNs).

Hajmeer et al. (1997) and Hajmeer and Basheer (2002, 2003a) describe the principles of ANNs and related technologies in the context of predictive microbiology, and numerous texts are dedicated to the subject but the following is largely drawn from the succinct and lucid description of Tu (1996).

Artificial neural networks were conceived decades ago by researchers attempting to reproduce the function of the human brain, i.e., its ability to learn and remember, but it was only in the 1980s that the "back-propagation" technique was rediscovered, enabling such computational systems to "learn" mathematical relationships between input and output variables.

Neural networks are effectively a series of mathematical relationships between predictor variables ("input nodes"), a series of hidden "nodes," and an output variable ("output node") (Figure 3.6). Each input node is related to each hidden node, and each hidden node is related to the output node, by some mathematical function. Each input is given a weight during the "training" routines, the value of each hidden node being the sum of a weighted linear combination of the input node values. In addition, bias values can be added to the weighted values of the inputs. These are analogous to the intercept in regression equations, while the weights are analogous to coefficients of the independent variables. The output node receives a weighted input from each of the hidden nodes, to which is often applied a logistic transformation or other function (the "activation function") to determine the overall output.

A set of input and corresponding output values is presented to the network, the error is evaluated, and the weights are then adjusted to minimize the difference between the predicted output and that which was observed. This process of adjustment of weights is the back-propagation step and involves algorithms based on complex equations. Input data are continuously presented to the neural network until

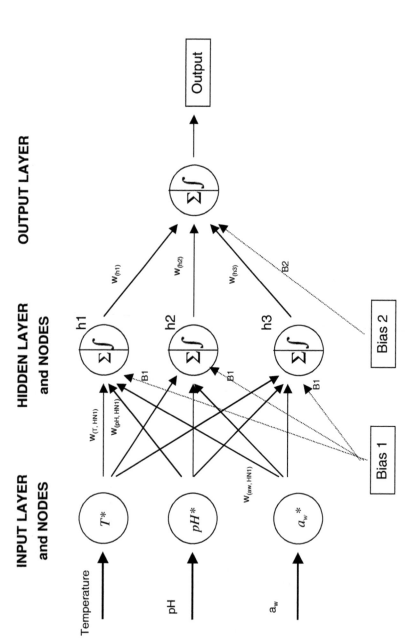

FIGURE 3.6 Diagram of an imaginary artificial neural network that might be used in predictive microbiology. The output is the response of the population of microorganisms to variations in the temperature, pH, and water activity of their growth medium. (The diagram is fully explained in the text.)

the overall error has been minimized, a process analogous to the iterative routines employed in nonlinear regression software. Optimal training algorithms can, at this time, only be determined empirically. Additionally, when using ANNs other elements of the modeling require experimentation, including the number of training cycles (too many can reduce predictive performance), the number of nodes in the hidden layer, and the ideal learning rate (the magnitude of change in the weights for each training case).

In Figure 3.6, the input, hidden, and output layers are shown, as well as the connections between them. Nodes are represented by circles. The $W_{(i,j)}$ terms indicate the weight applied to the inputs to hidden nodes. (Not all weights are represented in the diagram.) The hidden nodes have a transformation applied to them, e.g., a logistic function represented by the functions $h1$, $h2$, etc. Thus, in the example:

$$h1 = 1/(1+\exp(\text{Bias } 1 + W_{(T, HN1)} \times T^* + W_{(pH, HN1)} \times pH^* + W_{(aw, HN1)} \times a_w^*))$$

and, similarly:

$$\text{Output} = 1/(1+ \exp(B2 + W_{(h1)} + W_{(h2)}))$$

Sufficient data are required so that a subset of data can be used to train the ANN, while the remainder is used to test the predictive ability of the ANN. One complete cycle of the training data set is called an "epoch" and the duration of the training is often described as the number of epochs required to minimize the error in the training set.

Tu (1996) compared the advantages and disadvantages of the ANN approach to those of traditional statistical regression modeling, as summarized in Table 3.4. Evaluation of the approach as applied to predictive models for microbial growth is presented below, and in relation to growth limits models in Section 3.4.2. Further comment is provided in Chapter 4, Section 4.4.3.

The use of ANN in predictive growth modeling remains relatively little developed, and direct comparison of the performance of different ANN techniques is still lacking. To describe growth curves, Schepers et al. (2000) concluded ANN was less appropriate than classical nonlinear sigmoidal growth models. Cheroutre-Vialette and Lebert (2000), however, found a recurrent (i.e., back-propagation) ANN suitable to predict growth of *Listeria monocytogenes* under constant and fluctuating pH and NaCl conditions. As shown in Table 3.5 and Table 3.6, several secondary ANN models have been developed including models for *Aeromonas hydrophila*, *Brochothrix thermosphaca*, *Escherichia coli*, lactic acid bacteria, *Listeria monocytogenes*, and *Shigella flexneri*. These secondary ANN models have been compared with polynomial, square-root-type, and cardinal parameter models. The comparisons showed ANN typically fitted experimental data better and in most cases provided slightly more accurate predictions. Thus, in general, ANN may provide slightly improved predictions. Commercial neural network software is available and development of ANNs has become relatively easy. However, ANN is a data-driven approach and this could be a drawback because a secondary model that can be written as an equation with coefficients and parameters is not produced. The

TABLE 3.4
Advantages and Disadvantages of Neural Network Approaches to Modeling

Disadvantages	Advantages
Neural networks are a "black box" and have limited ability to specifically identify possible causal relationships between predictor and response variables	Neural networks require less formal statistical training to develop
Neural network models may be more difficult to use in the field	Neural network models can implicitly detect complex nonlinear relationships between predictor and response variables
Neural network modeling requires greater computational resources	Neural network models have the ability to detect all possible interactions between predictor variables
Neural network models are prone to overfitting	Neural networks can be developed using multiple different training algorithms
	Neural network model development is empirical, and many methodological issues remain to be resolved

Source: After Tu, J.V. *J. Clin. Epidemiol.*, 11, 1225–1231, 1996.

incorporation of classical secondary models in user-friendly application software has been essential for the usefulness of predictive microbiology in industry, teaching, and research. It remains to be demonstrated whether successful ANN models can, in a similar way, be communicated to and conveniently applied by wide groups of users within predictive microbiology.

3.3 SECONDARY MODELS FOR INACTIVATION

There are relatively few models that consider the effects of multiple environmental factors on the rate of death of microorganisms, and these are discussed in Chapter 2 and Chapter 5. Some available inactivation models are also summarized in Table 3.5 and Table 3.6.

3.4 PROBABILITY MODELS

3.4.1 INTRODUCTION

Models to predict the likelihood, as a function of intrinsic and extrinsic factors, that growth of a microorganism of concern could occur in a food were first explored in the 1970s. Those models were concerned with prediction of the probability of formation of staphylococcal enterotoxin or botulinum toxin within a specified period of time under defined conditions of storage and product composition (Genigeorgis, 1981; Gibson et al., 1987). Phenomena that have been modeled using this approach

TABLE 3.5
Examples of Secondary Models for Growth of Pathogenic and Indicator Microorganisms

Microorganisms and References	Type of Secondary Model	Response Variables	Independent Variables and Ranges
Aeromonas hydrophila			
Palumbo et al. (1992)[a]	Polynomial	GT,[b] lag	T (5–42°C); NaCl (0.5–4.5%); pH (5.0–7.3); Na-nitrite (0–200 ppm); anaerobic
McClure et al. (1994b)	Polynomial	GT, lag	T (3–20°C); NaCl (0.5–4.5%); pH (4.6–7.0); aerobic
Palumbo et al. (1996)[a]	Polynomial	GT, lag	T (5–42°C); NaCl (0.5–4.5%); pH (5.0–7.3); Na-nitrite (0–200 ppm); aerobic
Devlieghere et al. (2000a)	Square-root, polynomial	μ_{max}, lag	T (4–12°C); a_w (0.974–0.992); CO_2 (0–2403 ppm); pH 6.12; nitrite (22 ppm)
Jeyamkondan et al. (2001)	ANN	GT, lag	Data from McClure et al (1994b)
Aspergillus spp.			
Pitt (1995)	Kinetic with yield factors	Growth and aflatoxin formation	T; a_w; pH; and colony size: limits not specified in manuscript
Molina and Giannuzzi (1999)	Arrhenius	Colony growth	T (25–36°C); propionic acid (129–516 ppm)
Rosso and Robinson (2001)	CPM	Colony growth	T (25, 30, 37°C); a_w (0.83–0.99); pH (6.5); humectant: glucose/fructose
Sautour et al. (2001)	CPM	Colony growth	T (25°C); a_w (0.88–0.99)
Bacillus cereus			
Benedict et al. (1993)[a]	Polynomial	GT, lag	T (5–42°C); NaCl (0.5–5.0%); pH (4.5–7.5); Na-nitrite (0–200 ppm); aerobic
Sutherland et al. (1996)[c]	Polynomial	GT, lag	T (10–30°C); NaCl (0.5–10.5%); pH (4.5–7.0); CO_2 (10–80%)
Zwietering et al. (1996)	Gamma	μ_{max}	T (10–30°C); a_w (0.95–1.00); pH (4.9–6.6)
Chorin et al. (1997)	Polynomial	Growth rate, lag	T (7–30°C); a_w (0.95–0.991); pH (5–7.5); humectant: glycerol
Singaglia et al. (2002)	Polynomial	Spore germination	T (20–40°C); a_w (0.94–0.99); pH (4.5–6.5)
Clostridium botulinum			
Baker and Genigeorgis (1990)[a]	Polynomial	Time to toxin formation	T (4–30°C); initial spore conc. (–2 to +4 log cfu/g); initial aerobic plate count (–2 to +3 log cfu/g)
Graham et al. (1996)[c]	Polynomial	GT, time to toxin	T (4–30°C); NaCl (1.0–5.0%); pH (5.0–7.3)

TABLE 3.5 (Continued)
Examples of Secondary Models for Growth of Pathogenic and Indicator Microorganisms

Microorganisms and References	Type of Secondary Model	Response Variables	Independent Variables and Ranges
Whiting and Oriente (1997)[a]	Polynomial	Time to turbidity	T (4–28°C); NaCl (0–4%); pH (5–7); initial spore conc. (1–5 log cfu/g)
Chea et al. (2000)	Polynomial	Spore germination	T (15–30°C); NaCl (0.5–4.0%); pH (5.5–6.5)
Fernandez et al. (2001)	Polynomial	Time to turbidity	T (5–12°C); NaCl (0.5–2.5%); pH (5.5, 6.5); CO_2 (0–90%)
Clostridium perfringens			
Juneja et al. (1996)[a]	Polynomial	GT, lag	T (12–42°C); NaCl (0–3%); pH (5.5–7.0.); Na-pyrophosphate (0–3%)
Escherichia coli			
Buchanan and Bagi (1994)[a]	Polynomial	GT, lag	T (5–42°C); NaCl (0.5–5.0%); pH (4.5–8.5); Na-nitrite (0–200 ppm); aerobic and anaerobic
Sutherland et al. (1995, 1997)[c]	Polynomial	GT, lag	T (10–30°C); NaCl (0.5–6.5%); pH (4.0–7.0); Na-nitrite (0–200 ppm); aerobic
Rasch (2002)	Polynomial	Growth rate	T (10–30°C); NaCl (0.5–3.0%); pH (4.5–6.5); reuterin (0–4 AU/ml)
Ross et al. (2003)	Square-root	GT	T (7.6–47.4°C); a_w (0.951–0.999); pH (4.02–8.28); lactic acid (0–500 mM)
Skandamins et al. (2002)	Vitalistic	Time to decline in cell concentration	T (0–15°C); pH (4.0–5.0); oregano essential oil (0.0–2.1%)
Garcia-Gimeno et al. (2003)	ANN	Growth rate, lag time	T (9–21°C); NaCl (0–8%); pH (4.5–8.5); Na-nitrite (0–200 ppm)
Whiting and Golden (2003)	Polynomial	Time to decline in cell concentration	T (4–37°C); NaCl (0–15%); pH (3.5–7.0); Na-lactate (0–2%); Na-nitrite (0–75 ppm)
Listeria monocytogenes[d]			
Buchanan et al. (1997)	Polynomial	Time to decline in cell concentration	T (4–42°C); NaCl (0.5–19%); pH (3.2–7.3); lactic acid (0–2%); Na-nitrite (0–200 ppm)
Razavilar and Genigeorgis (1998)	Polynomial	Probability of growth	T (4–30°C); NaCl (0.5–12.5%); methyl paraben (0–0.2%); pH (~5.9); K-sorbate (0.3%); Na-propionate (0.1%); Na-benzoate (0.1%)
Cheroutre-Vialette and Lebert (2000)	ANN	Absorbance at 600 nm	T (20°C); pH (5.6–9.5); NaCl (0–8%)

TABLE 3.5 (Continued)
Examples of Secondary Models for Growth of Pathogenic and Indicator Microorganisms

Microorganisms and References	Type of Secondary Model	Response Variables	Independent Variables and Ranges
Bouttefroy et al. (2000)	Polynomial	Cell concentration	T (22°C); pH (5.0–8.2); NaCl (0–6%); curvaticin 13 (0–160 AU/ml)
Rodriguez et al. (2000)	Arrhenius	μ_{max}	T (4–20°C)
Augustin and Carlier (2000a,b)	CPM	μ_{max}, lag	T (−2.7 to −45.5°C); a_w (0.910–0.997); pH (4.55–9.61); acetic acid (0–20.1 mM); lactic acid (0–5.4 mM); citric acid (0–1.6 mM); Na-benzoate (0–0.7 mM); K-sorbate (0–5.1 mM); Na-nitrite (0–11.4 µM); glycerol monolaurin (0–118.5 ppm); butylated hydroxyanisole (0–254 ppm); butylated hydroxytoluene (0–48.7 ppm); $tert$-butylhydroquinone (0–1400 ppm); CO_2 (0–1.64 proportion); caffeine (0–10.8 g/l); phenol (0–12.5 ppm)
Ross et al. (in press)	Square-root	μ_{max}	T (3–40 °C); a_w (0.920–0.997); pH (4.0–7.8); lactic acid (0–450 mM); nitrite (0–150 ppm)
Buchanan and Phillips (2000)	Polynomial	GT, lag	T (4–37°C); pH (4.5–7.5); NaCl (0.5–10.5%); Na-nitrite (0–1000 ppm); aerobic
Buchanan and Phillips (2000)	Polynomial	GT, lag	T (4–37°C); pH (4.5–8.0); NaCl (0.5–5.0%); Na-nitrite (0–1000 ppm); anaerobic
Devlieghere et al. (2001)	Square-root, polynomial	μ_{max}, lag	T (4–12°C); a_w (0.9622–0.9883); Na-lactate (0–3.0%); Na-nitrite (20 ppm); pH (6.2)
Le Marc et al. (2002)	CPM	μ_{max}, lag	T (0.5–43°C); pH (4.5–9.4); acetic acid (16–64 mM); lactic acid (40–138 mM); propionic acid (18–55 mM)
Seman et al. (2002)	Polynomial	Growth rate	T (4°C); NaCl (0.8–3.6%); Na-diacetate (0.0–0.2%); K-lactate (0.15–5.6%); Na-erythrobate (317 ppm); Na-nitrite (97 ppm); Na-tripolyphosphate (0.276%)

TABLE 3.5 (Continued)
Examples of Secondary Models for Growth of Pathogenic and Indicator Microorganisms

Microorganisms and References	Type of Secondary Model	Response Variables	Independent Variables and Ranges
Gimenez and Dalgaard (in press)	Square-root	μ_{max}	T (4–10°C); %WPS (2–6%); smoke components/phenol (3–10 ppm); pH (5.9–6.3); lactic acid (0–20,000 ppm); interaction with lactic acid bacteria
Salmonella			
Gibson et al. (1988)[c]	Polynomial	GT, lag	T (10–30°C); NaCl (0.5–4.5%); pH (5.6–6.8); aerobic
Davey and Daughtry (1995)	Arrhenius	Growth rate, lag	Data from Gibson et al. (1988)
Koutsoumanis et al. (1998)	Polynomial	μ_{max}	T (22–42°C); pH (5.5–7.0); oleuropein (0–0.8%); aerobic
Oscar (1999)	Polynomial	μ_{max}, lag	T (15–40°C); pH (5.2–7.4); previous pH (5.7–8.6); aerobic
Oscar (2002)	Square-root, CPM	μ_{max}, lag	T (8–48°C); aerobic
Skandamins et al. (2002)	Vitalistic	Time to decline in cell concentration	T (5–20°C); pH (4.3–5.3); oregano essential oil (0.5–2.0%)
Shigella			
Zaika et al. (1994, 1998)[a]	Polynomial	GT, lag	T (10–37°C); NaCl (0.5–5.0%); pH (5.0–7.5); Na-nitrite (0–1000 ppm); aerobic
Jeyamkondan et al. (2001)	ANN	GT, lag	Data from Zaika et al. (1994)
Staphylococcus aureus			
Ross and McMeekin (1991)	Square-root	Growth rate	T (5–35°C); a_w (0.848–0.997)
Buchanan et al. (1993)[a]	Polynomial	GT, lag	T (12–45°C); NaCl (0.5–16.5%); pH (4.5–9.0); Na-nitrite (1–200 ppm); aerobic and anaerobic
Dengremont and Membré (1997)	Square-root	μ_{max}	T (10–37°C); NaCl (0–10%); pH (5–8)
Eifert et al. (1997)	Polynomial	Parameters in primary model	T (12–28°C); NaCl (0.5–8.5%); pH (5.0–7.0); acidulants HCl, acetic acid, or lactic acid; aerobic
Vibrio spp.			
Miles et al. (1997)	Square-root	GT	T (8–45°C); a_w (0.936–0.995); aerobic

TABLE 3.5 (Continued)
Examples of Secondary Models for Growth of Pathogenic and
Indicator Microorganisms

Microorganisms and References	Type of Secondary Model	Response Variables	Independent Variables and Ranges
Yersinia spp.			
Bhaduri et al. (1995)[a]	Polynomial	GT, lag	T (5–42°C); NaCl (0.5–5.0%); pH (4.5–8.5); Na-nitrite (0–200 ppm); aerobic
Sutherland and Bayliss (1994)[c]	Polynomial	GT, lag	T (5–30°C); NaCl (0.5–6.5%); pH (4.0–7.0); aerobic
Pin et al. (2000)	Polynomial	μ_{max}, lag	T (1–8°C); CO_2 (0–83%); O_2 (0–60%)
Wei et al. (2001)	Square-root	μ_{max}, lag	T (4–34°C); air; vacuum; CO_2 100%

[a] Models included in the Pathogen Modeling Program, which is available free of charge at www.arserrc.gov/mfs/PMP6_download.htm.

[b] Generation time = $\ln(2)/\mu_{max}$.

[c] Model included in Food MicroModel. The values of model parameter are not included in the manuscript.

[d] See Ross et al. (2000) for a list of *Listeria monocytogenes* growth models published prior to 2000.

include germination of spores, population growth, survival, and toxin formation. These types of models became known as "probability" models.

In the latter part of the 1990s it seemed that the only way to manage the risk to consumers from certain pathogens was to ensure that the organism was never present in foods, or to ensure that it was not able to grow in foods that could become contaminated. The latter imperative led to the re-development of "growth/no-growth boundary," or "interface" modeling.

This section is divided into three main parts. In the first, Section 3.4.2, "traditional" probability modeling is briefly discussed. Section 3.4.3 presents and discusses the newer growth/no-growth (G/NG) modeling approaches, while Section 3.4.4 considers methodological issues relevant to probability and G/NG modeling.

3.4.2 PROBABILITY MODELS

Several reviews of probability modeling were presented in the early 1990s (Baker, 1993; Baker and Genigeorgis, 1993; Dodds, 1993; Lund, 1993; Ross and McMeekin, 1994; Whiting, 1995) but, possibly because of the relative paucity of new publications in this field since then, there has been no more recent dedicated review. Whiting and Oriente (1997) and Zhao et al. (2001), however, provide succinct updates.

TABLE 3.6
Examples of Secondary Models for Growth of Spoilage Microorganisms

Microorganisms and References	Type of Model	Response Variables	Independent Variables and Ranges
Bacillus stearothermophilus			
Ng et al. (2002)	Polynomial	Growth rate, GOL[a]	T (45–60°C); NaCl (0–1.5%); pH (5.5–7.0)
Brochothrix thermosphacta			
McClure et al. (1993)	Polynomial	μ_{max}, lag	T (1–30°C); NaCl (0.5–8.0%); pH (5.6–6.8); aerobic
Abdullah et al. (1994)	Polynomial	GT, lag, MPD	T (–2 to –10°C); CO_2 (2–40%); diameter of meat particles (2–10 mm)
Geeraerd et al. (1998a)	ANN	μ_{max}, lag	Data from McClure et al. (1993)
Pin and Baranyi (1998)	Polynomial	μ_{max}, lag	T (2–11°C); pH (5.2–6.4); aerobic
Koutsoumanis et al. (2000)	Arrhenius	μ_{max}	T (0–20°C); CO_2 (0–100%)
Jeyamkondan et al. (2001)	ANN	GT, lag	Data from McClure et al. (1993)
Chryseomonas spp.			
Membré and Kubaczka (1998)	Square-root	μ_{max}	T (1.3–10°C); aerobic
Enterobacteriaceae			
Pin and Baranyi (1998)	Polynomial	μ_{max}, lag	T (2–11°C); pH (5.2–6.4); aerobic
Lactic acid bacteria			
Passos et al. (1993)	Kinetic	μ_{max}	pH (3.8–6.0); lactic acid (0–30 mM); acetic acid (0–40 mM); NaCl (0–9%); cucumber juice
Gänzle et al. (1998)	Square-root	μ_{max}	T (3–41°C); aerobic
Gänzle et al. (1998)	CPM	μ_{max}	pH (4.2–6.7); ionic strength (0.0–1.97); acetate (0–0.2 mM); aerobic
Devlieghere et al. (2000a,b)	Square-root, polynomial	μ_{max}, lag	T (4–12°C); a_w (0.962–0.9883); CO_2 (0–1986 ppm); Na-lactate (0.0–3.0 5); pH 6.2
Lou and Nakai (2001)	ANN	μ_{max}, lag	T (4–12°C); a_w (0.9 62–0.9883); CO_2 (0–2411 ppm); pH 6.2 Subset of data from Devlieghere et al. (2000a,b)
Wijtzes et al. (2001)	Square-root	μ_{max}	T (3–30°C); a_w (0.932–0.990); pH (5.0–7.5)

TABLE 3.6 (Continued)
Examples of Secondary Models for Growth of Spoilage Microorganisms

Microorganisms and References	Type of Model	Response Variables	Independent Variables and Ranges
Connil et al. (2002)	Polynomial	μ_{max}, lag	T (3–9°C); pH (2.5–6.5); glucose (0.2–0.6%)
Messens et al. (2002)	CPM	Growth, bacteriocin production	T (20–38°C); pH (4.8–7.0)
Leroy and De Vuyst (2003)	CPM	Growth, bacteriocin production	T (20–37°C); pH (4.5–6.5)
Garcia-Gimeno et al. (2002)	ANN	Growth rate, lag	T (20, 28°C); NaCl (0–6%); pH (4–7)
Messens et al. (2002)	CPM	Growth, bacteriocin production	T (20–38°C); pH (4.8–7.0)
Garcia-Gimeno et al. (2002)	ANN	Growth rate, lag	T (20, 28°C); NaCl (0–6%); pH (4–7)
Molds			
Gibson et al. (1994)	Polynomial	Colony growth	T (30°C); a_w (0.810–0.995)
Cuppers et al. (1997)	Square-root, CPM	Colony growth	T (5–37°C); NaCl (0–7%)
Valík et al. (1999)	Polynomial	Colony growth	T (25°C); a_w (0.87–0.995); aerobic
Battey et al. (2001)	Polynomial	Probability of growth	T (25°C); pH (2.8–3.8); titratable acidity (0.2–0.6%); sugar content (8–16°Brix); Na-benzoate (100–350 ppm); K-sorbate (100–350 ppm)
Panagou et al. (2003)	Polynomial, Arrhenius, CPM	Colony growth	T (20–40°C); NaCl (2–10%); pH (3.5–5.0)
Photobacterium phosphoreum			
Dalgaard et al. (1997)[b]	Polynomial, square-root	μ_{max}	T (0–15°C); CO_2 (0–100%)
Pseudomonas			
Membré and Burlot (1994)	Polynomial	μ_{max}, lag	T (4–30°C); pH (6–8); NaCl (0–5%)
Neumeyer et al. (1997)[c]	Square-root	GT	T (0–30°C); a_w (0.947–0.966)
Pin and Baranyi (1998)	Polynomial	μ_{max}, lag	T (2–11°C); pH (5.2–6.4); aerobic

TABLE 3.6 (Continued)
Examples of Secondary Models for Growth of Spoilage Microorganisms

Microorganisms and References	Type of Model	Response Variables	Independent Variables and Ranges
Koutsoumanis et al. (2000)	Arrhenius	μ_{max}	T (0–20°C); CO_2 (0–100%)
Koutsoumanis (2001)	Square-root	μ_{max}, lag	T (0–15°C)
Rasmussen et al. (2002)	Process risk model	GT	Data from Neumeyer et al. (1997)
Shewanella spp.			
Dalgaard (1993)[b]	Square-root	μ_{max}	T (0–35°C); aerobic, anaerobic
Koutsoumanis et al. (2000)	Arrhenius	μ_{max}	T (0–20°C); CO_2 (0–100%)
Yeasts			
Deak and Beuchat (1994)	Polynomial	Changes in conductance	T (10–30°C); a_w (0.93–0.99); pH (3.8–4.6); K-sorbate (0–0.06%)
Passos et al. (1997)	Kinetic with product inhibition	μ_{max}	T (30°C); pH (3.2–5.9); lactic acid (0–55 mM); acetic acid (0–35 mM); NaCl (0–6%); cucumber juice; aerobic and anaerobic
Gänzle et al. (1998)	Square-root	μ_{max}	T (8–36°C); aerobic
Gänzle et al. (1998)	CPM	μ_{max}	Ionic strength (0.0–3.2); acetic acid (0–90 mM); aerobic

[a] GOL = germination, outgrowth, and lag time.
[b] Model included in the Seafood Spoilage Predictor (SSP) software available free of charge at www.dfu.min.dk/micro/ssp/.
[c] Model included in the Food Spoilage Predictor (FSP) software.

3.4.2.1 Logistic Regression

Dodds (1993) explains that in relation to the hazard presented by *Clostridium botulinum* in foods, the detection of the toxin is often more important than growth and that while growth is continuous and fairly easily determined, the presence of detectable toxin was seen as an "all-or-none" response. This led workers to seek methods to predict the probability of production of detectable toxin levels in response to the independent variables.

In probability models in predictive microbiology the data are usually that the response (e.g., growth, detectable toxin production) is observed under the experimental conditions, or that it is not. Responses such as detectable toxin production can be coded as either 0 (response not observed) or 1 (response observed) or, if repeated observations have been made, as probability (between 0 and 1). The probability is related to potential predictor variables by some mathematical function using regression techniques.

Logistic regression is a widely used statistical modeling technique — and is the technique of choice — when the outcome of interest is dichotomous (i.e., has only two possible outcomes). It is widely used in medical research (e.g., Hosmer and Lemeshow, 1989). Because regression techniques do not exist for dichotomous data, the regression equation is usually related to the log odds, or *logit*, of the outcome of interest. This has the effect of transforming the response variable from a binary response to one that extends from $-\infty$ to $+\infty$ reflecting the possible ranges of the predictor variables, and has desirable mathematical features also (Hosmer and Lemeshow, 1989). The logit function is defined as:

$$\text{logit } P = \log(P/(1 - P)) \tag{3.42}$$

where P is the probability of the outcome of interest.
Logit P is commonly described as some function Y of the explanatory variables, i.e.:

$$\text{logit } P = Y \tag{3.43}$$

Equation 3.43 can be rearranged to:

$$1/(1 + e^{-Y}) = P$$

or

$$e^Y/(1 + e^Y) = P$$

where Y is the function describing the effects of the independent variables.

The latter parameterizations appear in some of the earlier probability modeling literature.

Zhao et al. (2001) assessed the performance of linear and logistic regression to model percentage data that are "bounded," and may be considered as rescaled probability values. It was confirmed that logistic regression provided a much more accurate description of percentage data than linear regression, which had the insurmountable problem of predicting values outside the range of the data (i.e., less than 0% or greater than 100%).

3.4.2.2 Confounding Factors

Probability modelers used logistic regression to define the probability that detectable toxins would be produced within a specified period of time and under specified product composition and storage conditions. Models were based on the idea that a product was safe/acceptable or that it was not. Nonetheless, the responses measured in "probability modeling" were related to a number of factors that were in turn related to the *growth* of the organism under study and, in some cases, also included elements of survival. This approach appears to have arisen from the ideas of Riemann

(1967) that the success of a preservation method with regard to *C. botulinum* is related to the probability that one spore will germinate and give rise to toxin in the finished product. In general, to assess the effect of preservation conditions on probability of toxin production, the probability of growth from a single cell is estimated as the number of spores able to initiate growth under the test conditions (usually determined by MPN [most probable number] methods) divided by the number originally inoculated (Lund, 1993). Often a series of increasingly dilute inocula are subjected to the test conditions to determine the minimum fraction able to initiate detectable growth under the test conditions.

It might be expected that the probability of detection would increase with time. Indeed, Lindroth and Genigeorgis (1986) recognized that the probability of growth detection was also dependent upon the lag time of the inoculum, its initial density, and the duration of the study. They introduced a modification to the logit model to specifically model these effects. That model was subsequently used in a number of other studies (Baker et al., 1990; Ikawa and Genigeorgis, 1987). Whiting and Call (1993) criticized earlier models for probability of *C. botulinum* outgrowth and toxin production because they did not specifically monitor the time at which growth/toxin formation was first detected, and specifically modeled the probability of formation of toxin as a function of time and storage conditions using the logistic function, i.e., the probability of detectable growth, when plotted as a function of time, is a sigmoid curve. That approach was further refined (Whiting and Oriente, 1997 ; Whiting and Strobaugh, 1998) by inclusion of the inoculum density as an independent variable in the model.

Clearly, the probability of the responses in many of these traditional probability models is strongly related to the growth rate of the organism under the experimental conditions, leading Ross and McMeekin (1994) to conclude that the distinction that had traditionally been made between probability and kinetic models was an artificial one. Similarly, Baker et al. (1990) noted that "The rate of *P* increase ... expresses the growth rate ... ."

However, under some experimental conditions *P* does not always reach an asymptote of 1. This is evident in the data of Whiting and colleagues (Whiting and Call, 1993; Whiting and Oriente, 1997; Whiting and Strobaugh, 1998), of Chea et al. (2000), and of Razavilar and Genigeorgis (1998). It had also been described earlier by Lund et al. (1987) who introduced to predictive microbiology a model that recognizes that under some conditions, no matter how long one waits, not all samples will show growth/toxigenesis.

While the above studies considered spores, Razavilar and Genigeorgis (1998) applied a logistic regression approach to the probability of growth initiation within 58 days of *Listeria monocytogenes* and other *Listeria* species in response to combinations of pH, salt, temperature, and methyl paraben, sodium propionate, sodium benzoate, and potassium sorbate (Table 3.5). Their results, also, suggested that under near-growth-limiting environmental conditions the asymptotic probability of growth (i.e., given infinite incubation time) was sometimes less than 1. Stewart et al. (2001) also commented that while kinetic models predict the mean growth rate, these estimates may be meaningless under stressed conditions owing to natural variability in biological responses. Similarly, Lund (1993) employed the

Gompertz model (see Chapter 2) to model the time-dependent probability of growth of *L. monocytogenes* Scott A as a function of environmental factors. Even at near-growth-limiting pH (4.3), however, the asymptote of the $\log(P_{growth})$ was still close to 1.

The above studies suggest that as environmental conditions become more inhibitory to growth, not only does the probability that growth will be observed during the course of the experiment decrease, but the probability that growth *is possible* also decreases. This may be because the generation or lag time of all cells within the inoculum becomes infinitely long. Under these conditions, one begins to identify the absolute limits to microbial growth under combined stresses, i.e., the G/NG interface.

3.4.3 GROWTH/NO GROWTH INTERFACE MODELS

Microbial growth is restricted to finite ranges for any environmental factor, with growth rate sometimes declining abruptly within a very small increment of change of environment. Individual factor limits have been determined and collated (e.g., ICMSF, 1996a). That the growth range of microorganisms for one factor is reduced when a second environmental factor is less than optimal is also well recognized, and underlies the Hurdle concept (Leistner et al., 1985) also known as (multiple) barrier technology, or combined processes (Gorris, 2000). While the physiological basis of this synergy remains incompletely understood, the ability to define the limits to growth under combined environmental factors has enormous practical application in maintaining the microbial safety and quality of foods. Whether pathogens grow at all and the position of the G/NG boundary are of more interest than their growth rate because any growth implies a potential to cause harm to consumers. Similarly, so-called shelf-stable foods are sold, stored, and consumed over long periods of time. Therefore, the ability of spoilage organisms to grow at all implies that they have the potential to multiply to sufficient numbers to cause spoilage (Jenkins et al., 2000).

In the early to mid-1990s, a vein of experimentation using logistic regression techniques was begun with the aim of developing models that could define *absolute* limits to microbial growth in multifactorial space, irrespective of time of incubation or number of cells in the inoculum. One impetus for this research was the problem of listeriosis (Parente et al., 1998; Tienungoon et al., 2000). Strategies proposed to control the threat of listeriosis included "zero tolerance" (i.e., not detectable in a 25-g sample) of the presence of *L. monocytogenes* in foods *that could support its growth*, or to limit levels of contamination *at the point of consumption* to less than 100 cfu/g. Thus, foods that did not support the growth of *L. monocytogenes* were considered to pose significantly less risk and to require much less regulatory "attention" and testing. It was, therefore, of great commercial interest to be able to predict, without the need for protracted and expensive challenge testing, the potential for growth of specific bacteria within a particular food or, equivalently, product formulation options that would preclude growth.

Models defining combinations of environmental conditions that *just* prevent growth have become known as "G/NG interface," "growth boundary," or more simply "growth limits" models. The importance of growth boundary models for the design of safe foods and setting of food safety regulations, for the design of shelf-stable foods, and as a means of empowering the Hurdle concept by allowing it to be applied quantitatively has been discussed by various authors (Masana and Baranyi, 2000b; McMeekin et al., 2000; Ratkowsky and Ross, 1995; Schaffner and Labuza, 1997). Various approaches have been suggested to define the G/NG boundary. For convenience, these are discussed below under three broad groupings:

1. Empirical, deterministic, approaches
2. Logistic regression techniques
3. Artificial neural networks

Table 3.7 provides an overview of G/NG models published since 1990.

3.4.3.1 Deterministic Approaches

The first explicit definition of a microbial G/NG interface appears to be Pitt (1992), who derived regression equations from published data to describe the tempera-ture/water activity interface for aflatoxin production and *Aspergillus* spp. growth. The equation used to describe the interaction between temperature and water activity limits for growth was:

$$T_g^{(min \cdot max)} = 29.27 \pm \sqrt{(856.71 - 2289 \times (1.172 - a_w))}$$

where $T_g^{(min \cdot max)}$ are the upper and lower temperature limits for growth at the specified water activity.

A similar equation was presented for aflatoxin production. The predicted inter-faces from both models are shown in Figure 3.7 .

To describe the pH/$a_{w(NaCl)}$ interface of the food spoilage organism *Brocothrix thermosphacta*, Masana and Baranyi (2000b) derived the midpoints of growth and no-growth observations by interpolation and fitted a polynomial function to those data. They noted that under some conditions, the interface was completely dominated by one factor or the other, so that their final model consisted of a pH vs. a_w parabolic curve and a NaCl-constant line. They also considered the effects of inoculum level on the interface, which was determined at 25°C for up to 24 days of incubation. Examples of the interface are shown in Figure 3.8.

Membré et al. (2001) estimated levels of sorbate that prevented growth of *Penicillium brevicompactum* in bakery products containing various levels of benzoate by extrapolation of kinetic data. Equations were derived to define growth-preventing combinations of sorbate and benzoate and were used to limit the range of predictions from the kinetic model they developed for *P. brevicompactum* growth rate.

Other workers have noted that the form and parameters of CPMs imply absolute limits to microbial growth, and suggested approaches to defining the G/NG interface based on estimates of cardinal parameters. In this vein Ratkowsky and Ross (1995), recognizing the relationship between absolute limits for each environmental factor and their relationship to the parameters of square-root-type models and CPMs, experimented with the use of a kinetic model as the basis of a growth boundary

TABLE 3.7
Summary of Published Growth Boundary Models

Reference	Organism	Strain	Medium	Environmental Factors	Ranges Lower	Ranges Upper	Levels	Replicates	Total Data Points	Measured by?	Time Limit	Other
Presser et al. (1998)	Escherichia coli	M23 (non-pathogenic)	Nutrient Broth	Temperature	10	37	6	1 to 4	627	OD Increase (confirmed as needed by culture)	50 days	Linear logistic regression, SAS PROCNONLIN
				a_w (NaCl)	0.955	0.995	4					
				pH	2.8	6.9	≥ 10					
				Lactic acid (mM)	0	500	6					
Parente et al. (1998)	Listeria monocytogenes	Scott A, V7, and L11	Tryptone Soy Broth + 0.6% Yeast Extract	Nisin (IU/ml)	1	2100					7 days (@30°C)	Logistic regression with polynomial using LOGIT 1.14 module of Systat
				Leucocidin F10 (AU/ml)	1	2100						
				pH	4.7	6.5						
				NaCl (% w/v)	0.7	4.5						
				EDTA (mmol)	0.1	0.9						
				Inoculum density	1.6×10^3	7.9×10^7						
		Validation Set 1		Nisin (IU/ml)	8	200		5			7 days (@30°C)	
				Leucocidin F10 (AU/ml)	8	200						
				pH	4.7	6.5						
				NaCl (% w/v)	0.7	4.5						
				EDTA (mmol)	0.08	4.72						
				Inoculum density	0.6×10^3	2.5×10^7						
		Validation Set 2		Nisin (IU/ml)	50	250		10			7 days (@30°C)	
				Leucocidin F10 (AU/ml)	1	250						
				pH	5.2	6						
				NaCl (% w/v)	1.8							
				EDTA (mmol)	0.2	0.6						
				Inoculum density	1×10^5							

TABLE 3.7 (Continued)
Summary of Published Growth Boundary Models

Reference	Organism	Strain	Medium	Environmental Factors	Ranges Lower	Ranges Upper	Levels	Replicates	Total Data Points	Measured by?	Time Limit	Other
Bolton and Frank (1999)	*Listeria monocytogenes*	Mixture (equal numbers) of Scott A, Brie 1, 71 Switzerland, 2379 LA	Soft fresh cheese (similar to "Mexican style" cheese)	Moisture (% w/w)	42	60	4	3	288	Viable count	21 and 42 days	Binary or "ordinal" logistic regression using SAS PROC LOGISTIC with link functions. For the latter, three responses: P of growth, stasis, or death (according to change in viable count; ± 0.5 log CFU) were modeled
				salt (% w/w)	2	8	4					
				pH	5	6.5	6					
Salter et al. (2000)	*Escherichia coli*	MR21 (STEC)	Nutrient Broth	Temperature	7.7	37	60	1–8, most 4	604	OD increase	50 days	Nonlinear logistic regression, SAS PROCLOGISTIC and PROCNONLIN
				a_w (NaCl)	0.943	0.987	28					
Jenkins et al. (2000)	*Zygosaccharomyces bailii*	4637, history unknown	Acidified yeast nitrogen broth	Salt (NaCl, %w/v)	2.6	4.2	3	3	243	OD increase	29 days	SAS LIFEREG
				Sugar (fructose, %w/v)	7	32	3					
				Total acetic acid (%v/v)	1.8	2.8	3					
				pH	3.5	4	3					

Reference	Organism	Strain/Model	Medium	Variable	Min	Max	Levels				Response	Time	Method
Tienungoon et al. (2000)	Listeria monocytogenes	(Scott A, L5 separate models)	TSB-YE	Temperature a_w (NaCl) pH Lactic acid (mM)	3.1 0.928 3.7 0	36.2 0.995 7.8 500	30 60 10 14	1 to 4	2839		OD increase	90 days	Nonlinear logistic regression, SAS PROCLOGISTIC and PROCNONLIN
López-Malo et al. (2000)	Saccharomyces cerevisiae	Not stated Model based on data of Cerruti et al. (1990)		a_w pH K-sorbate (ppm)	0.93 3 0	0.97 6 1000	3 4 6		72 ($\times 2$ observation times)		Viable count, including decrease in viable count	50 h or 350 h	Logistic regression with first-order polynomial using SPSS
Masana and Baranyi (2000b)	Brocothrix thermosphacta	MR 165	Tryptone Soya Broth	NaCl pH Inoculum (cfu/350 µl)	0.5 4.4 10	10 5.7 1 million	11 7 3				Viable count		Polynomial fitted to midpoints of data-pairs of adjacent growth and no growth observed combinations
		Fine grid experiments		NaCl pH	45 combinations at 5 levels of NaCl and 5 levels of pH close to the G/NG boundary								
				Inoculum (cfu/350 µl)	10	1000	2	10	450		Viable count		
Lanciotti et al. (2001)	Bacillus cereus	FG1	BHI Broth	Temperature	10	45	5	30 variables combinations over two independent trials for each organism	2 × 30 for each strain		OD increase (600nm)	2–7 days	Generalized linear logistic regression, Statistica (Statsoft) software
	Staphylococcus aureus	S33	BHI Broth	a_w (glycerol)	0.89	0.99	8						
	Salmonella enteritidis	B5	BHI Broth	pH	4	8	5						
				Ethanol (% v/v)	0	3	5						

TABLE 3.7 (Continued)
Summary of Published Growth Boundary Models

Reference	Organism	Strain	Medium	Environmental Factors	Ranges Lower	Ranges Upper	Levels	Replicates	Total Data Points	Measured by?	Time Limit	Other
Stewart et al. (2001)	*Staphylococcus aureus*	5 strain cocktail	BHI Broth	a_w (glycerol)	0.95	0.84	4	8	640	OD increase	168 days	Toxin assayed Modeled "time to growth" using LIFEREG
				Initial pH	4.5	7	4					
				K-sorbate (ppm) or								
				Ca-propionate (ppm)	0	1000	3					
				Temperature (°C)	37							
McKellar and Lu (2001)	*Escherichia coli* O15:H7	5 strain cocktail	TS Broth	Temperature (°C)	10	30	5	5	1820	Visible increase in turbidity	3 days	Linear logistic regression used (polynomial form)
				Acetic acid (modeled as undissociated form)	0	4%	8					
				NaCl	0.50%	16.50%	8					
				Sucrose	0	8%	3					
				pH	3.5	6.0	6					

Reference	Organism	Strain	Medium	Variable	Min	Max			n	Response	Time	Notes
Membré et al. (2001)	*Penicillium compactum*	Wild type from bakery products	MY50 agar	pH	2.5	7.5	± 6	1	76	Mycelial growth	75 days	Not directly modeled. growth limits due to sodium benzoate and sorbic acid at pH 5 were derived by extrapolation of growth rate data
				Sorbic acid			1					
				Propionic acid			1					
				Sodium benzoate			1					
				pH			1	5	122			
				Sorbic acid (mg l^{-1})	0.0	1000.0	6					
				Na-benzoate (mg l^{-1})	0.0	300.0	4					
				Commercial cakes				6	4			
Uljas et al. (2001)	*Escherichia coli* O15:H7	3 strain cocktail	Apple cider (juice)	pH	3.1	4.3	7	7	1600 × 3	Turbidity (growth within 48 h at 35°C) after dilution of treated sample	12 hours	SAS PROCLOGISTIC (dependency modeled as simple first order equations of predictor variables, no cross products)
				Temperature	5	35	4					
				Sodium benzoate	0	0.1%	3					
				Potassium sorbate	0	0.1%	3					
				Freeze–thaw	Not applied	Applied	3					
				Ciders type			3	756				

In this case the response modeled was P > 5 log inactivation after various treatment times

Reference	Organism	Strain	Medium	Variable	Min	Max			n	Response	Time	Notes
	Staphylococcus aureus	5 strain cocktail	BHI Broth	a_w (NaCl, or sucrose-fructose)	0.84	0.95	4	2 or 3	8	OD increase	168 days	Toxin assayed Modeled "time to growth" using LIFEREG
				pH	4.5	7	4		8			
				K-sorbate (ppm)	0	1000	2					

[Combined with data set of Stewart et al. (2001), 768 data]

Reference	Organism	Strain	Medium	Variable	Min	Max			n	Response	Time	Notes
	Listeria innocua	ATCC 33090	BHI Broth (+0.2% w/w glucose, +0.3% w/w yeast extract)	Temperature	0	43	16	(pH constant)	1792	Turbidimetry Viable count by culture	14 days / 1 month	Novel term based on relative inhibition of growth rate-affecting factors data generated for combined kinetic model that predicts no growth (NLINFIT in MATLAB 5.2)
				pH	4.5	9.4	15	(temperature constant)				
				Propionic acid (mM)	16	64	24	(pH varied from 5 to 7.5)				
				Lactic acid (mM)	20	138	27	(pH varied from 4.8 to 7.1)				

TABLE 3.7 (Continued)
Summary of Published Growth Boundary Models

Reference	Organism	Strain	Medium	Environmental Factors	Ranges Lower	Ranges Upper	Levels	Replicates	Total Data Points	Measured by?	Time Limit	Other
Battey and Schaffner (2001)	Spoilage bacteria: Acinetobacter calcoaceticus and Gluconobacter oxydans	2 strain cocktail	Cold filled, ready to drink, beverages	pH	2.8	3.8	3	84		Viable plate count	8 weeks at 25°C	Model is based on growth and inactivation rates. Included 14 duplicated validation trials (8 correctly predicted from model)
				Titratable acidity (%)	0.2	0.6	3					
				Sodium benzoate (ppm)	100	350	3					
				Sugar content (°C Brix)	8	16	3					
				Potassium sorbate (pp)	100	350	3					
Battey et al. (2002)	Spoilage yeasts: Saccharomyces cerevisiae, Zygosaccharomyces bailii, Candida lipolytica	3 strain cocktail	Cold filled, ready to drink, beverages	pH	2.8	3.8	3	84		Viable plate count	8 weeks at 25°C	Included 14 duplicated validation trials (all correctly predicted from model)
				Titratable acidity (%)	0.2	0.6	3					
				Sodium benzoate (ppm)	100	350	3					
				Sugar content (°C Brix)	8	16	3					
				Potassium sorbate (pp)	100	350	3					
Hajmeer and Basheer (2002, 2003a, 2003b)	Data of Salter et al. (2000), see above											Probabilistic Neural Network

model using linear logistic regression. This approach is discussed further below in Section 3.4.3.2.

The approaches of Augustin and Carlier (2000b) and Le Marc et al. (2002) were presented in Section 3.2.3. Essentially, these approaches are empirical. They are based on assumed interactions between factors and are not *fitted* to G/NG data. An example of the response predicted by these approaches is shown in Figure 3.11.

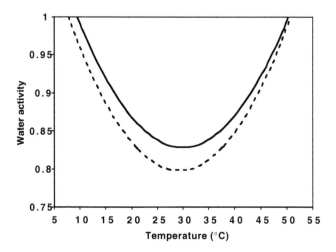

FIGURE 3.7 Predicted temperature–water activity interface for mold (*Aspergillus* spp.) growth (dashed line) and aflatoxin production (solid line) from the model of Pitt (1992).

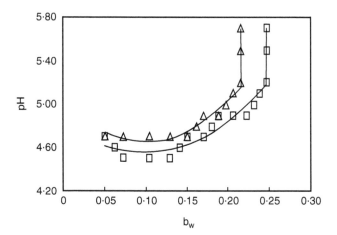

FIGURE 3.8 Data and modeled growth/no-growth boundary for *Brochothrix thermosphacta* in response to pH and water activity at 25°C. Water activity data were rescaled to $b_w = \sqrt{1 - a_w}$. The data are for an inoculum of 1.5×10^6 cells/well ($\square$), or for an inoculum of 1.5×10^1 and 1.5×10^3 cells/well ($\triangle$). (Reproduced from Masana, M.O. and Baranyi, J. *Food Microbiol.*, 17, 485–493, 2000b. With permission.)

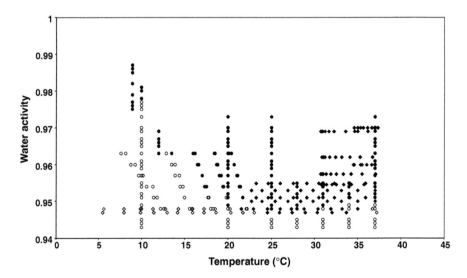

FIGURE 3.9 Data obtained from separate experiments for the growth/no-growth (G/NG) boundary of *Escherichia coli*. Data are from Salter et al. (2000) (circles) and from unpublished results of the authors (diamonds). Near the G/NG boundary, the data obtained from discrete experiments do not form a smooth (monotonic) boundary, suggesting that small differences in experimental procedures can significantly affect the position of the boundary. Open symbols denote no-growth conditions, and solid symbols indicate that growth was observed.

The above approaches can be considered to be deterministic; i.e., they predict only one position ($P_{growth} = 0.5$) for the boundary, although the position of boundaries can be adjusted by "weighting" data in the case of Masana and Baranyi (2000b) or by selecting an appropriate value for θ in the case of the Le Marc et al. (2002) approach (see Section 3.2.3). While the data of Masana and Baranyi (2000b) included tenfold replication, the midpoints of the *most* "extreme" conditions that *did* allow growth, and the *least* "extreme" conditions that *did not* allow growth were estimated by interpolation and considered to represent 50% probability of growth. Other workers have suggested that some problems require higher levels of confidence that growth will not occur, so that methods that enable definition of the interface at selected levels of statistical confidence may have greater utility.

Another approach that implicitly characterizes the G/NG interface is that of combined growth and death models in which the rate of growth and rate of death under specified conditions are estimated simultaneously. The G/NG interface can be inferred from those combinations of conditions where growth rate and death rate are equal (see, e.g., Jones and Walker, 1993; Jones et al., 1994). A similar approach is evident in Battey et al. (2001) who modeled both the rates of growth and rates of death of spoilage molds in ready to drink beverages. The G/NG interface was given, implicitly, as that set of conditions where the rate of growth was equal to the rate of death.

Ratkowsky et al. (1991) noted that as environmental conditions become more inhibitory to microbial growth the variability in growth rates increases widely, which implies that the probability that growth occurs at all becomes uncertain, because the left-hand tail of the growth rate distribution falls below zero. This is supported in the results of Whiting and colleagues (Whiting and Call, 1993; Whiting and Oriente, 1997; Whiting and Strobaugh, 1998), where P_{max} (the proportion of spores that successfully germinated and initiated growth) was shown to decline at increasingly stringent conditions. Conversely, Masana and Baranyi (2000b) observed, as have other workers (McKellar and Lu, 2001; Presser et al., 1998; Salter et al., 2000; Tienungoon et al., 2000), that the difference in conditions that allow growth and those that do not is usually abrupt, and often at or beyond the limits of resolution of instruments commonly used to measure such differences. Thus, Masana and Baranyi (2000b) questioned the need for approaches that model the transition between conditions leading to high probability of growth and those leading to low probabilities of growth. While this abrupt transition appears consistent *within* replicated experiments it is less certain, however, that the same consistency is true *between* experiments. Figure 3.9, showing experimental data, suggests that responses near the boundary may be inconsistent when data from several discrete experiments are combined. This may suggest subtle, but important, differences in response related to the physiology of the inoculum, or its concentration. Furthermore, it suggests that the ability to characterize probabilities of growth under specified sets of conditions may be an important element of growth boundary models and that the boundary may not be "absolute" but depend on the physiological state of the cell and, by inference, on the size of the inoculum. This will be discussed further in Section 3.4.4.

3.4.3.2 Logistic Regression

Ratkowsky and Ross (1995) and others (Bolton and Frank, 1999; Jenkins et al., 2000; Lanciotti et al., 2001; López-Malo et al., 2000; McKellar and Lu, 2001; Parente et al., 1998; Stewart et al., 2001, 2002; Uljas et al., 2001) reintroduced the use of logistic regression to model categorical data (i.e., growth or no growth) in predictive microbiology, enabling probabilistic determination of the G/NG boundary. Use of the logit function enabled the probability of growth under specific sets of conditions to be estimated, so that the G/NG boundary could be specified at selected levels of confidence.

Ratkowsky and Ross (1995) aimed to model absolute limits to growth in multifactorial space, but only had available data based on a 72-h observation period. While most other workers have preferred polynomial functions to describe the effect of independent variables on the logit function, in the former approach a square-root-type kinetic model was ln-transformed and used as the basis of the function relating the logit of probability of growth to independent variables, e.g., temperature, water activity, pH. This approach was adopted in an attempt to retain some level of biological interpretability of the models, a desire echoed by others (Augustin and Carlier, 2000a,b; Le Marc et al., 2002).

The form of the G/NG interface model of Presser et al. (1998) was derived from the kinetic model of Presser et al. (1997) for the growth rate of *E. coli* (see Equation

3.10). Novel data were generated specifically to assess the limits of *E. coli* growth under combinations of temperature, pH, a_w, and lactic acid. The corresponding G/NG model had the form:

$$\begin{aligned} \text{LogitP} = &\ 28.0 + 8.90 \ln(a_w - 0.943) \\ &+ 2.0\ln(T\text{-}4.00) + 4.59 \ln(1 - 10^{3.9\text{-pH}}) \\ &+ 6.96\ln[1 - LAC/(10.7 \times (1 + 10^{\text{pH-}3.86}))] \\ &+ 3.06\ln[1 - LAC/(823 \times (1 + 10^{3.86\text{-pH}}))] \end{aligned} \tag{3.44}$$

where all terms are as defined in Section 3.2.1.

Some parameters in that model had to be determined independently, i.e., were not determined in the regression, and were derived from the fitted values of square-root-type kinetic models. Essentially the same approach was adopted by Lanciotti et al. (2001) to develop G/NG models for *B. cereus, S. aureus,* and *Salmonella enteritidis.* Ratkowsky (2002) commented on the increased flexibility in being able to determine all of the parameters in the model during the regression, and subsequent studies developed the approach, eventually leading to a novel nonlinear logistic regression technique (Salter et al., 2000; Tienungoon et al., 2000). Ratkowsky (2002) pointed out that nonlinear logistic regression was a new statistical technique and discussed benefits and problems with that approach specifically in relation to growth limits modeling. A problem with models of the form of Equation 3.44 is that for conditions more extreme than the parameters corresponding to T_{min}, pH_{min}, $a_{w\ min}$, etc., and which are tested experimentally though not expected to permit growth, the terms containing those parameters would become negative. As all of those terms are associated with a logarithmic transformation, the expression cannot be calculated during the regression and such values are ignored in the model fitting process, or have to be eliminated from the data set before the fitting process begins. This, in turn, affects the values of the parameters of the fitted model. Ratkowsky (2002) comments that an objective method for selection and deletion of such data is necessary, but does not yet exist.

Bolton and Frank (1999) extended the binary logistic regression approach by recoding growth and no growth data to allow a third category: survival, or stasis. They termed this approach ordinal logistic regression. Parente et al. (1998) "reversed" the response variable, and applied logistic regression techniques to the probability of survival/no survival of *L. monocytogenes* in response to bacteriocins, pH, EDTA, and NaCl. Stewart et al. (2001) modeled the probability of growth of *S. aureus* within 6 months of incubation at 37°C, and at reduced water activity achieved by various humectants. They also compared the growth boundary with the boundary for enterotoxin production, and observed a close correlation between the two criteria.

Growth limits models have also been developed for spoilage organisms including *Saccharomyces cerevisiae* (López-Malo et al., 2000) and *Zygosaccharomyces bailii* (Jenkins et al., 2000) and cocktails of *Saccharomyces cerevisiae, Zygosaccharomyces bailii,* and *Candida lipolytica* (Battey et al., 2002). Interestingly, the study of Jenkins et al. (2000), while encompassing broader ranges of factor combinations, confirmed the simpler and earlier model of Tuynenburg Muys (1971). That model, which specifies combinations of molar salt plus sugar and percent undissociated acetic acid for stability of acidic sauces, still forms the basis of the industry standard for those products. This observation suggests that limits to growth under combined conditions can be highly reproducible.

3.4.3.3   Relationship to the Minimum Convex Polyhedron Approach

The concept of the MCP was introduced by Baranyi et al. (1996) (see Figure 3.10) to describe the multifactor "space" that just encloses the interpolation region of a predictive kinetic model. If the interpolation region exactly matched the growth region of the organism then the MCP would also describe the growth limits of the organism. In practice, however, it would be impossible to undertake sufficient measurements to completely define the MCP; i.e., the MCP has "sharp" edges because of the method of its calculation, whereas from available studies (see Figures 3.8 and 3.9) the G/NG interface forms a continuously curved surface. However, it might also be possible to use no-growth data to create a no-growth MCP and to combine the growth MCP and no-growth MCP to define a region within which the G/NG boundary must lie. This approach has been assessed and compared to a model of the form of Equation 3.43 by Le Marc and colleagues (Le Marc et al., 2003). These workers concluded that the logistic regression modeling approach produced a smoother response surface, more consistent with observations, but that the MCP approach had the advantage of being directly linked to observations and therefore was not a prediction from a model.

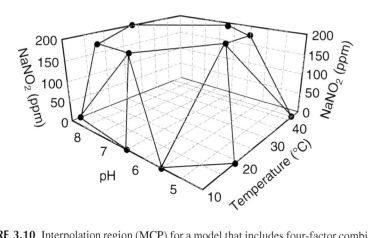

FIGURE 3.10 Interpolation region (MCP) for a model that includes four-factor combinations (T, pH, NaCl, NaNO₂). The interpolation region shown is that for NaCl = 0.5%, but is based on the complete data set. Solid circles indicate conditions under which observations have been made, while the lines represent the edges of the MCP. (From Masana, M.O. and Baranyi, J. *Food Microbiol.*, 17, 367–374, 2000a. With permission.)

3.4.3.4 Artificial Neural Networks

Recently, Hajmeer and Basheer (2002, 2003a,b) demonstrated the use of a Probabilistic Neural Network (PNN) approach to definition of the G/NG interface. PNNs
are a form of ANN (see Section 3.2.6). In a series of papers, based on modeling the
data of Salter et al. (2000) for the effects of temperature and water activity (due to
NaCl) on the growth limits of *E. coli*, Hajmeer and Basheer concluded that their
PNN models provided a better description of the data of Salter et al. (2000) than
did the nonlinear logistic regression method referred to above. Their conclusion is
considered in more detail in Section 3.4.3.5 below.

It should be noted that neither the logistic regression models described above,
nor the PNN, produce an equation that describes the interface. Rather, the output of
those models is the probability that a given set of conditions will allow growth. To
define the interface, it is necessary to rearrange the model for some selected value
of *P* to generate an equation that describes the G/NG boundary.

3.4.3.5 Evaluation of Goodness of Fit and Comparison
of Models

Methods for evaluation of performance of logistic regression models include the
receiver operating curve (ROC; also referred to as the concordance rate), the Hosmer–Lemeshow goodness-of-fit statistic, and the maximum rescaled R^2 statistic.
These are considered in greater detail in Tienungoon et al. (2000).

Briefly, the ROC is obtained from the proportion of events that were correctly
predicted compared to the proportion of nonevents that were correctly predicted.
The closer the value is to 1, the better the level of discrimination. In epidemiological
studies, ROC values > 0.8 are considered excellent. ROC values for G/NG models
are typically much higher.

The Hosmer–Lemeshow index involves grouping objects into a contingency
table and calculating a Pearson chi-square statistic. Small values of the index indicate
a good fit of the model.

The maximum rescaled R^2 value is proposed for use with binomial error as an
analogy to the R^2 value used with normally distributed error. The closer the value
is to 1, the greater is the success of the model in predicting the observed outcome
from the independent variables. Zhao et al. (2001) cite the deviance test and graphical
tools such as the index plot and half normal plot as methods for determining goodness
of fit of linear logistic regression models.

Other methods based on calculation of indices from the "confusion matrix"
(Hajmeer and Basheer, 2002, 2003b) or the equivalent "contingency matrix"
(Hajmeer and Basheer, 2003a) were used to compare performance between models
derived from different approaches and applied to the same data.

Another method of evaluation is to compare the fitted model to independent data
sets (Bolton and Frank, 1999; Masana and Baranyi, 2000b; Tienungoon et al., 2000)
although, generally, such data are not readily available (see, e.g., McKellar and Lu,
2001). The model of Tienungoon et al. (2000) for *L. monocytogenes* growth boundaries showed very good agreement with the data of McClure et al. (1989) and George

et al. (1988) despite that different strains were involved. There is also a remarkable level of similarity between the observations of Tienungoon et al. (2000) and the observations of Le Marc et al. (2002) on growth limits of *L. innocua*. Several publications, however, report growth of *L. monocytogenes* at temperatures lower than the minimum growth limit predicted by the Tienungoon et al. (2000) model, possibly indicating strain variation or that the experimental design failed to recognize important elements that facilitate *L. monocytogenes* growth at temperatures < 3°C, i.e., that an inappropriate growth substrate was used. Similarly, McKellar and Lu (2001) reported that their model failed to predict growth of *E. coli* O157:H7 under conditions where it had been previously reported, although it should be noted that their model was limited to observation of growth within 72 h. Bolton and Frank (1999) compared the predictions of their growth limits models for *L. monocytogenes* in cheese to the data of Genigeorgis et al. (1991) for *L. monocytogenes* growth in market cheese. The models predicted correctly in 65% of trials (42-day model) and 81% of trials (21-day model).

Given the diversity of approaches, it is pertinent to ask: does one method for defining the G/NG interface perform better than another? As with kinetic models, the ability to describe a specific experimental data set does not necessarily reflect the ability to predict accurately the probability of growth under novel sets of conditions. While measures of performance of logistic regression models are available, they can be readily affected by the data set used. Perfect agreement between observed and modeled data responses may not be possible if there are anomalies in the data. Figure 3.11 provides a clear example. Nonetheless, for many growth limits models high rates of concordance (typically >90%) have been reported. As noted earlier, in epidemiological logistic regression modeling, rates higher than 70% are considered to represent good fits to the data, implying that the limits to microbial growth are highly reproducible when well-controlled experiments are conducted.

To date, only one direct comparison of G/NG modeling approaches has been presented (Hajmeer and Basheer, 2002, 2003a,b) but this was based on one data set only, i.e., that of Salter et al. (2000) for the growth limits of *E. coli* in temperature/water activity space. Only by comparing the performance of different modeling approaches applied to multiple data sets does an appreciation of overall model performance emerge. Nonetheless, to illustrate differences between models and give some appreciation of their overall performance we compare several models using the data of Salter et al. (2000) for the growth limits of *E. coli* R31 in response to temperature and water activity. The model types compared are:

1. The PNN of Hajmeer and Basheer (2003a), which those authors were able to summarize as a relatively simple equation
2. A model of the type of Equation 3.44 fitted to a subset of the Salter et al. (2000) data set by Hajmeer and Basheer (2003a) (It should be noted that, contrary to what is stated in that publication, the model presented by Hajmeer and Basheer was not generated by nonlinear logistic regression but by a two-step linear logistic regression as described in Presser et al., 1998)

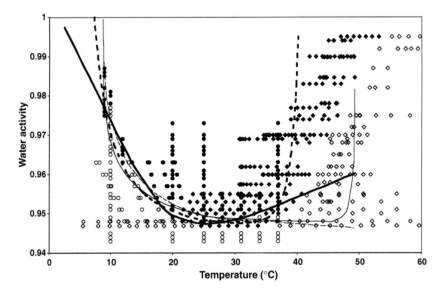

FIGURE 3.11 Comparison of predicted no growth boundaries for four modeling approaches applied to the data of Salter et al. (2000) (circles) for the growth limits of *Escherichia coli* R31 in response to temperature and water activity (NaCl) combinations. Approach of Hajmeer and Basheer PNN (heavy solid line); Linear Logistic /Equation 3.44 (heavy dashed line); Le Marc et al. (2002) (light solid line); Augustin and Carlier (2000a) (light dashed line). The data set was subsequently augmented with new data (diamonds), which reveals that none of the models extrapolate reliably. (Solid symbols: growth; open symbols: no growth.)

3. A model of the type of Le Marc et al. (2002; Equation 3.25 to Equation 3.27), where $T_{max} = 49.23°C$ (to be consistent with the logistic regression model parameter), $a_{w\ min} = 0.948$, and $T_{min} = 8.8°C$, based on the minimum water activity and temperature, respectively, at which growth were observed

4. A model of the type of Augustin and Carlier (2000b; Equation 3.24) assuming that $T_{min} = 8.8°C$ and $a_{w\ min} = 0.948$, consistent with the parameter values used for the Le Marc et al. (2002) model

The predicted interfaces are shown in Figure 3.11, together with the data used to generate the models. (Note that the subsets of 143 of the 179 data of Salter et al. (2000) used by Hajmeer and Basheer (2003a) to fit the PNN and the Equation 3.44 type model were not identified.)

When compared to the full data set, the level of misprediction ranged from ~15 to 20 of the 179 data points for each of the models, suggesting that the level of performance was not greatly different despite very different modeling approaches.* A complication in the comparison of G/NG model performance is that most of the

---

* It should be noted that this analysis disagrees with the results of Hajmeer and Basheer (2002) who reported only two to four mispredictions for the total (i.e., training and validation) data set.

data are readily predicted, e.g., those that fall outside the known limits for growth for individual environmental factors when all other factors are optimal. Such data can "overwhelm" the data in which one is most interested, i.e., in the relatively narrow region where factors interact to reduce the biokinetic ranges and, yet more specifically, where the probability of growth rapidly changes from "growth is very likely" to "growth is very unlikely." These data define the interface and, consequently, data closest to the interface are more important when comparing model performance. This has implications for experimental design, as discussed in Section 3.4.4 below.

To assess whether one model might be preferred on theoretical grounds, as adjudged by its ability to extrapolate reliably, the predictions of all models in the temperature range above 35°C can be compared to data subsequently generated, shown in Figure 3.11, and not used to generate the models. Clearly, none of the models extrapolate well.

From the above comparison, it appears that despite very different modeling approaches and degrees of complexity of modeling, there is currently little to differentiate those approaches on the basis of their ability to describe the G/NG interface or on their ability to predict outside the interpolation region.

3.4.4 EXPERIMENTAL METHODS AND DESIGN CONSIDERATIONS

As suggested above, currently there is little mechanistic understanding of how environmental factors interact to prevent bacterial growth and it must be recognized as a possibility that there is no single, common mechanism underlying the observed boundaries for different factor combinations. Consequently, it is not possible from first principles to design the optimal experiment that captures the essential information that will characterize the response and lead to reliable models. Instead, at this time, experimental methods must be focused toward gaining enough data in the interface region to be able to describe empirically the limits to growth.

First of all, two approaches may be distinguished that could affect the experimental methods chosen. In one, the interest is in whether growth/toxin production, etc. is possible within some specified time limit, which may be related to the shelf life of the product. In other approaches, the objective is to define absolute limits to growth, i.e., the most extreme combinations of factors that just allow growth. McKellar and Lu (2001) argue that there is always a time limit imposed on G/NG modeling studies. Strategies exist, however, that provide greater confidence that the "absolute" limits to growth are being measured. Some of these are discussed below.

3.4.4.1 Measuring Both Growth and Inactivation

Several groups have assessed both growth and inactivation in their experimental treatments (McKellar and Lu, 2001; Parente et al., 1998; Presser et al., 1999; Razavilar and Genigeorgis, 1998). In this way the position of the boundary is inferred from two "directions." If growth is not observed, an observer cannot be sure whether growth is not possible or has not occurred *yet*. If it is known that at some more extreme condition inactivation occurs, it can be inferred that the G/NG boundary lies between those two sets of conditions.

A potential problem with this strategy is that cultures can initially display some loss of viability, but with survivors eventually initiating growth; i.e., population decline cannot unambiguously be interpreted as "growth is not possible." Numerous studies (e.g., Mellefont et al., 2003) have demonstrated that rapid transfer of a culture from one set of conditions to another that is more stressful can induce injury and death, but that survivors will eventually adjust and be able to grow. This has been termed the Phoenix phenomenon (Shoemaker and Pierson, 1976). Such regrowth has been reported in the context of G/NG modeling (Bolton and Frank, 1999; Masana and Baranyi, 2000b; Parente et al., 1998; Tienungoon et al., 2000).

Clearly, an experimenter interested in determining the "absolute" G/NG boundary will need to maximize the resistance of the inoculum to stress on exposure to the new, more stressful, environment. The use of stationary phase cultures as inocula would seem to be a minimum requirement. It may be necessary to habituate cultures to the test conditions (e.g., growth at conditions just less harsh) prior to inoculation into the test conditions to maximize the chance that growth, if possible, will be observed. One way to maximize the likelihood of observing the most extreme growth limits would be to use cultures growing at the apparent limits as inocula into slightly more stringent conditions. This also has the advantage of minimizing growth lags on inoculation into a harsher environment.

3.4.4.2 Inoculum Size

Masana and Baranyi (2000b) indicated that inoculum size affected the position of the boundary. Robinson et al. (1998) reported similar effects of inoculum density on bacterial lag times. While it is clear that time to detection would depend on inoculum density, growth detection methods were not cell-density-dependent in those studies. Parente et al. (1998) also reported that a decrease in inoculum size decreased the probability of survival. If the shock of transfer is known to inactivate a fixed proportion of the cells in the inoculum, to develop a robust model it will be necessary to use an inoculum that ensures that even after inactivation there is a high probability that at least one cell will survive.

The above observations lend support to the hypothesis that it is the distribution of physiological states of readiness to survive and multiply in a new environment that determines the position of the G/NG boundary, i.e., all other things being equal, the more cells in the inoculum the more likely it is that there is one cell that has the capacity to survive and grow. This also reinforces the equivalence between G/NG boundary modeling and the modeling of conditions that lead to infinite lag times. The importance of the distribution of lag times on the observed lag times of whole populations has been discussed by Baranyi (1998).

There may be more involved reasons for inoculum density-dependent responses also, such as chemical messaging between cells (see, e.g., Miller and Bassler, 2001; Winans and Bassler, 2000).

In conclusion, if the aim is to determine absolute limits to growth, a higher number of cells is preferable. Masana and Baranyi (2000b) stated that growth boundaries "represented the chance of growth for each sample; therefore, to assure a low probability of growth in many samples, it will be more relevant to consider boundaries for high inoculum levels." Equally, as noted above, steps to maximize the cell's chances of survival and growth in the environment are also recommended.

There is potentially a caveat, however, that needs to be applied. Maximum population densities of batch cultures are reported to decline under increasingly harsh growth conditions. Thus, the use of high inocula may mask the true position of the G/NG boundary if the inoculum used is already denser than the MPD of the organism in a very stressful test environment.

3.4.4.3 Are There Absolute Limits to Microbial Growth?

In the above discussion it has been implicitly assumed that there are absolute limits to microbial growth under combined environmental stresses. It is pertinent to examine this assumption.

Numerous authors have noted that, within an experiment, the transition between conditions that allow growth, and those that do not, is abrupt and that usually all replicates at the last growth condition grow, while all the replicates at the first-growth-preventing condition do not (Masana and Baranyi, 2000; Presser et al., 1998; Tienungoon et al., 2000). McKellar and Xu (2001), for example, reported that of 1820 conditions tested, all five replicates of each condition either grew or did not grow. This abruptness, however, is not always evident in the modeled results (Tienungoon et al., 2000).

Conversely, between experiments by the same researcher, using the same methods and the same strain, results are not always reproducible. Figure 3.9 provides an example and Masana and Baranyi (2000b) make the same observation of their data for *Brochothrix thermosphacta*. At the same time, however, there is evidence of excellent reproducibility of boundaries between independent workers, using different strains, and different methods in different locations. The results of Tienungoon et al. (2000) were highly consistent between two strains tested, and more notably, with those of George et al. (1988) and Cole et al. (1990) presented a decade earlier, including different strains in one case. There is also a remarkable similarity between the pH/temperature G/NG interface of *Listeria innocua* reported by Le Marc et al. (2002) and the same interface for two species of *L. monocytogenes* presented in Tienungoon et al. (2000).

Jenkins et al. (2000) noted that the boundary they derived for the growth limits of *Zygosaccharomyces baillii* in beverages was very consistent with a model developed 30 years earlier for the stability of acidic sauces.

Stewart et al. (2002) noted that with *S. aureus*, as conditions became increasingly unfavorable for growth, the contour lines (P_{growth}) they generated drew closer and closer together, suggesting that conditions were approaching absolute limits that do not allow growth. Conversely, there are examples where one group's observations do not agree well with another's for an analogous organisms/environmental pair (e.g., Bolton and Frank, 1999; McKellar and Xu, 2001). Delignette-Muller and Rosso (2000) reported strain variability in the minimum temperature for growth.

While the above discussion points to heterogeneity in the physiological readiness of bacteria to grow in a new environment, Masana and Baranyi (2000b) also infer that differences in microenvironments, particularly within foods, could also be a source of heterogeneity in observed growth limits.

In conclusion, there is a body of experimental evidence that suggests that growth boundaries, if carefully determined, might be highly reproducible. Conversely, counterexamples exist. It remains to be determined whether the incongruous results arise from significant and measurable differences in methodology, e.g., the detection time used in the respective studies, or are due to uncontrollable sources (Table 3.7).

3.4.4.4 Experimental Design

As noted above, it is not possible from first principles to design the optimal experiment that captures the information to characterize the G/NG boundary. Various authors have suggested physiological interpretations (Battey et al, 2001; Battey and Schaffner, 2001; Jenkins et al., 2000; López-Malo et al., 2000; McMeekin et al., 2000) but none have yet been experimentally tested.

Thus, an empirical approach that aims to collect as much information in the region of most interest, i.e., the G/NG interface, is recommended by most workers. Several groups of researchers have indicated that they use a two-stage modeling process. The first uses a coarse grid of conditions of variables to roughly establish the position of the boundary. The second phase monitors responses at conditions near the boundary and at close intervals of the environmental parameters. Variable combinations far from the interface, at which growth is either highly likely or highly unlikely, do not provide much information to the modeling process, which seeks to define the interface with a high level of precision. Equally, it is ideal to use a design that gives roughly equal numbers of conditions where growth is, and growth is not, observed (Jenkins et al., 2000; Legan et al., 2002; Masana and Baranyi, 2000b, Uljas et al., 2001). Pragmatically, Legan et al. (2002) recommend setting up "marginal" and "no-growth" treatments first because these treatments will run for the longest time (possibly several months). Those conditions in which growth is expected to be relatively quick can be set up last because they only need monitoring until growth is detected.

The nature of these studies necessarily involves long incubation times. Legan et al. (2002) noted that particular care must be taken to ensure that the initial conditions do not change over time solely as a result of an uncontrolled interaction with the laboratory environment. Prevention of dessication or uptake of water vapor requires particular attention. Changes resulting from microbial activity may, however, be an important part of the mechanism leading to growth initiation and should not be stabilized at the expense of growth that would naturally occur in a food. Legan et al. (2002) comment that, for example, maintaining the initial pH over time is typically neither possible nor practical, even in buffered media, and that allowing a change in pH due to growth of the organism more closely mimics what would happen in a food product than maintaining the initial pH over time.

3.4.4.5 Conclusion

From the above discussion, unambiguous definition of the G/NG boundary of an organism in multidimensional space presents several paradoxical challenges. While an experimenter will do well to remember these considerations in the interpretation of his/her results, it seems probable that methods that have been used to date will have come close to identifying the "true" G/NG boundary, and that the position of the boundary will move only slightly if an experimenter acts to control all of the above variables and to maximize the potential for the observation of growth in the chosen experimental system.

While the discussion has not focused specifically on appropriate methods for probability of growth within a defined time, many of the same principles and considerations will apply.

Moreover, the field of growth limits modeling, while having an equally long history as kinetic modeling, now seems to be quite disjointed, with little rigorous comparison of approaches, let alone agreement on the most appropriate model structures or experimental methods. In particular, the earlier work in probability modeling seems to have been ignored by some more contemporary workers, without reasons being indicated.

The results of G/NG studies are clearly of great interest to food producers and food safety managers. It is perhaps time, then, that the G/NG modeling community seeks to find common ground and to begin to develop a rigorous framework for the development, and interpretation, of growth limits studies.

APPENDIX A3.1 — CHARACTERIZATION OF ENVIRONMENTAL PARAMETERS AFFECTING MICROBIAL KINETICS IN FOODS

A3.1.1 TEMPERATURE

In most situations, temperature is the major environmental parameter influencing kinetics of microorganisms in food and its effect is included in most predictive microbiology models. During processing, storage, and distribution the temperature of foods can vary substantially, frequently including periods of temperature abuse for chilled foods (see, e.g., Audits International, 1999; James and Evans, 1990; O'Brien, 1997; Sergelidis et al., 1997). Thus, it is an important property of secondary models that they can predict the effect of changing temperatures on microbial kinetics and application of these models relies on information about product temperature and its possible variation over time. Numerous types of thermometers, temperature probes, and data loggers are available (McMeekin et al., 1993, pp. 257–269; seagrant.oregonstate.edu/extension/fisheng/loggers.html) to measure the temperature of foods or food processing equipment. Infrared non-contact thermometers are often appropriate for foods but their use is limited for process equipment with stainless surfaces.

A3.1.2 STORAGE ATMOSPHERE

Foods are typically stored aerobically, vacuum packed, or by using modified atmosphere packing (MAP). "Controlled atmosphere packaging" can be considered a special case of MAP. MAP foods are exposed to an atmosphere different from both air and vacuum packed usually involving mixtures of the gasses carbon dioxide (CO_2), nitrogen (N_2), and oxygen (O_2).

O_2 and CO_2 influence growth of most microorganisms and secondary predictive models must take their effect into account. The solubility of O_2 in water, and thereby into the water phase of foods, is low (~0.03 l/l) but it can be important for growth and metabolism of microorganisms in both aerobic and MAP-stored products (Dainty and Mackey, 1992). Numerous techniques and instruments are available to determine O_2 in the gas phase or dissolved in food. Microelectrodes to determine gradients of dissolved O_2 in foods are available (www.instechlabs.com/oxygen.html; www.microelectrodes.com/) but models to predict the effect of such gradients remain to be developed. To account for the effect of aerobic or vacuum packed storage of foods a categorical approach has been used within predictive microbiology. For aerobic conditions growth media with access to air have been agitated. For vacuum packed foods microorganisms typically have been grown under 100% N_2.

CO_2 inhibits growth of some microorganisms substantially and, to predict microbial growth in MAP foods, it is important to determine the equilibrium concentration in the gas phase or the concentration of CO_2 dissolved into the foods water phase. At equilibrium, the concentration of CO_2 dissolved into the water phase of foods is proportional to the partial pressure of CO_2 in the atmosphere surrounding the product. Henry's law (Equation A3.1) provides a good approximation for the solubility of CO_2.

$$\left[CO_2\right]_{aqueous}^{Equilibrium} = K_H \cdot \rho CO_2 \qquad (A3.1)$$

In Equation A3.1, K_H is Henry's constant (mg/l/atm) and ρCO_2 is the partial pressure (atm) of CO_2. Between 0 and 160°C the temperature dependence of the Henry's constant can be predicted by Equation A3.2:

$$K_H(mg \cdot l^{-1} \cdot atm^{-1}) =$$

$$\frac{101325 \cdot 2.4429}{\exp(-6.8346 + 1.2817 \cdot 10^4 / K - 3.7668 \cdot 10^6 / K^2 + 2.997 \cdot 10^3 / K^3)} \qquad (A3.2)$$

where K is the absolute temperature (Carroll et al., 1991). Those authors expressed K_H as MPa/mole fraction. In Equation A3.2 the constants 101,325 Pa/atm and 2.4429 was used to convert this unit into mg CO_2/l H_2O/atm.

For MAP foods in flexible packaging the partial pressure of CO_2 is conveniently determined from the percentage of CO_2 inside the pack. A range of analytical methods is available to determine CO_2 concentration in gas mixtures or concentrations of dissolved CO_2 (Dixon and Kell, 1989; www.pbi-dansensor.com/Food.htm).

As shown from Equation A3.1 and Equation A3.2, the concentration of CO_2 dissolved in the water phase of a MAP food with 50% CO_2 in the headspace gas at equilibrium is 1.67 g/l at 0°C and 1.26 g/l at 8°C. Because of the high solubility of

CO_2 in water the gas composition in the headspace of MAP foods changes after packaging. The equilibrium gas composition is influenced by several factors, e.g., the percentage of CO_2 in the initial headspace gas ($\%CO_2^{Initial}$), the initial gas/product volume ratio (G/P), temperature, pH, lipids in the food, respiration of the food, and, of course, permeability of the packing film. Different mass-balance equations to predict the rate of adsorption and solubility of CO_2 have been suggested (Devlieghere et al., 1998; Dixon and Kell, 1989; Gill, 1988; Löwenadler and Rönner, 1994; Simpson et al., 2001a,b; Zhao et al., 1995). In chilled foods the rate of absorption of CO_2 is rapid compared to growth of microorganisms. Therefore, to predict microbial growth in these MAP foods it is sufficient to take into account the equilibrium concentration of CO_2.

Devlieghere et al. (1998) suggested Equation A3.3 to predict the concentration of CO_2 in the water phase as a function of $\%CO_2^{Initial}$ and G/P. In Equation A3.3, d_{CO2} is the density of CO_2 (1.976 g/l).

$$[CO_2]_{aqueous}^{Equilibrium} = \frac{\left(\frac{G}{P}\cdot dCO_2 + K_H\right) - \sqrt{\left(\frac{G}{P}\cdot dCO_2 + K_H\right)^2 - \left(\frac{4}{100}\cdot K_H\cdot\frac{G}{P}\cdot \%CO_2^{Initial}\cdot dCO_2\right)}}{2} \qquad (A3.3)$$

Equation A3.3 does not take into account the effect of the storage temperature and Devlieghere et al. (1998) developed a polynomial model to predict the concentration of dissolved CO_2 as a function of $\%CO_2^{Initial}$, G/P, and temperature. If, for example, $\%CO_2^{Initial}$ is 25, the polynomial model predicts that a G/P ratio of three results in higher concentration of dissolved CO_2 than does a G/P ratio of 4, which is not logical. In contrast we have found that the combined use of Equation A3.2 and Equation A3.3 provides realistic predictions for concentrations of dissolved CO_2. It also seems relevant to include the effect of product pH on dissolved CO_2, and thereby the equilibrium concentration of CO_2 in the gas phase of MAP foods.

A3.1.3 SALT, WATER-PHASE SALT, AND WATER ACTIVITY

While temperature is the single most important *storage* condition influencing growth of microorganisms in foods, NaCl is the most important *product* characteristic in many foods. The concentration of NaCl in foods can be determined as chloride by titration (Anon., 1995a). Instruments to determine NaCl indirectly from conductivity measurements are available but extensive calibration for particular types of products may be required. In fresh and intermediate moisture foods, NaCl is dissolved in the water phase of the products.

To predict the effect of NaCl on growth of microorganisms in these products the concentration of water-phase salt (WPS) or relative humidity, i.e., the water activity (a_w) must be determined (Equation A3.4 to Equation A3.7).

Water-phase salt can be calculated from Equation A3.4:

$$\% \text{ Water phase salt} =$$
$$\%NaCl \text{ (w/v)} \times 100/(100 - \% \text{ dry matter} +\%NaCl \text{ (w/v)}) \qquad \text{(A3.4)}$$

Water activity is a fundamental property of aqueous solutions and is defined as:

$$a_w = \frac{\rho}{\rho_o} \qquad \text{(A3.5)}$$

where ρ is the vapor pressure of the solution and ρ_0 is the vapor pressure of the pure water under the same conditions of temperature, etc.

For mixtures of NaCl and water there is a direct relation between the WPS content and a_w (Resnik and Chirife, 1988; Equation A3.6 and Equation A3.7). For cured foods where NaCl is the only major humectant these relations are valid as documented, e.g., for cold-smoked salmon (Jørgensen et al., 2000) and processed "delicatessen" meats (Ross and Shadbolt, 2001). To determine water activity of foods, instruments relying on the dew point method are now widely used because of their speed (providing results within a few minutes), robustness, and reliability but other methods and instruments are available (Mathlouthi, 2001).

$$a_w = 1 - 0.0052471 \cdot \%WPS - 0.00012206 \cdot \%WPS^2 \qquad \text{(A3.6)}$$

$$\% WPS = 8 - 140.07 \cdot (a_w - 0.95) - 405.12 \cdot (a_w - 0.95)^2 \qquad \text{(A3.7)}$$

A3.1.4 pH

For many microorganisms, small pH variations in the pH range ~6 to ~7 have very little or no effect on population kinetics. In more acidic foods, however, pH *per se* can greatly influence microbial kinetics but can also accentuate the effect of other added preservative compounds. The pH of solid foods is often determined by homogenizing 10 g of a sample with 10 to 20 ml of distilled water and measuring the pH of the suspension using a standard combined electrode.

A3.1.5 ADDED PRESERVATIVES INCLUDING ORGANIC ACIDS, NITRATE, AND SPICES

High concentrations of organic acids occur naturally in some foods and various organic acids including acetic acid, ascorbic acid, benzoic acid, citric acid, lactic acid, and sorbic acid are frequently added to foods. Organic acids can inhibit growth of microorganisms markedly and secondary models to predict their inhibitory effect are frequently needed. As for NaCl the secondary models must take into account the concentration of organic acids in the water phase of products. In addition, secondary models may need to describe the combined effect of organic acids and other environmental parameters particularly the pH.

In solution, organic acids exist either as the dissociated (ionized) or undissociated species. The Henderson–Hasselbalch equation (Equation A3.8) relates the proportion of undissociated and dissociated forms of organic acid to pH and pK_a according to the following expression:

$$[A^-]/[HA] = 10^{pH-pK_a} \tag{A3.8}$$

where [HA] is the concentration of undissociated form of the acid, [A$^-$] the concentration of dissociated (ionized) form of the acid, and pK_a is the pH at which the concentrations of the two forms are equal.

While both the dissociated and the undissociated forms of organic acids have inhibitory effects on bacterial growth the undissociated form is more inhibitory, usually by two to three orders of magnitude, than the dissociated form (Eklund, 1989).

Cross-multiplying and rearranging Equation A3.8 to solve for [HA] gives:

$$[HA] = [LAC]/(1 + 10^{pH-pK_a}) \tag{A3.9}$$

where [LAC] is the total lactic acid concentration and all other terms are as previously defined.

As the concentration of an undissociated acid increases the growth rate of microorganisms decreases, eventually ceasing completely at a level described as the MIC. This behavior, and its dependence on the interaction of pH and total organic acid concentration, is included explicitly in several secondary models (Augustin and Carlier, 2000a; Presser et al., 1997).

Simple enzyme kits are available to determine several of the organic acids that are important in foods. Simultaneous determination of a range of organic acids is possible by HPLC analysis and is often an appropriate method to use (Dalgaard and Jørgensen, 2000; Pecina et al., 1984).

Nitrite can be added to some types of meat products and its concentration in the water phase of products must be taken into account when secondary predictive models for these products are developed. Colorimetric methods are available to measure the concentration of nitrite in foods (Anon., 1995b; Karl, 1992).

Spices and herbs can have substantial antimicrobial activity and appropriate terms may need to be included in secondary models (Koutsoumanis et al., 1999; Skandamis and Nychas, 2000). The concentration of active antimicrobial components in spices, herbs, and essential oils can vary substantially as a function, e.g., of geographical region and season (Nychas and Tassou, 2000; Sofos et al., 1998). Therefore, the development of accurate secondary predictive models most likely will have to rely on the concentration of their active antimicrobial components. Recently, Lambert et al. (2001) showed the antimicrobial effect of the oregano essential oil quantitatively corresponded to the effect of its two active components, i.e., thymol and carvacrol. To quantitatively determine active components in spices, herbs, and essential oils appropriate extracts can be analyzed by GC/MS techniques (Cosentino et al., 1999; Cowan, 1999).

A3.1.6 SMOKE COMPONENTS

It has long been known that high concentrations of smoke components have strong antimicrobial activity (Shewan, 1949). Today many meat and seafood products are smoked but typically less intensively than some decades ago. However, even moderate concentrations of smoke components can influence growth rates, growth limits, and rates of death/inactivation of microorganisms in foods (Leroi et al., 2000; Leroi and Joffraud, 2000; Ross et al., 2000b; Suñen, 1998; Thurette et al., 1998). Thus, to obtain accurate prediction of microbial kinetics in smoked foods it is important to include terms for the effect of smoke components in secondary models. Phenols are important antimicrobials in wood smoke, or in liquid smokes, and a few secondary models include the total phenol concentration as an environmental parameter (Augustin and Carlier 2000a,b; Giménez and Dalgaard, in press; Membré et al., 1997).

Classical colorimetric methods can be used to determine the total concentration of phenols in smoked foods. These methods rely on formation of colored complexes, e.g., between phenols and Gibb's reagent (2,6-dichloroquinone-4-chloroimide) or 4-aminoantipyrine (Leroi et al., 1998; Tucker, 1942). The total phenol concentration is a crude measure of how intensely foods have been smoked. By using GC/MS techniques more detailed information about specific smoke components can be obtained (Guillén and Errecalde, 2002; McGill et al., 1985; Tóth and Potthast, 1984). In the future, secondary models may be developed to include the effect of specific phenols, other specific smoke components, and possibly their interaction with NaCl. During the smoking of foods, phenols and other smoke components are mainly deposited in the outer 0.5 cm of the product (Chan et al., 1975). Modeling the effect of the spatial distribution in foods is another challenge.

A3.1.7 OTHER ENVIRONMENTAL PARAMETERS

The environmental parameters discussed above include those that are of major importance in traditional methods of food preservation. Many modern methods of food preservation also rely on combinations of these environmental parameters. However, the effect of a few well-known and several emerging food processing technologies relies on the antimicrobial effect of other environmental parameters, e.g., bacteriocins, gamma irradiation, high electric field pulses, high pressure, and UV light. Secondary models for the effect of some of these environmental parameters have been developed but will not be discussed here in detail. Other environmental parameters related to food structure and to the effect of microbial metabolism on changes in environmental parameters are discussed in Chapter 5 whereas the effect of time-varying environmental parameters is discussed in Chapter 7.

REFERENCES

Abdullah, B., Gardini, F., Paparella, A., and Guerzoni, M.E. Growth modelling of the predominant microbial groups in hamburgers in relation to modulation of atmosphere composition, storage temperature, and diameter of meat particle. *Meat Sci.*, 38, 511–526, 1994.

Adair, C., Kilsby, D.C., and Whittall, P.T. Comparison of the Schoolfield (non-linear Arrhenius) model and the square root model for predicting bacterial growth in foods. *Food Microbiol.*, 6, 7–18, 1989.

Adams, M.R., Little, C.L., and Easter, M.C. Modelling the effect of pH, acidulant and temperature on growth of *Yersinia enterocolitica*. *J. Appl. Bacteriol.*, 71, 65–71, 1991.

Anon. Salt (chlorine as sodium chloride) in seafood. Volumetric method. In *Official Methods of Analysis*, AOAC, Arlington, VA, 1995a, p. 937.09.

Anon. Nitrites in cured meat. Colorimetric method. In *Official Methods of Analysis*, AOAC, Arlington, VA, 1995b, p. 937.31.

Audits International, US Food Temperature Evaluation. Audits International, Northbrook, IL, 1999. http://www.foodriskclearinghouse.umd.edu/audits_international.htm. February, 2002.

Augustin, J.-C. and Carlier, V. Mathematical modelling of the growth rate and lag time for *Listeria monocytogenes*. *Int. J. Food Microbiol.*, 56, 29–51, 2000a.

Augustin, J.-C. and Carlier, V. Modelling the growth of *Listeria monocytogenes* with a multiplicative type model including interactions between environmental factors. *Int. J. Food Microbiol.*, 56, 53–70, 2000b.

Augustin, J.-C., Rosso, L., and Carlier, V. A model describing the effect of temperature history on lag time for *Listeria monocytogenes*. *Int. J. Food Microbiol.*, 57, 169–181, 2000.

Bajard, S., Rosso, L., Fardel, G., and Flandrois, J.P. The particular behaviour of *Listeria monocytogenes* under sub-optimal conditions. *Int. J. Food Microbiol.*, 29, 201–211, 1996.

Baker, D.A. Probability models to assess the safety of foods with respect to *Clostridium botulinum*. *J. Ind. Microbiol.*, 12, 156–161, 1993.

Baker, D.A. and Genigeorgis, C. Predicting the safe storage of fresh fish under modified atmospheres with respect to *Clostridium botulinum* toxigenesis by modelling length of the lag phase of growth. *J. Food Prot.*, 53, 131–140, 1990.

Baker, D.A. and Genigeorgis, C. Predictive modeling. In *Clostridium botulinum Ecology and Control in Foods*, Eds. A.W.H. Hauschild and K.L. Dodds. Marcel Dekker, New York, 1993, 343–406.

Baker, D., Genigeorgis, C., Glover, J., and Razavilar, V. Growth and toxigenesis of *C. botulinum* type E in fishes packaged under modified atmospheres. *Int. J. Food Microbiol.*, 10, 269–290, 1990.

Baranyi, J. Comparison of stochastic and deterministic concepts of bacterial lag. *J. Theor. Biol.*, 192, 403–408, 1998.

Baranyi, J. and Roberts, T.A. A dynamic approach to predicting bacterial growth in food. *Int. J. Food Microbiol.*, 23, 277–294, 1994.

Baranyi, J. and Roberts, T.A. Mathematics of predictive food microbiology. *Int. J. Food Microbiol.*, 26, 199–218, 1995.

Baranyi, J., Ross, T., McMeekin, T.A., and Roberts, T.A. Effects of parametrization on the performance of empirical models used in "predictive microbiology." *Food Microbiol.*, 13, 83 –91, 1996.

Battey, A.S. and Schaffner, D.W. Modelling bacterial spoilage in cold-filled ready to drink beverages by *Acinetobacter calcoaceticus* and *Gluconobacter oxydans*. *J. Appl. Microbiol.*, 91, 237–247, 2001.

Battey, A.S., Duffy, S., and Schaffner, D.W. Modelling mould spoilage on cold-filled ready-to-drink beverages by *Aspergillus niger* and *Penicillium spinulosum*. *Food Microbiol.*, 18, 521–529, 2001.

Battey, A.S., Duffy, S., and Schaffner, D.W. Modeling yeast spoilage in cold-filled ready-to-drink beverages with *Saccharomyces cerevisiae, Zygosaccharomyces bailii*, and *Candida lipolytica*. *Appl. Environ. Microbiol.*, 68, 1901–1906, 2002.

Benedict, R.C., Pertridge, T., Wells, D., and Buchanan, R.L. *Bacillus cereus*: aerobic growth kinetics. *J. Food Prot.,* 56, 211–214, 1993.

Bernaerts, K., Versyck, K.J., and van Impe, J.F. On the design of optimal experiments for parameter estimation of a Ratkowsky-type growth kinetic at suboptimal temperatures. *Int. J. Food Microbiol.*, 54, 27–38, 2000.

Bhaduri, S., Buchanan, R.L., and Phillips, J.G. Expanded response surface model for predicting the effects of temperature, pH, sodium chloride contents and sodium nitrite concentration on the growth rate of *Yersinia enterocolitica*. *J. Appl. Bacteriol.*, 79, 163–170, 1995.

Bolton, L.F. and Frank, J.F. Defining the growth/no growth interface for *Listeria monocytogenes* in Mexican-style cheese based on salt, pH and moisture content. *J. Food Prot.*, 62, 601–609, 1999.

Bouttefroy, A., Mansour, M., Linder, M., and Milliere, J.B. Inhibitory combinations of nisin, sodium chloride, and pH on *Listeria monocytogenes* ATCC 15313 in broth by an experimental design approach. *Int. J. Food Microbiol.,* 54, 109–115, 2000.

Broughall, J.M. and Brown, C. Hazard analysis applied to microbial growth in foods: development and application of three-dimensional models to predict bacterial growth. *Food Microbiol.*, 1, 13–22, 1984.

Broughall, J.M., Anslow, P., and Kilsby, D.C. Hazard analysis applied to microbial growth in foods: development of mathematical models describing the effect of water activity. *J. Appl. Bacteriol.*, 55, 101–110, 1983.

Buchanan, R.L. Developing and distributing user-friendly application software. *J. Ind. Microbiol.*, 12, 251–255, 1993a.

Buchanan, R.L. Predictive food microbiology. *Trends Food Sci. Technol.*, 41, 6–11, 1993b.

Buchanan, R.L. and Bagi, K. Expansion of response surface models for the growth of *Escherichia coli* O157:H7 to include sodium nitrite as a variable. *Int. J. Food Microbiol.*, 23, 317–332, 1994.

Buchanan, R.L. and Phillips, J.G. Updated models for the effects of temperature, initial pH, NaCl, and $NaNO_2$ on aerobic and anaerobic growth of *Listeria monocytogenes*. *Quant. Microbiol.*, 2, 103–128, 2000.

Buchanan, R.L., Smith, J.L., McColgan, C., Marmer, B.S., Golden, M., and Dell, B. Response surface models for the effects of temperature, pH, sodium chloride, and sodium nitrite on the aerobic and anaerobic growth of *Staphylococcus aureus* 196E. *J. Food Saf.*, 13, 159–175, 1993.

Buchanan, R.L., Golden, M.H., and Phillips, J.G. Expanded models for the non-thermal inactivation of *Listeria monocytogenes*. *J. Appl. Microbiol.,* 82, 567–577, 1997.

Budavari, S. *The Merck Index: An Encyclopaedia of Chemicals, Drugs, and Biologicals,* 11th ed., Merck and Co., Inc., Rahway, NJ, 1989.

Carroll, J.J., Slupsky, J.D., and Mather, A.E. The solubility of carbon dioxide in water at low pressure. *J. Phys. Chem. Ref. Data,* 20, 1201–1209, 1991.

Chan, W.S., Toledo, R.T., and Deng, J. Effect of smokehouse temperature, humidity and air flow on smoke penetration into fish muscle. *J. Food Sci.*, 40, 240–243, 1975.

Chandler, R.E. and McMeekin, T.A. Combined effect of temperature and salt concentration/water activity on growth rate of *Halobacterium* spp. *J. Appl. Bacteriol.*, 67, 71–76, 1989.

Chea, F.P., Chen, Y., Montville, T.J., and Schaffner, D.W. Modelling the germination kinetics of *Clostridium botulinum* 56A spores as affected by temperature, pH, and sodium chloride. *J. Food Prot.*, 63, 1071–1079, 2000.

Cheroutre-Vialette, M. and Lebert, A. Modelling the growth of *Listeria monocytogenes* in dynamic conditions. *Int. J. Food Microbiol.*, 55, 201–207, 2000.

Chorin, E., Thuault, D., Cleret, J.-J., and Bourgeois, C.-M. Modelling *Bacillus cereus* growth. *Int. J. Food Microbiol.*, 38, 229–234, 1997.

Cole, M.B., Jones, M.V., and Holyoak, C. The effect of pH, salt concentration and temperature on the survival and growth of *Listeria monocytogenes*. *J. Appl. Bacteriol.*, 69, 63–72, 1990.

Connil, N., Plissoneau, L., Onno, B., Pilet, M.-F., Prévost, H., and Dousset, X. Growth of *Carnobacterium divergens* V41 and production of biogenic amines and divergicin in sterile cold-smoked salmon extract at varying temperatures, NaCl levels, and glucose concentrations. *J. Food Prot.*, 65, 333–338, 2002.

Cosentino, S., Tuberoso, C.I.G., Pisano, B., Satta, M., Mascia, V., Arzedi, E., and Palmas, F. *In-vitro* antimicrobial activity and chemical composition of Sardinian *thymus* essential oils. *Lett. Appl. Microbiol.*, 29, 130–135, 1999.

Cowan, M.M. Plant products as antimicrobial agents. *Clin. Microbiol. Rev.*, 12, 564–582, 1999.

Cuppers, H.G.A., Oomes, S., and Brul, S. A model for the combined effects of temperature and salt concentration on growth rate of food spoilage molds. *Appl. Environ. Microbiol.*, 63, 3764–3769, 1997.

Dainty, R.H. and Mackey, B.M. The relationship between the phenotypic properties of bacteria from chill-stored meat and spoilage processes. *J. Appl. Bacteriol.*, 73, 103S–114S, 1992.

Dalgaard, P. *Evaluation and Prediction of Microbial Fish Spoilage*, Ph.D. thesis, Technological Laboratory, Danish Ministry of Fisheries, Lynby, Denmark, 1993.

Dalgaard, P. Modelling of microbial activity and prediction of shelf life for packed fresh fish. *Int. J. Food Microbiol.*, 26, 305–317, 1995.

Dalgaard, P. Modelling and prediction the shelf-life of seafood. In *Safety and Quality Issues in Fish Processing*, Bremner, H.A. (Ed.), Woodhead Publishing, Cambridge, England, 2002, pp. 191–219.

Dalgaard, P. and Jørgensen, L.V. Cooked and brined shrimps packed in a modified atmosphere have a shelf-life of >7 months at 0°C, but spoil at 4–6 days at 25°C. *Int. J. Food Sci. Technol.*, 35, 431–442, 2000.

Dalgaard, P., Ross, T., Kamperman, L., Neumeyer, K., and McMeekin, T.A. Estimation of bacterial growth rates from turbidimetric and viable count data. *Int. J. Food Microbiol.*, 23, 391–404, 1994.

Dalgaard, P., Mejlholm, O., and Huss, H.H. Application of an iterative approach for development of a microbial model predicting the shelf-life of packed fish. *Int. J. Food Microbiol.*, 38, 169–179, 1997.

Dalgaard, P., Cowan, B. J., Heilmann, J., and Silberg, S. The seafood spoilage and safety predictor (SSSP). In *Predictive Modeling in Foods — Conference Proceedings*, Van Impe, J.F.M., Geeraerd, A.H., Leguerinel, I., and Mafart, P. (Eds.), Katholieke Universiteit Leuven/BioTec, Belgium, 2003, pp. 256–258 .

Dantigny, P. Dimensionless analysis of the microbial growth rate dependence on sub-optimal temperatures. *J. Ind. Microbiol. Biotechnol.*, 21, 215–218, 1998.

Dantigny, P. and Molin, P. Influence of the modelling approach on the estimation of the minimum temperature for growth in Belehradek-type models. *Food Microbiol.*, 17, 597–604, 2000.

Daughtry, B.J., Davey, K.R., and King, K.D. Temperature dependence of growth kinetics of food bacteria. *Food Microbiol.*, 14, 21–30, 1997.

Davey, K.R. A predictive model for combined temperature and water activity on microbial growth during the growth phase. *J. Appl. Bacteriol.*, 67, 483–488, 1989.

Davey, K.R. Applicability of the Davey (linear Arrhenius) predictive model to the lag phase of microbial growth. *J. Appl. Bacteriol.*, 70, 253–257, 1991.

Davey, K.R. Modeling the combined effect of temperature and pH on the rate coefficient for bacterial growth. *Int. J. Food Microbiol.*, 23, 295–303, 1994.

Davey, K.R. Belehradek models: application to chilled foods preservation. *Refrig. Sci. Technol. Proc.*, EUR 18816, 37–47, 1999.

Davey, K.R. Models for predicting the combined effect of environmental process factors on the exponential and lag phases of bacterial growth: development and application and an unexpected correlation. In *Proceedings of the 6th World Conference of Chemical Engineering*, Melbourne, 2001.

Davey, K.R. and Daughtry, B.J. Validation of a model for predicting the combined effect of 3 environmental-factors on both exponential and lag phases of bacterial-growth — temperature, salt concentration and pH. *Food Res. Int.*, 28, 233–237, 1995.

Davies, K.W. Design of experiments for predictive microbial modeling. *J. Ind. Microbiol.*, 12, 295–300, 1993.

Deak, T. and Beuchat, L.R. Use of indirect conductimetry to predict the growth of spoilage yeasts, with special consideration of *Zagosaccharomyces bailii*. *Int. J. Food Microbiol.*, 23, 405–417, 1994.

Delignette-Muller, M.L. Relation between the generation time and the lag time of bacterial growth kinetics. *Int. J. Food Microbiol.*, 43, 97–104, 1998.

Delignette-Muller, M.L. and Rosso, L. Biological variability and exposure assessment. *Int. J. Food Microbiol.*, 58, 203–212, 2000.

Delignette-Muller, M.L., Baty, F., Cornu, M., and Bergis, H. The effect of the temperature shifts on the lag phase of *Listeria monocytogenes*. In *Predictive Modeling in Foods— Conference Proceedings*, Van Impe, J.F.M., Geeraerd, A.H., Leguerinel, I., and Mafart, P. (Eds.), Katholieke Universiteit Leuven/BioTec, Belgium, 2003, pp. 209–211.

Dengremont, E. and Membré, J.M. Statistical approach for comparison of the growth rates of the strains of *Staphylococcus aureus*. *Appl. Environ. Microbiol.*, 61, 4389–4395, 1997.

Devlieghere, F., Debevere, J., and van Impe, J. Concentration of carbon dioxide in the water-phase as a parameter to model the effect of a modified atmosphere on microorganisms. *Int. J. Food Microbiol.*, 43, 105–113, 1998.

Devlieghere, F., Geeraerd, A.H., Versyck, K.J., Bernaert, H., van Impe, J., and Debevere, J. Shelf life of modified atmosphere packed cooked meat products: addition of Na-lactate as a fourth shelf life determinative factor in a model and product validation. *Int. J. Food Microbiol.*, 58, 93–106, 2000a.

Devlieghere, F., Lefevre, I., Magnin, A., and Debevere, J. Growth of *Aeromonas hydrophila* in modified-atmosphere-packed cooked meat products. *Food Microbiol.*, 17, 185–196, 2000b.

Devlieghere, F., Geeraerd, A.H., Versyck, K.J., Vandewaetere, B., van Impe, J., and Debevere, J. Growth of *Listeria monocytogenes* in modified atmosphere packed cooked meat products: a predictive model. *Food Microbiol.*, 18, 53–66, 2001.

Dixon, N.M. and Kell, D.B. The control and measurement of "CO_2" during fermentations. *J. Microbiol. Methods*, 10, 155–176, 1989.

Dodds, K.L. An introduction to predictive microbiology and the development and use of probability models with *Clostridium botulinum*. *J. Ind. Microbiol.*, 12, 139–143, 1993.

Doe, P.E. and Heruwati, E. A model for the prediction of the microbial spoilage of sun-dried tropical fish. *J. Food Eng.*, 8, 47–72, 1988.

Draper, N.R. and Smith, H. *Applied Regression Analysis*, 2nd ed., John Wiley & Sons, New York, 1981, 709 pp.

Eifert, J.D., Hackney, C.R., Pierson, M.D., Duncan, S.E., and Eigel, W.N. Acetic, lactic, and hydrochloric acid effects on *Staphylococcus aureus* 196E growth based on a predictive model. *J. Food Sci.*, 62, 174–178, 1997.

Eklund, T. Organic acids and esters. In *Mechanisms of Action of Food Preservation Procedures,* Gould, G.W. (Ed.), Elsevier Science Publishers, London, 1989, pp. 160–200.

Eyring, H. The activated complex in chemical reactions. *J. Chem. Phys.*, 3, 107–115, 1935.

Farber, J. M. Predictive modeling of food deterioration and safety. In *Foodborne Microorganisms and Their Toxins: Developing Methodology,* Pierson, M.D. and Stern, N.J. (Eds.), Marcel Dekker, New York, 1986, pp. 57–90.

Fernandez, P.S., Baranyi, J., and Peck, M.W. A predictive model of growth from spores of non-proteolytic *Clostridium botulinum* in the presence of different CO_2 concentrations as influenced by chill temperature, pH and NaCl. *Food Microbiol.*, 18, 453–461, 2001.

Gänzle, M.G., Ehmann, M., and Hammes, W.P. Modeling of growth of *Lactobacillus sanfranciscensis* and *Candida milleri* in response to process parameters of sourdough fermentation. *Appl. Environ. Microbiol.*, 64, 2616–2623, 1998.

Garcia-Gimeno, R.M., Hervás-Martinez, C., and de Silóniz, M.I. Improving artificial neural networks with a pruning methodology and genetic algorithms for their application in microbial growth predictions in food. *Int. J. Food Microbiol.*, 72, 19–30, 2002.

Garcia-Gimeno, R.M., Hervás-Martinez, C., Barco-Alcaia, E., Zurera-Cosano, G., and Sanz-Tapia, E. An artificial neural network approach to *Escherichia coli* O157:H7 growth estimation. *J. Food Sci.*, 68, 639–645, 2003.

Geeraerd, A.H., Herremans. C.H., Cenens, C., and Van Impe J.F. Application of artificial neural networks as a non-linear modular modeling technique to describe bacterial growth in chilled food products. *Int. J. Food Microbiol.*, 44, 49–68, 1998a.

Geeraerd, A.H., Herremans, C.H., Ludikhuyze, L.R., Hendrickx, M.E., and Van Impe, J.F. Modeling the kinetics of isobaric–isothermal inactivation of *Bacillus subtilis* alpha-amylase with artificial neural networks. *J. Food Eng.*, 36, 263–279, 1998b.

Geeraerd, A.H., Valdramidis, V.P., Devlieghere, F., Bernaert, H., Debevere, J., and van Impe, J. Development of a novel modelling methodology by incorporating *a priori* microbiological knowledge in a black box approach. *Int. J. Food Microbiol.*, in press.

Genigeorgis, C.A. Factors affecting the probability of growth of pathogenic microorganisms in foods. *J. Am. Vet. Med. Assoc.*, 179, 1410–1417, 1981.

Genigeorgis, C., Carniciu, M., Dutulescu, D., and Farver, T.B. Growth and survival of *Listeria monocytogenes* in market cheeses stored at 4°C to 30°C. *J. Food Protect.*, 54, 662–668, 1991.

George, S.M., Lund, B.M., and Brocklehurst, T.F. The effect of pH and temperature on initiation of growth of *Listeria monocytogenes*. *Lett. Appl. Microbiol.*, 6, 153–156, 1988.

Gibson, A.M., Bratchell, N., and Roberts, T.A. The effect of sodium chloride and temperature on the rate and extent of growth of *Clostridium botulinum* type A in pasteurized pork slurry. *J. Appl. Bacteriol.*, 62, 479–490, 1987.

Gibson, A.M., Bratchell, N., and Roberts, T.A. Predicting microbial growth: growth responses of salmonellae in a laboratory medium as affected by pH, sodium chloride and storage temperature. *Int. J. Food Microbiol.*, 6, 155–178, 1988.

Gibson, A.M., Baranyi, J., Pitt, J.I., Eyles, M.J., and Roberts, T.A. Predicting fungal growth: the effect of water activity on *Aspergillus flavus* and related species. *Int. J. Food Microbiol.*, 23, 419–431, 1994.

Gill, C.O. The solubility of carbon dioxide in meat. *Meat Sci.*, 22, 65–71, 1988.

Giménez, B.C. and Dalgaard, P. Modelling and predicting the simultaneous growth of *Listeria monocytogenes* and spoilage microorganisms in cold-smoked salmon. *J. Appl. Microbiol.*, in press.

Gorris, L. Hurdle technology. In *Encyclopaedia of Food Microbiology*, Robinson, R., Batt, C.A., and Patel, P. (Eds.), Academic Press, London, 2000, pp. 1071–1076.

Graham, A.F., Mason, D.R., and Peck, M.W. Predictive model of the effect of temperature, pH and sodium chloride on growth from spores of non-proteolytic *Clostridium botulinum*. *Int. J. Food Microbiol.*, 31, 69–85, 1996.

Guillén, M.D. and Errecalde, C. Volatile components of raw and smoked black bream (*Brama rail*) and rainbow trout (*Oncorhynchus mykiss*) studied by means of solid phase microextraction and gas chromatography/mass spectrometry. *J. Sci. Food Agric.*, 82, 945–952, 2002.

Hajmeer, M.N. and Basheer, I. A probabilistic neural network approach for modeling and classification of bacterial growth/no-growth data. *J. Microbiol. Methods*, 51, 217–226, 2002.

Hajmeer, M.N. and Basheer, I.A. A hybrid Bayesian-neural network approach for probabilistic modeling of bacterial growth/no-growth interface. *Int. J. Food Microbiol.*, 82, 233–243, 2003a.

Hajmeer, M.N. and Basheer, I. Comparison of logistic regression and neural network-based classifiers for bacterial growth. *Food Microbiol.*, 20, 43–55, 2003b.

Hajmeer, M.N., Basheer, I.A., and Najjar, Y.M. Computational neural networks for predictive microbiology. 2. Application to microbial growth. *Int. J. Food Microbiol.*, 34, 51–66, 1997.

Heitzer, A., Kohler, H.E., Reichert, P., and Hamer, G. Utility of phenomenological models for describing temperature dependence of bacterial growth. *Appl. Environ. Microbiol.*, 57, 2656–2665, 1991.

Hinshelwood, C.N. Influence of temperature on the growth of bacteria. In *The Chemical Kinetics of the Bacterial Cell*, Clarendon Press, Oxford, 1946, 254–257.

Hosmer, D.W. and Lemeshow, S. *Applied Logistic Regression*, John Wiley & Sons, New York, 1989.

Hultin, E. The influence of temperature on the rate of enzymic processes. *Acta Chem. Scand.*, 9, 1700–1710, 1955.

ICMSF (International Commission on Microbiological Specifications for Foods). *Microorganisms in Foods. 5. Microbiological Specifications of Food Pathogens,* Blackie Academic & Professional, London, 1996a.

ICMSF. Modelling microbial responses in foods. In *Microorganisms in Foods. 5. Microbiological Specifications of Food Pathogens,* Roberts, T.A., Baird-Parker, A.C., and Tompkin, R.B. (Eds.), Blackie Academic & Professional, London, 1996b, pp. 493–500.

Ikawa, J.Y. and Genigeorgis, C. Probability of growth and toxin production by nonproteolytic *Clostridium botulinum* in rockfish fillets stored under modified atmospheres. *Int. J. Food Microbiol.*, 4, 167–181, 1987.

James, S. and Evans, J. Performance of domestic refrigerators and retail display cabinets for chilled products. *Refrig. Sci. Technol. Proc.*, 1990–1994, 401–409, 1990.

Jenkins, P., Poulos, P.G., Cole, M.B., Vandeven, M.H., and Legan, J.D. The boundary for growth of *Zygosaccharomyces bailii* in acidified products described by models for time to growth and probability of growth. *J. Food Prot.*, 63, 222–230, 2000.

Jeyamkondan, S., Jayas, D.S., and Holley, R.A. Microbial growth modelling with artificial neural networks. *Int. J. Food Microbiol.*, 64, 343–354, 2001.

Johnson, F.H. and Lewin, I. The growth rate of *E. coli* in relation to temperature, quinine and coenzyme. *J. Cell. Comp. Physiol.*, 28, 47–75, 1946.

Jones, J.E. and Walker, S. Advances in modeling microbial growth. *J. Ind. Microbiol.*, 12, 200–205, 1993.

Jones, J.E., Walker, S.J., Sutherland, J.P., Peck, M.W., and Little, C.L. Mathematical modelling of the growth, survival and death of *Yersinia enterocolitica*. *Int. J. Food Microbiol.*, 23, 433–437, 1994.

Jørgensen, L.V., Dalgaard, P., and Huss, H.H. Multiple compound quality index for cold-smoked salmon (*Salmo salar*) developed by multivariate regression of biogenic amines and pH. *J. Agric. Food Chem.*, 48, 2448–2453, 2000.

Juneja, V.K., Marmer, B.S., Phillips, J.G., and Palumbo, S.A. Interactive effects of temperature, initial pH, sodium chloride, and sodium pyrophosphate on the growth kinetics of *Clostridium perfringens*. *J. Food Prot.*, 59, 963–968, 1996.

Kalina, V. Dynamics of microbial growth and metabolic activity and their control by aeration. *Antonie Van Leeuwenhoek*, 63, 353–373, 1993.

Karl, H. Bestimmung des nitritgehaltes in rucherfischen und anden fischprodukten. *Deutsche Lebensmittel-Rundschau*, 2, 41–45, 1992.

Koutsoumanis, K. Predictive modeling of the shelf life of fish under nonisothermal conditions. *Appl. Environ. Microbiol.*, 67, 1821–1829, 2001.

Koutsoumanis, K. and Nychas, G.-J.E. Application of a systematic experimental procedure to develop a microbial model for rapid fish shelf life prediction. *Int. J. Food Microbiol.*, 60, 171–184, 2000.

Koutsoumanis, K., Tassou, C.C., Taoukis, P.S., and Nychas, G.-J.E. Modelling the effectiveness of a natural antimicrobial on *Salmonella enteritidis* as a function of concentration, temperature and pH, using conductance measurements. *J. Appl. Microbiol.*, 84, 981–987, 1998.

Koutsoumanis, K., Lambropoulou, K., and Nychas, G.-J.E. A predictive model for the non-thermal inactivation of *Salmonella enteritidis* on a food model system supplemented with a natural antimicrobial. *Int. J. Food Microbiol.*, 49, 63–74, 1999.

Koutsoumanis, K.P., Taoukis, P.S., Drosinos, E.H., and Nychas, G.-J.E. Application of an Arrhenius model for the combined effect of temperature and CO_2 packaging on the spoilage microflora of fish. *Appl. Environ. Microbiol.*, 66, 3528–3534, 2000.

Lambert, R.J.W., Skandamis, P.N., Coote, P.J., and Nychas, G.-J.E. A study of the minimum inhibitory concentration and mode of action of oregano essential oil, thymol and carvacrol. *J. Appl. Microbiol.*, 91, 453–462, 2001.

Lanciotti, R., Sinigaglia, M., Guardini, F., Vannini, L, and Guerzoni, M.E. Growth/no growth interface of *Bacillus cereus, Staphylococcus aureus* and *Salmonella enteritidis* in model systems based on water activity, pH, temperature and ethanol concentration. *Food Microbiol.*, 18, 659–668, 2001.

Legan, D., Vandeven, M., Stewart, C., and Cole, M.B. Modelling the growth, survival and death of bacterial pathogens in food. In *Foodborne Pathogens: Hazards, Risk Analysis, and Control,* Blackburn, C. de W. and McClure, P.J. (Eds.), Woodhead Publishing, Cambridge, England/CRC Press, Boca Raton, FL, 2002, pp. 53–95.

Leistner, L. Hurdle technology applied to meat products of the shelf stable and intermediate moisture food types. In *Properties of Water in Foods in Relation to Quality and Stability,* Simatos, D. and Multon, J.L., Eds., Martinus Nijhoff Publishers, Dordrecht, 1985.

Le Marc, Y., Huchet, V., Bourgeois, C.M., Guyonnet, J.P., Mafart, P., and Thuault, D. Modelling the growth kinetics of *Listeria* as a function of temperature, pH and organic acid concentration. *Int. J. Food Microbiol.,* 73, 219–237, 2002.

Le Marc, Y., Pin, C., and Baranyi, J. Methods to determine the growth domain in a multidimensional environmental space. In *Predictive Modeling in Foods—Conference Proceedings,* Van Impe, J.F.M., Geeraerd, A.H., Leguerinel, I., and Mafart, P. (Eds.), Katholieke Universiteit Leuven/BioTec, Belgium, 2003, pp. 34–36.

Leroi, F. and Joffraud, J.J. Salt and smoke simultaneously affect chemical and sensory quality of cold-smoked salmon during 5°C storage predicted using factorial design. *J. Food Prot.,* 63, 1222–1227, 2000.

Leroi, F., Joffraud, J.-J., Chevalier, F., and Cardinal, M. Study of the microbial ecology of cold-smoked salmon during storage at 8°C. *Int. J. Food Microbiol.,* 39, 111–121, 1998.

Leroi, F., Joffraud, J.J., and Chevalier, F. Effect of salt and smoke on the microbiological quality of cold-smoked salmon during storage at 5°C as estimated by the factorial design method. *J. Food Prot.,* 63, 502–508, 2000.

Leroy, F. and de Vuyst, L. A combined model to predict the functionality of the bacteriocin-producing *Lactobacillus sakei* strain CTC 494. *Appl. Environ. Microbiol.,* 69, 1093–1099, 2003.

Levenspiel, O. *Chemical Reaction Engineering,* 2nd ed., John Wiley & Sons, New York, 1972.

Lindroth, S. and Genigeorgis, C. Probability of growth and toxin production by nonproteolytic *Clostridium botulinum* in rock fish stored under modified atmospheres. *Int. J. Food Microbiol.,* 3, 167–181, 1986.

López-Malo, A., Guerrero, S., and Alzamora, S.M. Probabilistic modelling of *Saccharomyces cerevisiae* inhibition under the effects of water activity, pH and potassium sorbate concentration. *J. Food Prot.,* 63, 91–95, 2000.

Lou, W. and Nakai, S. Application of artificial neural networks for predicting the thermal inactivation of bacteria: a combined effect of temperature, pH and water activity. *Food Res. Int.,* 34, 573–579, 2001.

Löwenadler, J. and Rönner, U. Determination of dissolved carbon dioxide by colorimetric titration in modified atmosphere systems. *Lett. Appl. Microbiol.,* 18, 285–288, 1994.

Lund, B.M. Quantification of factors affecting the probability of development of pathogenic bacteria, in particular clostridium botulinum, in foods. *J. Ind. Microbiol.,* 12, 144–155, 1993.

Lund, B.M., Graham, S.M., and Franklin, J.G. The effect of acid pH on the probability of growth of proteolytic strains of *Clostridium botulinum. Int. J. Food Microbiol.,* 4, 215–226, 1987.

Masana, M.O. and Baranyi, J. Adding new factors to predictive models: the effect on the risk of extrapolation. *Food Microbiol.,* 17, 367–374, 2000a.

Masana, M.O. and Baranyi, J. Growth/no growth interface of *Brochothrix thermosphacta* as a function of pH and water activity. *Food Microbiol.,* 17, 485–493, 2000b.

Mathlouthi, M. Water content, water activity, water structure and the stability of foodstuffs. *Food Control*, 12, 409–417, 2001.

McClure, P.J., Roberts, T.A., and Otto Oguru, P. Comparison of the effects of sodium chloride, pH and temperature on the growth of *Listeria monocytogenes* on gradient plates and in liquid medium. *Lett. Appl. Microbiol.*, 9, 95–99, 1989.

McClure, P.J., Baranyi, J., Boogard, E., Kelly, T.M., and Roberts, T.A. A predictive model for the combined effect of pH, sodium chloride and storage temperature on the growth of *Brochotrix thermosphacta*. *Int. J. Food Microbiol.*, 19, 161–178, 1993.

McClure, P.J., Blackburn, C. de W., Cole, M.B., Curtis, P.S., Jones, J.E., Legan, J.D., Ogden, I.D., and Peck, M.W. Modelling the growth, survival and death of microorganisms in foods: the UK Food Micromodel approach. *Int. J. Food Microbiol.*, 23, 265–275, 1994a.

McClure, P.J., Cole, M.B., and Davies, K.W. An example of the stages in the development of a predictive mathematical model for microbial growth: the effect of NaCl, pH and temperature on the growth of *Aeromonas hydrophila*. *Int. J. Food Microbiol.*, 23, 359–375, 1994b.

McDonald, K. and Sun, D.W. Predictive food microbiology for the meat industry: a review. *Int. J. Food Microbiol.*, 52, 1–27, 1999.

McGill, A.S., Murray, J., and Hardy, R. Some observations on the determination of phenols in smoked fish by solvent extraction: difficulties and improvements in the methodology. *Z. Lebens. Unters. Forsch.*, 181, 363–369, 1985.

McKellar, R.C. and Lu, X.W. A probability of growth model for *Escherichia coli* O157:H7 as a function of temperature, pH, acetic acid, and salt. *J. Food Prot.*, 64, 1922–1928, 2001.

McMeekin, T.A., Chandler, R.E., Doe, P.E., Garland, C.D., Olley, J., Putro, S., and Ratkowsky, D.A. Model for the combined effect of temperature and salt concentration/water activity on growth rate of *Staphylococcus xylosus*. *J. Appl. Bacteriol.*, 62, 543–550, 1987.

McMeekin. T.A., Olley, J., Ross, T., and Ratkowsky, D.A. *Predictive Microbiology. Theory and Application*, Research Studies Press, Taunton, U.K., 1993.

McMeekin, T.A., Presser, K., Ratkowsky, D., Ross, T., Salter, M., and Tienungoon, S. Quantifying the hurdle concept by modelling the bacterial growth/no growth interface. *Int. J. Food Microbiol.*, 55, 93–98, 2000.

Mellefont, L.A. and Ross, T. The effect of abrupt shifts in temperature on the lag phase duration of *Escherichia coli* and *Klebsiella oxytoca*. *Int. J. Food Microbiol.*, 83, 295–305, 2003.

Mellefont, L.A., McMeekin, T.A., and Ross, T. The effect of abrupt osmotic shifts on the lag phase duration of foodborne bacteria. *Int. J. Food Microbiol.*, 83, 281–293, 2003.

Mellefont, L.A., McMeekin, T.A., and Ross, T. The effect of abrupt osmotic shifts on the lag phase duration of physiologically distinct populations of *Salmonella typhimurium*. *Int. J. Food Microbiol.*, in press.

Membré, J.-M. and Burlot, P.M. Effect of temperature, pH, and NaCl on growth and pectinolytic activity of *Pseudomonas marginalis*. *Appl. Environ. Microbiol.*, 60, 2017–2022, 1994.

Membré, J.M. and Kubaczka M. Degradation of pectic compounds during pasteurised vegetable juice spoilage by *Chryseomonas luteola*: a predictive microbiology approach. *Int. J. Food Microbiol.*, 42, 159–166, 1998.

Membré, J.M., Thurette, J., and Catteau, M. Modelling the growth, survival and death of *Listeria monocytogenes*. *J. Appl. Bacteriol.*, 82, 345–350, 1997.

Membré, J-M., Ross, T., and McMeekin, T.A. Behaviour of *Listeria monocytogenes* under combined chilling processes. *Lett. Appl. Bacteriol.*, 28, 216–220, 1999.

Membré, J.M., Kubaczka, M., and Chene, C. Growth rate and growth–no-growth interface of *Penicillium brevicompactum* as functions of pH and preservative acids. *Food Microbiol.*, 18, 531–538, 2001.

Membré, J.-M., Leporq, B., Vilette, M., Mettler, E., Perrier, L., and Zwietering, M. Experimental protocols and strain variability of cardinal values (pH and a_w) of bacteria using Bioscreen C: microbial and statistical aspects. In *Microbial Adaptation to Changing Environments, Proceedings and Abstracts of the 18th Symposium of the International Committee on Food Microbiology and Hygiene (ICFMH)*, Axelson, L., Tronrud, E.S., and Merok, K.J. (Eds.), MATFORSK, Norwegian Food Research Institute, Ås, Norway, 2002, pp. 143–146.

Messens, W., Verluyten, J., Leroy, F., and De Vurst, L. Modelling growth and bacteriocin production by *Lactobacillus curvatus* LTH 1174 in response to temperature and pH values used for European sausage fermentation processes. *Int. J. Food Microbiol.*, 81, 41–52, 2002.

Miles, D.W., Ross, T., Olley, J., and McMeekin, T.A. Development and evaluation of a predictive model for the effect of temperature and water activity on the growth rate of *Vibrio parahaemolyticus*. *Int. J. Food Microbiol.*, 38, 133–142, 1997.

Miller, M.B. and Bassler, B.L. Quorum sensing in bacteria. *Ann. Rev. Microbiol.*, 55, 165–199, 2001.

Molina, M. and Giannuzzi, L. Combined effect of temperature and propionic acid concentration on the growth of *Aspergillus parasiticus*. *Food Res. Int.*, 32, 677–682, 1999.

Najjar, Y.M., Basheer, I.A., and Hajmeer, M.N. Computational neural networks for predictive microbiology. 1. Methodology. *Int. J. Food Microbiol.*, 34, 27–49, 1997.

Neumeyer, K., Ross, T., and McMeekin, T.A. Development of a predictive model to describe the effect of temperature and water activity on the growth of spoilage pseudomonads. *Int. J. Food Microbiol.*, 38, 45–54, 1997.

Ng, T.M., Viard, E., Caipo, M.L., Duffy, S., and Schaffner, D.W. Expansion and validation of a predictive model for the growth of *Bacillus stearothermophilus* in military rations. *J. Food Sci.*, 67, 1872–1878, 2002.

Nichols, D.S., Olley, J., Garda, H., Brenner, R.R., and McMeekin, T.A. Effect of temperature and salinity stress on growth and lipid composition of *Shewanella gelidimarina*. *Appl. Environ. Microbiol.*, 66, 2422–2429, 2002a.

Nichols, D.S., Presser, K.A., Olley, J., Ross, T., and McMeekin, T.A. Variation of branched-chain fatty acids marks the normal physiological range for growth in *Listeria monocytogenes*. *Appl. Environ. Microbiol.*, 68, 2809–2813, 2002b.

Nychas, G.-J.E. and Tassou, C.C. Traditional preservatives: oils and spices. In *Encyclopaedia of Food Microbiology*, Robinson, R.K., Batt, C.A., and Patel, P.D. (Eds.), Academic Press, San Diego, 2000, pp. 1717–1722.

O'Brien, G.D. Domestic refrigerator air temperatures and the public's awareness of refrigerator use. *Int. J. Environ. Health Res.*, 7, 41–148, 1997.

Oscar, T.P. Response surface models for effects of temperature, pH, and previous growth pH on growth of *Salmonella typhimurium* in brain heart infusion broth. *J. Food Prot.*, 62, 106–111, 1999.

Oscar, T.P. Development and validation of a tertiary simulation model for predicting the potential growth of *Salmonella typhimurium* on cooked chicken. *Int. J. Food Microbiol.*, 76, 177–190, 2002.

Palumbo, S.A., Williams, A.C., Buchanan, R.L., and Phillips, J.G. Model for the anaerobic growth of *Aeromonas hydrophila* K144. *J. Food Prot.*, 55, 260–265, 1992.

Palumbo, S.A., Williams, A.C., Buchanan, R.L., Call, J.C., and Phillips, J.G. Expanded model for the aerobic growth of *Aeromonas hydrophila. J. Food Saf.*, 16, 1–13, 1996.

Panagou, E.Z., Skandamis, P.N., and Nychas, G.-J.E. Modelling the combined effect of temperature, pH and a_w on growth of *Monascus ruber*, a heat-resistant fungus isolated from green table olives. *J. Appl. Microbiol.*, 94, 146–156, 2003.

Parente, E., Giglio, M.A., Ricciardi, A., and Clementi, F. The combined effect of nisin, leucocin F10, pH, NaCl and EDTA on the survival of *Listeria monocytogenes* in broth. *Int. J. Food Microbiol.*, 40, 65–74, 1998.

Passos, F.V., Fleming, H.P., Felder, R.M., and Ollis, D.F. Modeling growth of *Saccharomyces rosei* in cucumber fermentation. *Food Microbiol.*, 14, 533–542, 1997.

Passos, F.V., Fleming, H.P., Ollis, D.F., Hassan, H.M., and Felder, R.M. Modeling the specific growth rate of *Lactobacillus plantarum* in cucumber extract. *Appl. Microbiol. Biotechnol.*, 40, 143–150, 1993.

Pecina, R., Bonn, G., and Burtscher, O. High-performance liquid chromatography elution behaviour of alcohols, aldehydes, ketones, organic acids and carbohydrates on a strong cation-exchange stationary phase. *J. Chromatogr.*, 287, 245–258, 1984.

Pin, C. and Baranyi, J. Predictive models as means to quantify the interactions of spoilage organisms. *Int. J. Food Microbiol.*, 41, 59–72, 1998.

Pin, C., Baranyi, J., and de Fernando, G.G. Predictive model for the growth of *Yersinia enterocolitica* under modified atmospheres. *J. Appl. Microbiol.*, 88, 521–530, 2000.

Pitt, R.E. A descriptive model of mold growth and aflatoxin formation as affected by environmental conditions. *J. Food Prot.*, 56, 139–146, 1992.

Pitt, R.E. A model of aflatoxin formation in stored products. *Trans. ASAE*, 38, 1445–1453, 1995.

Pouillot, R., Albert, I., Cornu, M., and Denis, J.-B. Estimation of uncertainty and variability in bacterial growth using Bayesian inference. Application to *Listeria monocytogenes. Int. J. Food Microbiol.*, 81, 87–104, 2003.

Presser, K., Ratkowsky, D.A., and Ross, T. Modelling the growth rate of *Escherichia coli* as a function of pH and lactic acid concentration. *Appl. Environ. Microbiol.*, 63, 2355–2360, 1997.

Presser, K.A., Ross, T., and Ratkowsky, D.A. Modelling the growth limits (growth/no growth interface) of *E. coli* as a function of pH, lactic acid and temperature. *Appl. Environ. Microbiol.*, 64, 1773–1779, 1998.

Presser, K.A., Salter, M.A., Ratkowsky, D.A., and Ross, T. (1999). Development of growth limits (growth/no growth interface) modelling and its application to predictive food microbiology. *Recent Res. Dev. Microbiol.*, 3, 535–549, 1999.

Rasch, M. The influence of temperature, salt and pH on the inhibitory effect of reuterin on *Escherichia coli. Int. J. Food Microbiol.*, 72, 225–231, 2002.

Rasmussen, S.K.J., Ross, T., Olley, J., and McMeekin, T. A process risk model for the shelf-life of Atlantic salmon fillets. *Int. J. Food Microbiol.*, 73, 47–60, 2002.

Ratkowsky, D.A. *Nonlinear Regression Modeling: A Unified Practical Approach*, Marcel Dekker, New York, 1983, 276 pp.

Ratkowsky, D.A. Principles of nonlinear-regression modeling. *J. Ind. Microbiol.*, 12, 195–199, 1993.

Ratkowsky, D.A. Some examples of, and problems with, the use of nonlinear logistic regression in predictive food microbiology. *Int. J. Food Microbiol.*, 73, 119–125, 2002.

Ratkowsky, D.A. and Ross, T. Modelling the bacterial growth/no growth interface. *Lett. Appl. Microbiol.*, 20, 29–33, 1995.

Ratkowsky, D.A., Olley, J., McMeekin, T.A., and Ball, A. Relation between temperature and growth rate of bacterial cultures. *J. Bacteriol.*, 149, 1–5, 1982.

Ratkowsky, D.A., Lowry, R.K., McMeekin, T.A., Stokes, A.N., and Chandler, R.E. Model for bacterial culture growth rate throughout the entire biokinetic temperature range. *J. Bacteriol.*, 154, 1222–1226, 1983.

Ratkowsky, D.A., Ross, T., McMeekin, T.A., and Olley, J. Comparison of Arrhenius-type and Bêlehrádek-type models for prediction of bacterial growth in foods. *J. Appl. Bacteriol.*, 71, 452–459, 1991.

Razavilar, V. and Genigeorgis, C. Prediction of *Listeria* spp. growth as affected by various levels of chemicals, pH, temperatures and storage time in a model broth. *Int. J. Food Microbiol.*, 40, 149–157, 1998.

Resnik, S.L. and Chirife, J. Proposed theoretical water activity values at various temperatures for selected solutions to be used as reference sources in the range of microbial growth. *J. Food Prot.*, 51, 419–423, 1988.

Riemann, H. The effect of the number of spores on growth and toxin formation by *C. botulinum* type E in inhibitory environments. In *Botulism, 1966.* Proceedings of the 5th International Symposium on Food Microbiology, Moscow, July 1966, Ingram, M. and Roberts, T.A., Eds., Chapman and Hall, London, 1967, 150–169.

Roberts, T.A. and Jarvis, B. Predictive modelling of food safety with particular reference to *Clostridium botulinum* in model cured meat systems. In *Food Microbiology: Advances and Prospects,* Roberts, T.A. and Skinner, F.A. (Eds.), Academic Press, New York, 1983, pp. 85–95.

Robinson, T.P., Ocio, M.J., Kaloti, A., and Mackey, B.M. The effect of growth environment on the lag phase of *Listeria monocytogenes. Int. J. Food Microbiol.,* 44, 83–92, 1998.

Rodriguez, A.M.C., Alcalá, E.B., Gimeno, R.M.G., and Cosano, G.Z. Growth modelling of *Listeria monocytogenes* in packed fresh green asparagus. *Food Microbiol.*, 17, 421–427, 2000.

Ross, T. *A Philosophy for the Development of Kinetic Models in Predictive Microbiology,* Ph.D. thesis, University of Tasmania, Hobart, Australia, 1993a.

Ross, T. Belehrádek-type models. *J. Ind. Microbiol.,* 12, 180–189, 1993b.

Ross, T. *Predictive Food Microbiology Models in the Meat Industry,* Meat and Livestock Australia, Sydney, 1999a, 196 pp.

Ross, T. Assessment of a theoretical model for the effects of temperature on bacterial growth rate. *Refrig. Sci. Technol. Proc.,* EUR 18816, 64–71, 1999b.

Ross, T. and McMeekin, T.A. Predictive microbiology: a review. *Int. J. Food Microbiol.,* 23, 241–264, 1994.

Ross, T. and Shadbolt, C.T. *Predicting Escherichia coli Inactivation in Uncooked Comminuted Fermented Meat Products,* Meat and Livestock Australia, Sydney, 2001, 66 pp.

Ross, T., Fazil, A., and Paoli, G. *Listeria monocytogenes* in Australian processed meat products: risks and their management. *Meat and Livestock Australia,* Sydney, in press.

Ross, T., Baranyi, J., and McMeekin, T.A. Predictive microbiology and food safety. In *Encyclopaedia of Food Microbiology,* Robinson, R., Batt, C.A., and Patel, P. (Eds.), Academic Press, London. 2000a, pp. 1699–1710.

Ross, T., Dalgaard, P., and Tienungoon, S. Predictive modelling of the growth and survival of *Listeria* in fishery products. *Int. J. Food Microbiol.,* 62, 231–245, 2000b.

Ross, T., Ratkowsky, D.A., Mellefont, L.A., and McMeekin, T.A. Modelling the effects of temperature, water activity, pH and lactic acid concentration on the growth rate of *Escherichia coli. Int. J. Food Microbiol.,* 82, 33–43, 2003.

Rosso, L. *Modélisation et Microbiologie Prévissionnelle: Élaboration d'un nouvel outil pour l'Agro-alimentaire,* Université Claude Bernard, Lyon, 1995.

Rosso, L. Models using cardinal values. *Refrig. Sci. Technol. Proc.,* EUR 18816, 48–55, 1999.

Rosso, L., Lobry, R., and Flandrois, J.P. An unexpected correlation between cardinal temperatures of microbial growth highlighted by a new model. *J. Theor. Biol.*, 162, 447–463, 1993.

Rosso, L., Lobry, J.R., Bajard, S., and Flandrois, J.P. Convenient model to describe the combined effects of temperature and pH on microbial growth. *Appl. Environ. Microbiol.*, 61, 610–616, 1995.

Rosso, L. and Robinson, T.P. A cardinal model to describe the effect of water activity on the growth of moulds. *Int. J. Food Microbiol.*, 63, 265–273, 2001.

Salter, M., Ross, T., and McMeekin, T.A. Applicability of a model for non-pathogenic *Escherichia coli* for predicting the growth of pathogenic *Escherichia coli*. *J. Appl. Microbiol.*, 85, 357–384, 1998.

Salter, M.A., Ratkowsky, D.A., Ross, T., and McMeekin, T.A. Modelling the combined temperature and salt (NaCl) limits for growth of a pathogenic *Escherichia coli* strain using generalised nonlinear regression. *Int. J. Food Microbiol.*, 61, 159–167, 2000.

Sautour, M., Dantigny, P., Divies, C., and Bensoussan, M. A temperature-type model for describing the relationship between fungal growth and water activity. *Int. J. Food Microbiol.*, 67, 63–69, 2001.

Schaffner, D.W. and Labuza, T.P. Predictive microbiology: where are we, and where are we going? *Food Technol.*, 51, 95–99, 1997.

Schepers, A.W., Thibault, J., and Lacroix, C. Comparison of simple neural networks and nonlinear regression models for descriptive modelling of *Lactobacillus helveticus* growth in pH-controlled batch cultures. *Enzyme Microb. Technol.*, 26, 431–445, 2000.

Schoolfield, R.M., Sharpe, P.J.H., and Magnuson, C.E. Non-linear regression of biological temperature-dependent rate models based on absolute reaction-rate theory. *J. Theor. Biol.*, 88, 719–731, 1981.

Seman, D.L., Borger, A.C., Meyer, J.D., Hall, P.A., and Milkowski, A.L. Modeling the growth of *Listeria monocytogenes* in cured ready-to-eat processed meat products by manipulation of sodium chloride, sodium diacetate, potassium lactate, and product moisture content. *J. Food Prot.*, 65, 651–658, 2002.

Sergelidis, D., Abrahim, A., Sarimvei, A., Panoulis, C., Karaioannoglou, P., and Genigeorgis, C. Temperature distribution and prevalence of *Listeria spp.* in domestic, retail, and industrial refrigerators in Greece. *Int. J. Food Microbiol.*, 34, 171–177, 1997.

Sharpe, P.J.H. and DeMichele, D.W. Reaction kinetics and poikilotherm development. *J. Theor. Biol.*, 64, 649–670, 1977.

Shewan, J.M. The biological stability of smoked and salted fish. *Chem. Ind.*, July, 501–505, 1949.

Shoemaker, S.P. and Pierson, M.D. "Phoenix phenomenon" in the growth of *Clostridium perfringens*. *Appl. Environ. Microbiol.*, 32, 803–806, 1976.

Simpson, R., Almonacid, S., and Acevedo, C. Development of a mathematical model for MAP systems applied to nonrespiring foods. *J. Food Sci.*, 66, 561–567, 2001a.

Simpson, R., Almonacid, S., and Acevedo, C. Mass transfer in Pacific hake (*Merluccius australis*) packed in refrigerated modified atmosphere. *J. Food Process. Eng.*, 24, 405–421, 2001b.

Sinigaglia, M., Corbo, M.R., Altieri, C., and Massa, S. Response surface model for effects of temperature, water activity and pH on germination of *Bacillus cereus* spores. *J. Food Prot.*, 22, 121–133, 2002.

Skandamis, P.N. and Nychas, G.-J.E. Development and evaluation of a model predicting the survival of *Escherichia coli* O157:H7 NCTC 12900 in homemade eggplant at various temperatures, pHs, and oregano essential oil concentrations. *Appl. Environ. Microbiol.*, 66, 1646–1653, 2000.

Skandamis, P.N., Davies, K.W., McClure, P.J., Koutsoumanis, K., and Tassou, C. A vialistic approach for non-thermal inactivation of pathogens in traditional Greek salads. *Food Microbiol.,* 19, 405–421, 2002.

Skinner, G.E., Lerkin, J.W., and Rhodehamel, E.J. Mathematical modeling of microbial growth: a review. *J. Food Saf.,* 14, 175–217, 1994.

Smith, M.G. The generation time, lag time and minimum temperature of growth of coliform organisms on meat, and the implications for codes of practice in abattoirs. *J. Hyg. (Camb.),* 94, 289–300, 1985.

Sofos, J.N., Beuchat, L.R., Davidson, P.M., and Johnson, E.A. Naturally Occurring Antimicrobials in Food, Task Force Report 132, Council of Agricultural Science and Technology, Ames, IA, 1998.

Stewart, C.M., Cole, M.B., Legan, J.D., Slade, L., Vandeven, M.H., and Schaffner, D.W. Modeling the growth boundary of *Staphylococcus aureus* for risk assessment purposes. *J. Food Prot.,* 64, 51–57, 2001.

Stewart, C.M., Cole, M.B., Legan, J.D., Slade, L., Vandeven, M.H., and Schaffner, D.W. *Staphylococcus aureus* growth boundaries: moving towards mechanistic predictive models based on solute specific effects. *Appl. Environ. Microbiol.,* 68,1864–1871, 2002.

Suñen, E. Minimum inhibitory concentration of smoke wood extracts against spoilage and pathogenic micro-organisms associated with foods. *Lett. Appl. Microbiol.,* 27, 45–48, 1998.

Sutherland, J.P. and Bayliss, A.J. Predictive modelling of growth of *Yersinia enterocolitica*: the effects of temperature, pH and sodium chloride. *Int. J. Food Microbiol.,* 21, 197–215, 1994.

Sutherland, J.P., Aherne, A., and Beaumont, A.L. Preparation and validation of a growth model for *Bacillus cereus*: the effects of temperature, pH, sodium chloride and carbon dioxide. *Int. J. Food Microbiol.,* 30, 359–372, 1996.

Sutherland, J.P., Bayliss, A.J., and Braxton, D.S. Predictive modelling of growth of *Escherichia coli* O157:H7: the effects of temperature, pH and sodium chloride. *Int. J. Food Microbiol.,* 25, 29–49, 1995.

Sutherland, J.P., Bayliss, A.J., Braxton, D.S., and Beaumont, A.L. Predictive modelling of *Escherichia coli* O157-H7 — inclusion of carbon dioxide as a fourth factor in a preexisting model. *Int. J. Food Microbiol.,* 37, 113–120, 1997.

Thurette, J., Membre, J.M., Han Ching, L., Tailliez, R., and Catteau, M. Behavior of *Listeria* spp. in smoked fish products affected by liquid smoke, NaCl concentration, and temperature. *J. Food Prot.,* 61, 1475–1479, 1998.

Tienungoon, S. *Some Aspects of the Ecology of Listeria monocytogenes in Salmonoid Aquaculture,* Ph.D. thesis. University of Tasmania, Hobart, Australia, 1988.

Tienungoon, S., Ratkowsky, D.A., McMeekin, T.A., and Ross, T. Growth limits of *Listeria monocytogenes* as a function of temperature, pH, NaCl and lactic acid. *Appl. Environ. Microbiol.,* 66, 4979–4987, 2000.

Tóth, L. and Potthast, K. Chemical aspects of the smoking of meat and meat products. *Adv. Food Res.,* 29, 87–158, 1984.

Tu, J.V. Advantages and disadvantages of using artificial neural networks versus logistic regression for predicting medical outcomes. *J. Clin. Epidemiol.,* 11, 1225–1231, 1996.

Tucker, I.W. Estimation of phenols in meat and fat. *Assoc. Off. Agric. Chem.,* 25, 779–782, 1942.

Tuynenberg Muys, G. Microbial safety in emulsions. *Process Biochem.,* 6, 25–28, 1971.

Uljas, H.E., Schaffner, D.W., Duffy, S., Zhao, L., and Ingham, S.C. Modeling of combined steps for reducing *Escherichia coli* O157:H7 populations in apple cider. *Appl. Environ. Microbiol.,* 67, 133–141, 2001.

Valík, L., Baranyi, J., and Görner, F. Predicting fungal growth: the effect of water activity on *Penicillium rocqueforti*. *Int. J. Food Microbiol.*, 47, 141–146, 1999.

Wei, Q.K., Fang, T.J., and Chan, W.C. Development and validation of growth model for *Yersinia enterocolitica* in cooked chicken meats packed under various atmosphere packaging and stored at different temperatures. *J. Food Prot.*, 64, 987–993, 2001.

Whiting, R.C. Microbial modeling in foods. *Crit. Rev. Food Sci. Nutr.*, 35, 467–494, 1995.

Whiting, R.C. and Bagi, L.K. Modeling the lag phase of *Listeria monocytogenes*. *Int. J. Food Microbiol.*, 73, 291–295, 2002.

Whiting, R.C. and Buchanan, R.L. Letter to the editor: a classification of models in predictive microbiology. A reply to K.R. Davey. *Food Microbiol.*, 10, 175–177, 1993.

Whiting, R.C. and Call, J.E. Time of growth model for proteolytic *Clostridium botulinum*. *Food Microbiol.*, 10, 295–301, 1993.

Whiting, R.C. and Cygnarowic-Provost, M. A quantitative model for bacterial growth and decline. *Food Microbiol.*, 9, 269–277, 1992.

Whiting, R.C. and Golden, M.H. Modeling temperature, pH, NaCl, nitrite and lactate on the survival of *Escherichia coli* O157:H7 in broth. *J. Food Safety*, 23, 61–74, 2003.

Whiting, R.C. and Oriente, J.C. Time-to-turbidity model for non-proteolytic type B *Clostridium botulinum*. *Int. J. Food Microbiol.*, 36, 49–60, 1997.

Whiting, R.C. and Strobaugh, T.P. Expansion of the time-to-turbidity model for proteolytic *Clostridium botulinum* to include spore numbers. *Food Microbiol.*, 15, 449–453, 1998.

WHO/FAO (World Health Organisation and Food and Agriculture Organisation). Risk Assessment of *Listeria monocytogenes* in ready-to-eat foods. Food and Agriculture Organisation, Rome, in press.

Wijtzes, T., De Wit, J.C., Huis in't Veld, J.H., van't Riet, K., and Zwietering, M.H. Modelling bacterial growth of *Lactobacillus curvatus* as a function of acidity and temperature. *Appl. Environ. Microbiol.*, 61, 2533–2539, 1995.

Wijtzes, T., van't Riet, K., Huis in't Veld, J.H., and Zwietering, M.H. A decision support system for the prediction of microbial food safety and food quality. *Int. J. Food Microbiol.*, 42, 79–90, 1998.

Wijtzes, T., Rombouts, F.M., Kant-Muermans, M.L., van't Riet, K., and Zwietering, M. Development and validation of a combined temperature, water activity, pH model for bacterial growth rate of *Lactobacillus curvatus*. *Int. J. Food Microbiol.*, 63, 57–64, 2001.

Winans, S.C. and Bassler, B.L. Mob psychology. *J. Bacteriol.*, 184, 873–888, 2000.

Yan, W. and Hunt, L.A. An equation for modelling the temperature response of plants using only the cardinal temperatures. *Ann. Bot.*, 84, 607–614, 1999.

Yin, W. and Wallace, D.H. A nonlinear model for crop development as a function of temperature. *Agric. Forest Meteorol.*, 77, 1–16, 1995.

Zaika, L.L., Moulden, E., Weimer, L., Phillips, J.G., and Buchanan, R.L. Model for the combined effects of temperature, initial pH, sodium chloride and sodium nitrite concentrations on anaerobic growth of *Shigella flexneri*. *Int. J. Food Microbiol.*, 23, 345–358, 1994.

Zaika, L.L., Phillips, J.G., Fanelli, J.S., and Scullen, O.J. Revisited model for aerobic growth of *Shigella flexneri* to extend the validity of predictions at temperatures between 10 and 19°C. *Int. J. Food Microbiol.*, 41, 9–19, 1998.

Zhao, Y., Wells, J.H., and McMillin, K.W. Dynamic changes of headspace gases in CO_2 and N_2 packed fresh beef. *J. Food Sci.*, 60, 571–575, 591, 1995.

Zhao, L., Chen, Y., and Schaffner, D.W. Comparison of logistic regression and linear regression on modelling percentage data. *Appl. Environ. Microbiol.*, 67, 2129–2135, 2001.

Zwietering, M. Microbial and engineering characteristic numbers for process evaluation. *Refrig. Sci. Technol. Proc.*, EUR 18816, 272–279, 1999.

Zwietering, M.H., Cuppers, H.G.A.M., De Wit, J.C., and van't Riet, K. Evaluation of data transformations and validation of models for the effect of temperature on bacterial growth. *Appl. Environ. Microbiol.*, 60, 195–203, 1994.

Zwietering, M.H., de Koos, J.T., Hasenack, B.E., De Wit, J.C., and van't Riet, K. Modeling of bacterial growth as a function of temperature. *Appl. Environ. Microbiol.*, 57, 1094–1101, 1991.

Zwietering, M.H., De Wit, J.C., and Notermans, S. Application of predictive microbiology to estimate the number of *Bacillus cereus* in pasteurised milk at the time of consumption. *Int. J. Food Microbiol.*, 30, 55–70, 1996.

Zwietering, M.H., Jongenburger, I., Rombouts, F.M., and van't Reit, K. Modeling of the bacterial growth curve. *Appl. Environ. Microbiol.*, 56, 1875–1881, 1990.

Zwietering, M.H., Wijtzes, T., De Wit, J.C., and van't Riet, K. A decision support system for prediction of the microbial spoilage in foods. *J. Food Prot.*, 55, 973–979, 1992.

4 Model Fitting and Uncertainty

David A. Ratkowsky

CONTENTS

0-8493-1237-X/04/$0.00+$1.50
© 2004

4.1 OVERVIEW

This chapter is divided into two main sections, viz. (1) model fitting, featuring the principles of, and examples of, the use of regression models, especially nonlinear regression models, for data sets in food preservation and safety, and comparisons of various modeling approaches and (2) the consequences of uncertainty, i.e., variation in the measurements, and its implications for product shelf life. These sections are designated as 4.2 and 4.3, respectively. The chapter concludes with an epilogue, in which the author raises a few additional issues (Section 4.4).

4.2 MODEL FITTING

This section examines various models used in predictive microbiology, focusing attention upon those criteria and factors that have to be taken into account when the models are being fitted. Also, criteria for assessing goodness-of-fit are presented.

4.2.1 THE MODELS

Three groups of models will be considered in this chapter, which serve to illustrate the various facets of modeling in predictive microbiology. The first group of data (see Table A4.1) and their associated models involve lag time as a function of temperature, and were examined recently by Oscar (2002). The lag time is usually determined experimentally by fitting a primary model, or by noting the time taken before perceptible growth of a bacterial culture is observed. Primary modeling is the subject of Chapter 2 of this book.

Denoting the lag time by λ, the models considered are:
Hyperbola model

$$\lambda = \exp[p/(T - q)] \tag{4.1}$$

Extended hyperbola model

$$\lambda = [p/(T - q)]^m \tag{4.2}$$

Linear Arrhenius model (Davey, 1989)

$$\lambda = \exp[-(A + B/T + C/T^2)] \tag{4.3}$$

Simple square-root model (Ratkowsky et al., 1983)

$$\lambda = 1/\{[b(T - T_{min})]^2\} \tag{4.4}$$

The second group of data (see Table A4.2) was obtained from experiments conducted by students in the on-going predictive microbiology research program at the University of Tasmania, on the maximum specific growth rate constant μ for three species of microorganism as a function of temperature throughout the entire biokinetic temperature range. The models for fitting and predicting μ are confined here to just two models, viz. the four-parameter square-root model of Ratkowsky et al. (1983)

$$\sqrt{\mu} = b(T - T_{min})\{1 - \exp[c(T - T_{max})]\} \tag{4.5}$$

and the cardinal temperature model of Rosso et al. (1993)

$$\mu = \mu_{opt}(T - T_{max})(T - T_{min})^2 / [(T_{opt} - T_{min})^2(T - T_{opt}) -$$
$$(T_{opt} - T_{min})(T_{opt} - T_{max})(T_{opt} + T_{min} - 2T)] \tag{4.6}$$

In each of the above models, T represents temperature in degrees absolute, although the only model in which it is essential that degrees absolute be employed is the Linear Arrhenius model (4.3). The other models all involve differences between temperatures, and therefore other temperature scales are acceptable, since they result in equivalent answers. T_{min} and T_{max} represent notional minimum and maximum temperatures, respectively, the term "notional" meaning that these temperatures are not to be interpreted as "true" minimum and maximum temperatures, although various authors have mistakenly or misguidedly given them this interpretation (e.g., Dantigny and Molin, 2000). In 4.5 and 4.6, they are nothing more than intercepts on the rate (μ) axis, i.e., the temperatures at which the rate equals zero when a graph of μ vs. T is extrapolated outside the range of the observed data. In 4.6, the additional cardinal temperature T_{opt} represents the temperature at which growth is optimal (i.e., μ is greatest), and the fourth parameter μ_{opt} is the maximum specific growth rate corresponding to T_{opt}. Thus, the cardinal temperature model (4.6) is the only one in which all its parameters can be considered to be biologically interpretable, although not necessarily achievable (i.e., T_{min} and T_{max}). All other models contain arbitrary constants, viz. p, q, m, A, B, C, b, and c, which are devoid of biological meaning. They are simply parameters included in the model to enhance the curve-fitting prospects of the model.

It should also be noted that 4.5 and 4.6 apply only in the range $T_{min} < T < T_{max}$, and that outside these ranges, i.e., for $T < T_{min}$ and $T > T_{max}$, the rate μ is zero. To be mathematically correct, these bounds should be stated along with the equation definitions, but they are omitted here for simplicity, and it should be understood that nonzero rates can only apply at temperatures between T_{min} and T_{max}. It should also be self-evident that when modeling data, using either of the above models, only nonzero rates should be employed. Data corresponding to temperatures at which the observed rates are zero need to be discarded when curve fitting.

4.2.2 STOCHASTIC ASSUMPTIONS

Equation 4.1 to Equation 4.6 are nonlinear regression models with two to four parameters, which may be estimated using nonlinear regression modeling. Some of the equations can be transformed by rearranging terms, thereby linearizing them.

For example, the reason why 4.3 is called a "Linear Arrhenius" model can be seen by rewriting it as:

$$\ln \text{rate} = \ln(1/\lambda) = A + B/T + C/T^2$$

The result follows from the fact that a time, whether it is a lag time, a generation time, or some other time-based variable, may be expressed as a rate by taking its reciprocal. The right-hand side is a quadratic polynomial in $1/T$, the reciprocal of absolute temperature, a term that is often seen in Arrhenius-type models. Similarly, Equation 4.4 may be rearranged as:

$$\sqrt{\text{rate}} = \sqrt{(1/\lambda)} = b(T - T_{\min})$$

which shows that the model is in reality the simple square-root model of Ratkowsky et al. (1982). Whether one should or should not transform response variables in this manner is decided by the so-called *stochastic assumption*, i.e., the assumption that one makes about how the response variable, the lag time λ or the specific growth rate constant μ, varies with change in the explanatory variable, the temperature T. The lag time λ will almost never have a homogeneous variance, as the lag time in the suboptimal range tends to be much more variable at low temperatures where growth rates are slow, than near the optimal temperature T_{opt}. Therefore, for modeling 4.1 to 4.6, careful consideration has to be given to the form in which these models are fitted, to reflect the stochastic assumption made.

For the specific growth rate constant μ, Ratkowsky et al. (1983), in the paper in which the four-parameter square-root model (4.5) first appeared, assumed that the variance was homogeneous in $\sqrt{\mu}$; that is, the transformed response $\sqrt{\mu}$ was assumed to have the same variance at each temperature T. This implies that the variance of the untransformed μ is a function of T, the variance increasing as μ increases. The near constancy of the variance of $\sqrt{\mu}$ has previously been demonstrated by R.K. Lowry (unpublished data) on a variety of data sets when the square-root model was first developed for suboptimal data sets (Ratkowsky et al., 1982).

On the other hand, Rosso et al. (1993) implicitly assumed that the variance of μ was homogeneous (i.e., unchanging with T). This assumption results in a different set of parameter estimates from what is obtained by assuming that $\sqrt{\mu}$ is homogeneous in T. An alternative assumption, also frequently used in the predictive microbiology literature (e.g., see Schaffner, 1998), is that $\ln \mu$ is homogeneous in μ, where $\ln \mu$ is the natural logarithm of the rate constant μ. (One may use "base 10" logarithms but mathematicians prefer the "base e" natural logarithms.) Incorporation of the stochastic assumption is most easily done by applying the transformation to both the left-hand side and the right-hand side of the expression. The result is a proliferation of forms in which the same basic equation may appear, each of which depends upon the stochastic assumption. The equations that follow express the other alternative forms in which the models used in this chapter may appear.

Hyperbola model
  Log assumption

$$\ln(1/\lambda) = -p/(T-q) \tag{4.1a}$$

Square-root assumption

$$\sqrt{(1/\lambda)} = \sqrt{\exp[-p/(T-q)]} \tag{4.1b}$$

Extended hyperbola model
  Log assumption

$$\ln(1/\lambda) = -m\ln p + m\ln(T-q) \tag{4.2a}$$

Square-root assumption

$$\sqrt{(1/\lambda)} = [p/(T-q)]^{-m/2} \tag{4.2b}$$

Linear Arrhenius model
  Log assumption

$$\ln(1/\lambda) = A + B/T + C/T^2 \tag{4.3a}$$

Square-root assumption

$$\sqrt{(1/\lambda)} = \sqrt{\exp(A + B/T + C/T^2)} \tag{4.3b}$$

Simple square-root model
  Square-root assumption

$$\sqrt{(1/\lambda)} = b(T - T_{min}) \tag{4.4a}$$

Log assumption

$$\ln(1/\lambda) = 2\ln b + 2\ln(T - T_{min}) \tag{4.4b}$$

Four-parameter square-root model
  Rate assumption

$$\mu = b^2(T - T_{min})^2\{1 - \exp[c(T - T_{max})]\}^2 \tag{4.5a}$$

Log assumption

$$\ln \mu = 2\log b + 2\log(T - T_{min}) + 2\log\{1 - \exp[c(T - T_{max})]\} \qquad (4.5b)$$

Cardinal temperature model
Square-root assumption

$$\sqrt{\mu} =$$

$$\sqrt{\{\mu_{opt}(T - T_{max})(T - T_{min})^2 / [(T_{opt} - T_{min})^2(T - T_{opt}) - (T_{opt} - T_{min})(T_{opt} - T_{max})(T_{opt} + T_{min} - 2T)]\}}$$

$$(4.6a)$$

Log assumption

$$\ln \mu = \ln \mu_{opt} + \ln(T - T_{max}) + 2\ln(T - T_{min}) -$$
$$\ln[(T_{opt} - T_{min})^2(T - T_{opt}) - (T_{opt} - T_{min})(T_{opt} - T_{max})(T_{opt} + T_{min} - 2T)] \quad (4.6b)$$

4.2.3 DATA SETS AND SOFTWARE

To illustrate the methodology, and show the effects of the various stochastic assumptions, several groups of data are employed. The first group is given in Table A4.1, and involves the lag time for the growth of *Salmonella typhimurium* on autoclaved, and therefore sterile, ground chicken breast and thigh burgers at 2°C intervals from 8 to 48°C. Colonies were counted on inverted spiral plates after incubation at 30°C for 24 h. The two sets of data (breast vs. thigh) enable a comparison of regressions to be made to test whether the same model can successfully fit both data sets.

The second group of data is for the specific growth rate constant μ vs. temperature, these being the same three data sets as used by Lowry and Ratkowsky (1983). They involve an *Alteromonas* sp. (CLD38), the temperatures ranging between 1.3 and 29.9°C, a *Pseudomonas* Group I species (16L16) in the range of 0 to 31.6°C, and a mesophilic species *Morganella morganii* (M68) (formerly *Proteus morganii*), with data in the range of 19 to 41.5°C. Unlike the data from Oscar (2002), turbidimetric measurements, rather than plate counts, were used. Table A4.2 lists these data sets, which are used to compare the four-parameter square-root and cardinal temperature models (4.5 and 4.6, respectively).

The third group of data is for the growth of *Listeria monocytogenes* and involves five complete replicates of growth data throughout the entire biokinetic range for temperature. Four of these replicates were used in a recently published study on variation of branched-chain fatty acids (Nichols et al., 2002), with a fifth set of data being added, which was not used in that study because it lacked fatty acid compositions. Different temperatures were obtained using a temperature gradient incubator and growth was monitored by measuring the percentage of transmittance of light at a wavelength of 540 nm. The data are given in Table A4.3, and are expressed as the square root of rate vs. temperature in degree Celsius. Note that there are a few zero rates at some low and some high temperatures. As indicated in Section 4.2.1, these data points have to be discarded before modeling can begin. These data sets will be fitted using the square-root model (4.5) and are used here to illustrate methodology for the examination of residuals and the effect of replication.

Nonlinear regression modeling was carried out using the SAS© statistical software, Version 8.2, PROC NLIN. The Gauss–Newton method was chosen as the fitting option. No derivatives need to be supplied, as the procedure computes them automatically. A measure of nonlinear behavior of the parameter estimators, the Hougaard measure of skewness (see Ratkowsky, 1990, pp. 27–28), was calculated using the option Hougaard. Even when the regression model was linear, e.g., 4.3a, PROC NLIN was still employed, as the Gauss–Newton method converges to the correct least-squares estimates in a single iteration, irrespective of the initial parameter values.

4.2.4   RESULTS OF MODEL FITTING

4.2.4.1   Lag Time Modeling

Oscar (2002) concluded that lag times were similar for breast and thigh meat for all temperatures in the data set of Table A4.1, probably as a consequence of the autoclaving process, and combined the individual data sets ($n = 21$ data points for each) into a single data set ($n = 42$). The graphs shown in Figure 4.1a (time scale) and Figure 4.1b (ln rate scale) visually indicate that the data sets are similar, and this will be confirmed by formal testing later in this chapter (see Section 4.2.5.2). We will use the combined data set to test the efficacy of the models 4.1 to 4.4 in the paragraphs that follow.

Table 4.1 presents parameter estimates and their asymptotic standard errors obtained by fitting models 4.1 to 4.4 to the data in Table A4.1 using the lag time λ as the response variable, and also from models 4.1a to 4.4b, which incorporate the logarithmic and the square-root transformations, respectively, applied after converting lag time into rate by taking the reciprocal of λ. Table 4.2 presents the residual mean squares corresponding to these models.

Superficially, from Table 4.2 it appears that the extended hyperbola model (4.2) is best, having a smaller residual mean square than the three alternative models when lag time λ is used as the response variable and also using the "log rate" stochastic assumption, but the Linear Arrhenius model (4.3) has a slightly smaller error mean square using the "square root of rate" assumption. Irrespective of assumption, the hyperbola model (4.1) performs badly and the simple square-root model (4.4) is intermediate. The parameter estimates given in Table 4.1 show a strong dependence upon the stochastic assumption for all models. The reasons for this will be explored in subsequent sections of this chapter.

4.2.4.2   Modeling μ

Table 4.3 lists the parameter estimates obtained from models 4.5 and 4.6 for the data in Table A4.2 using the rate assumption i.e., with μ as the response variable, and also with the square-root assumption (i.e., with $\sqrt{\mu}$ as the response) and the log assumption (i.e., with $\ln \mu$ as the response). Table 4.4 presents the residual mean squares corresponding to these models.

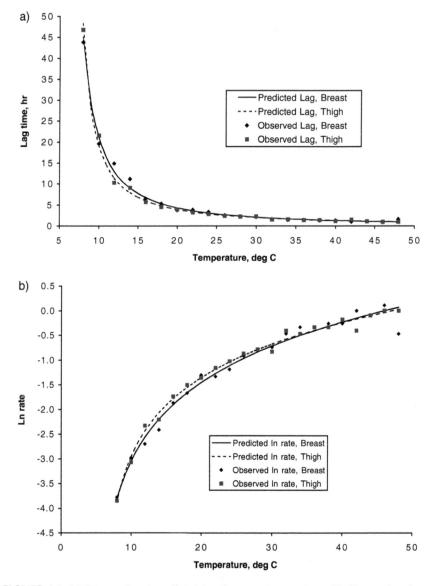

FIGURE 4.1 (a) Observed and predicted lag times vs. temperature. (b) Observed and predicted ln rates vs. temperature. Predicted values were obtained using the extended hyperbola model (4.2) and the log rate assumption on the breast and thigh data separately.

   The parameter estimates in Table 4.3 indicate that they are not strongly dependent upon the stochastic assumption. For example, the maximum range of the estimates of T_{min} or T_{max} from either model is 1.2 degree and is less than half a degree for T_{opt} in the cardinal temperature model. We have seen from the results for lag time modeling in Table 4.1 that the stochastic assumption in regression modeling can have a big impact on the magnitude and the precision of the estimates. That it

TABLE 4.1
Parameter Estimates and Their Asymptotic Standard Errors for Models 4.1 to 4.4 in Their Original and Transformed Forms, Data of Table A4.1 for the Different Stochastic Assumptions

Model	Parameter	Assumption		
		Untransformed	Log Rate	Square-Root Rate
(4.1) Hyperbola	p	28.9 ± 1.02	20.9 ± 1.51	15.5 ± 1.51
	q	0.434 ± 0.278	2.846 ± 0.516	5.69 ± 0.89
(4.2) Extended hyperbola	p	40.6 ± 2.81	41.0 ± 0.96	41.1 ± 1.11
	q	5.23 ± 0.32	5.66 ± 0.41	6.48 ± 0.87
	m	1.42 ± 0.09	1.34 ± 0.07	1.23 ± 0.09
(4.3) Linear Arrhenius	A	-540.5 ± 37.1	-274.5 ± 19.8	-214.2 ± 20.7
	B	3.52 ± 0.25	1.74 ± 0.13	1.34 ± 0.14
	C	-0.0057 ± 0.0004	-0.0028 ± 0.0002	-0.0021 ± 0.0002
(4.4) Simple square-root	B	0.0312 ± 0.0009	0.0236 ± 0.0007	0.0207 ± 0.0008
	T_{min}	3.22 ± 0.16	0.60 ± 0.51	-2.93 ± 1.27

TABLE 4.2
Residual Mean Squares for the Four Models 4.1 to 4.4 in Their Original and Transformed Forms, Data of Table A4.1 for the Different Stochastic Assumptions

Model	Assumption		
	Untransformed	Log Rate	Square-Root Rate
(4.1) Hyperbola	1.0896	0.0855	0.00910
(4.2) Extended hyperbola	0.6787	0.0185	0.00251
(4.3) Linear Arrhenius (T in Kelvin)	2.4674	0.0343	0.00245
(4.4) Simple square-root	1.0090	0.0394	0.00388

has not for the rate data is probably a reflection of the fact that the data fit each of the models well, as evidenced by the low residual mean squares in Table 4.4. The better the fit of the data to the model, the lesser the importance of the stochastic assumption. In the limit, a perfect fit would result in a fitted model that is independent of the error assumption.

Other differences may be observed from the examination of the estimates. Estimates of T_{min} from the square-root model are consistently lower than those from the cardinal temperature model, whereas estimates of T_{max} from the square-root model are consistently higher than those from the cardinal temperature model. This means that the predicted "biokinetic range," the temperature range at which nonzero growth is predicted, is always higher when estimated from the square-root model than when estimated from the cardinal temperature model. Despite the differences

TABLE 4.3
Parameter Estimates from Models 4.5 and 4.6 Fitted to the Data of Table A4.2 for the Different Stochastic Assumptions

Model	Parameter	Assumption		
		Square-Root Rate	Rate	Log Rate
CLD38 (Temperature Estimates in Kelvin)				
(4.5) 4-Parameter square root	T_{min}	266.9	267.2	266.7
	T_{max}	309.3	309.5	309.0
	b	0.0100	0.0102	0.0099
	c	0.1817	0.1732	0.1929
(4.6) Cardinal temperature	T_{min}	267.6	268.1	267.2
	T_{max}	306.2	306.5	305.8
	T_{opt}	299.1	299.0	299.3
	μ_{opt}	0.0742	0.0738	0.0747
16L16 (Temperature Estimates in Kelvin)				
(4.5) 4-Parameter square root	T_{min}	266.2	266.6	265.9
	T_{max}	310.4	310.9	309.7
	b	0.0107	0.0110	0.0105
	c	0.3096	0.2773	0.3572
(4.6) Cardinal temperature	T_{min}	266.6	267.3	266.1
	T_{max}	306.9	307.5	306.3
	T_{opt}	302.7	302.5	302.8
	μ_{opt}	0.1274	0.1263	0.1293
M68 (Temperature Estimates in Kelvin)				
(4.5) 4-Parameter square root	T_{min}	272.1	272.0	272.1
	T_{max}	317.5	317.6	317.5
	b	0.00227	0.00227	0.00227
	c	0.3397	0.3390	0.3420
(4.6) Cardinal temperature	T_{min}	274.8	275.1	274.4
	T_{max}	315.9	316.0	315.9
	T_{opt}	310.0	310.0	310.1
	μ_{opt}	0.00623	0.00624	0.00622

when these models are extrapolated, both models closely fit the data within the observed range of the data (see Figure 4.2a to Figure 4.2c). There does not appear to be any systematic departure of either model from the data. From the residual mean squares from each model in Table 4.4, it can be seen that the square-root model fits better, regardless of the stochastic assumption, for two of the three data sets, but the cardinal temperature model fits slightly better, depending upon the stochastic assumption, for the third data set. More data sets are needed to see if there is a consistent pattern. These limited results suggest that there is little to choose between the two models in terms of their ability to fit data over the whole of the temperature range. Further examination of goodness-of-fit will be made in the following sections.

TABLE 4.4
Residual Mean Squares for Models 4.5 and 4.6 Fitted to the Data of Table A.2 for the Different Stochastic Assumptions

	Assumption		
Model	Square-Root Rate	Rate	Log Rate
	CLD38		
(4.5) 4-Parameter square root	5.924×10^{-6}	0.735×10^{-6}	9.82×10^{-4}
(4.6) Cardinal temperature	8.536×10^{-6}	1.091×10^{-6}	13.0×10^{-4}
	16L16		
(4.5) 4-Parameter square root	8.75×10^{-6}	1.73×10^{-6}	10.6×10^{-4}
(4.6) Cardinal temperature	13.5×10^{-6}	2.69×10^{-6}	14.1×10^{-4}
	M68		
(4.5) 4-Parameter square root	1.17×10^{-6}	2.35×10^{-8}	10.4×10^{-4}
(4.6) Cardinal temperature	1.16×10^{-6}	2.14×10^{-8}	11.2×10^{-4}

4.2.5 Measures of Goodness-of-Fit

4.2.5.1 R^2 and "Adjusted R^2" Are Not Appropriate in Nonlinear Regression

The residual mean squares presented in Table 4.2 to Table 4.4 are part of the process of assessing the goodness-of-fit of a regression model. An oft-used criterion appearing in the scientific literature for judging whether or not a regression model fits well is R^2, the proportion of "explained variation" based upon the sum of squares; that is, it is nothing more than the ratio of the sum of squares due to regression to that of the total sum of squares of the response variable around its mean. As such, it purports to indicate how much of the total variation in the response variable, ignoring the regression, is explained by the regression model, i.e., by the inclusion of terms which are introduced to help explain the variation in the response variable. Similarly, another goodness-of-fit measure, the so-called "adjusted R^2" or "percent variance accounted for," is based upon the variances (i.e., the mean squares) rather than upon the sum of squares. The use of either of these measures for nonlinear regression is inappropriate, usually leading to a rather overoptimistic view of the success of the modeling process. We now look into these measures, and some alternatives to them, in some detail.

In linear regression models, where the model contains an intercept, as in the simple straight-line model

$$Y = \alpha + \beta X \tag{4.7}$$

where X is an explanatory variable and Y a response variable, or in the multiple regression model

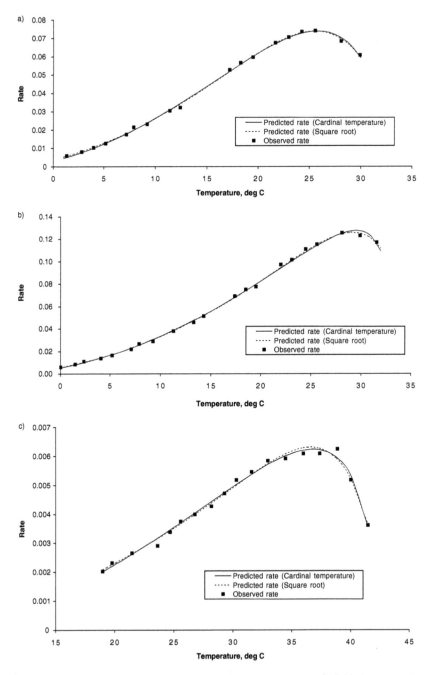

FIGURE 4.2 Observed and predicted rates vs. temperature for (a) CLD38 data, (b) 16L16 data, and (c) M68 data. Predicted rates were obtained using the cardinal temperature and square-root models.

$$Y = \alpha + \beta_1 X_1 + \beta_2 X_2 + \cdots\cdots + \beta_p X_p \qquad (4.8)$$

where X_1, X_2, ..., X_p are explanatory variables and Y a response variable, the parameters α, β, β_1, β_2, ..., β_p are estimated using the criterion of least squares. This widely used criterion finds a set of parameter estimates that minimizes the sum of squares of the differences between the observed and the fitted points, these differences being referred to as the residuals. The ratio of the sum of squares of the residuals to the corrected sum of squares of the response variable Y (the denominator being the sum of squares of the observed Ys around its mean) is the complement of R^2, also known as the "coefficient of determination," being the proportion of the total variation of Y (about its mean) that is explained by the regression.

For a linear regression model with an intercept (e.g., A in Equation 4.3a), the use of R^2 as a measure of goodness-of-fit seems sensible, but even there it may be misleading. As pointed out by Sen and Srivastava (1990, p.14), R^2 depends not only on the sum of squares of the residuals, as one would wish, but also on the corrected sum of squares of the response variable about its mean, and increasing the latter, which has nothing to do with goodness-of-fit, can also increase R^2. For example, the explanatory variables may be chosen such that half of them are in one closely spaced group and the other half in another closely spaced group, with the two groups spaced widely apart. This disposition of the Xs tends to make the denominator of R^2 large, while having no effect whatsoever upon how well the regression model fits the observed data.

For linear regression models *without* an intercept, such as

$$Y = \beta X \qquad (4.9)$$

or

$$Y = \beta_1 X_1 + \beta_2 X_2 + \cdots\cdots + \beta_p X_p \qquad (4.10)$$

R^2 cannot function as a goodness-of-fit criterion without modification. The modification is usually made by defining R^2 to be the complement of the ratio of the sum of squares of the residuals to the *uncorrected* sum of squares, where the least-squares regression line is determined by forcing the line to pass through the origin, i.e., the point $(0,0)$ on the (X,Y) axis.

Attempts have been made to generalize R^2 by correcting it in such a way that it becomes appropriate for nonlinear regression as well as for models without an intercept, including models with stochastic assumptions other than the normal (Nagelkerke, 1991). Rather than looking at the question from a theoretical point of view, which involves complicated mathematics, we will look at a practical example. Consider the data set for CLD38 using the square-root model with the stochastic assumption that the variance is homogeneous in $\sqrt{\mu}$. From Table 4.4, the residual mean square was 5.924×10^{-6} for this data set. Carrying out the least-squares regression analysis in the usual way, and presenting the results in the form of an analysis of variance (ANOVA) table, leads to the following tabulation.

Source of Variation	df	Sum of Squares	Mean Squares	Approx. F	Pr > F
Regression	4	0.7458	0.1864	31474	<0.0001
Residual	14	0.000083	5.924E-6		
Corrected total	17	0.0826			
Uncorrected total	18	0.7459			

If one were to define the coefficient of determination in the usual way as the complement of the residual sum of squares divided by the sum of squares corrected for the mean, one would obtain $R^2 = 1 - 0.000083/0.0826 = 0.999$. This would suggest that 99.9% of the "explainable" variation in the response variable $\sqrt{\mu}$ has been accounted for by the square-root nonlinear regression model, a suspiciously high figure. The inappropriateness of R^2 for nonlinear regression models, whether they have an intercept or not, has been dealt with previously (Ratkowsky, 1983, 1990). In the above calculation, the corrected total sum of squares is the sum of squares of the response variable adjusted for its mean. Although centering around the mean is justifiable when the model is a straight line or a plane that passes through the mean, there is no heuristic reason why it should be a sensible procedure when the model is nonlinear. Despite warnings of its potentially misleading nature, R^2 continues to be misused not only in the predictive microbiology literature but also wherever nonlinear regression models are employed.

Another widely used criterion is the "percentage variance accounted for" or "adjusted R^2," which differs from the traditional R^2 in being based upon the mean squares rather than the sum of squares; that is, this alternative measure is defined as the complement of the ratio of the residual mean square to a mean square based upon the corrected total. In the above example, this adjusted R^2 would be

$$\text{Adjusted } R^2 = 1 - 0.000005924/(0.0826/17) = 0.9988$$

a closely similar result to that for the traditional R^2. This too is inappropriate for a nonlinear regression model. This criterion was used by Davey and Amos (2002), but its use was appropriate, because there the model in question was a linear regression model with an intercept.

Another widely used but inappropriate indicator of goodness-of-fit both for linear and nonlinear regression models that often appears in the predictive microbiology literature is a plot of predicted response vs. observed response. Although it might appear that predicted responses are "close" superficially to the 45° line, there may also be a clear pattern of discrepancy manifested by long runs of like-signed residuals, which are the differences between the observed and the predicted values. Misleading inferences may easily be made by the indiscriminate use of such graphs.

4.2.5.2 Root Mean Square Error

Probably the most simple and the most informative measure of goodness-of-fit for regression models, both linear and nonlinear, is the root mean square error (RMSE), defined as the square root of the residual mean square. The RMSE may be viewed as the "average" discrepancy between the observed data, transformed if necessary, and their predicted values. Hence, its magnitude, especially when one also considers the precision in the original data, is useful in assessing whether a given model truly fits the data well. For the data sets from Table A4.1 and Table A4.2, the RMSEs are simply the square roots of the entries in Table 4.2 and Table 4.4, respectively.

The most desirable situation occurs when each experimental condition in the whole experiment is replicated, as that will enable one to calculate a measure of precision from the contributions of the replications to the residual variance. If the variances are of similar magnitude for each experimental condition, a pooled variance may be calculated. If the magnitudes of the residual mean square and the pooled variance are similar, this suggests that the model fits the data well. If the residual variance is much larger than the pooled variance, improvements to the model should be sought. When experiments are not replicated, the data required to calculate the pooled variance are not available. For the data given in Table A4.1, a pair of data sets is available, and if it can be shown that the lag times obtained from the breast data are not significantly different from those obtained from the thigh data, the two data sets may be pooled. We will now formally carry out the tests of significance to test the null hypothesis that the breast and thigh data sets are closely similar.

Since the extended hyperbola model, coupled with the log rate stochastic assumption as model (4.2a), seemed to be best (from the residual mean squares of Table 4.2), we will use that model to illustrate the procedure for testing whether the breast and thigh data may be pooled. Fitting 4.2a to the data sets separately produces regression results of which the following ANOVA table may be extracted:

Source of Variation	df	Sum of Squares	Mean Squares
Breast Meat Data ($n = 21$)			
Due to regression	3	49.8649	16.6216
Residual	18	0.4937	0.0274
Corrected total	20	23.3117	
Uncorrected total	21	50.3585	
Thigh Meat Data ($n = 21$)			
Due to regression	3	46.8351	15.6117
Residual	18	0.1696	0.0094
Corrected total	20	21.8457	
Uncorrected total	21	47.0047	

It is clear from the high ratio of regression to residual mean squares that the model fits the data well for each data set. The two sets of data are now combined into a single pooled data set of size $n = 42$, and model 4.2a fitted to the combined set. The following ANOVA table extract is obtained.

Source of Variation	df	Sum of Squares	Mean Squares
Combined Data ($n = 42$)			
Due to regression	3	96.6415	32.2138
Residual	39	0.7217	0.0185
Corrected total	41	45.1745	
Uncorrected total	42	97.3633	

We can now compare the residual sum of squares of 0.7217 with 39 df to the "pooled" residual sum of squares of 0.4937 + 0.1696 = 0.6633 with 18 + 18 = 36 df. The difference between these two sums of squares is 0.7217 − 0.6633 = 0.0584 with 39 − 36 = 3 df. This leads to the following variance ratio test:

$$F_{3,36} = \frac{0.0584 / 3}{0.6633 / 36} = 1.057$$

This variance ratio of 1.057 has an F distribution with 3 and 36 df and is clearly nonsignificant; hence, the breast and thigh data sets may be combined. We note that the residual mean square for the combined data, 0.0185, is almost identical to the pooled residual mean square of 0.0184, so that we have no hesitation in pooling these two sets of data into a single combined set.

Even if the entire experiment cannot be replicated, there is merit in trying to replicate some of the experimental conditions in one's experiment. Doing so provides one with a pooled error against which the residual mean square may be formally tested using the variance ratio test.

4.2.5.3 Examination of Residuals

Examination of the residuals is an important component of the evaluation of regression models, enabling the user to assess whether the model fits the data adequately. A residual is defined as the difference between the observation and the fitted or predicted value,

$$r_i = y_i - \hat{y}_i$$

where r_i is the residual corresponding to the ith observation y_i, and $\hat{y}_i$ is the corresponding predicted value. Commonly used techniques for examining residuals include plots of residuals vs. predicted values, normal probability plots, and calculating measures of influence. These procedures are described in books such as those by Mendenhall and Sincich (1996) and Fox (1991), and are carried out by software packages such as SAS (1990). Employing plots of residuals vs. predicted values and normal probability plots and associated tests enables the modeler to examine the assumptions inherent in regression analysis, such as normality of the residuals and equality of the error variance. In particular, they readily identify outlying observations, some of which may be data entry errors. Measures of influence extend the examination further, shedding further light on unusual observations.

4.2.5.3.1 The Runs Test

A simple first step in the examination of residuals is to order the residuals so that they are arranged according to increasing order of the explanatory variable X (also referred to as the "independent" or "regressor" variable), and then count the number of runs of like-signed residuals. The more runs there are, the more the fitted model tends to be centrally located within the set of data points, and thus the better the

TABLE 4.5
Number of Runs of Like-Signed Residuals[a]

Model	Assumption		
	Untransformed	Log Rate	Square-Root Rate
(4.1) Hyperbola	9	7	7
(4.2) Extended hyperbola	20	20	20
(4.3) Linear Arrhenius (T in Kelvin)	8	20	16
(4.4) Simple square root	10	13	15

[a] Data of Table A4.1.

goodness-of-fit. The runs test was used by Oscar (2002), who failed to mention that an ambiguity arises when there is more than one observation at an X value. For the data in Table A4.1, there are two observations at each of the 21 temperatures, if, as appears justified for these data, the data for breast lag time are pooled with the data for thigh lag time. For want of a better procedure, we arrange the data in such a way so that we start with the breast measurement at 8°C, follow it with the thigh measurement at 8°C, and continue alternating breast and thigh measurements until the thigh measurement at 48°C is reached.

Results of applying the runs test to the fitted models for the data of Table A4.1 are presented in Table 4.5. The model with a consistently poor fit is the "hyperbola" model (4.1), which has few (range 7 to 9) runs irrespective of the stochastic assumption, and that with a consistently good fit is the "extended hyperbola" model (4.2), with 20 runs of like-signed residuals for each error assumption. The figures for the "Linear Arrhenius" model (4.3) are interesting. When the response variable is untransformed, so that lag time λ is modeled directly, the fit is poor (8 runs), but when log rate is used, as intended by Davey (1989, 1991), the fit is excellent (20 runs). It is less good (16 runs) when the square-root transformation is applied to the rate instead of log rate. The simple square-root model (4.4) occupies an intermediate position. Although at its best when used with the square-root transformation of the rate, its 15 runs of like-signed residuals do not compete with models 4.2 or 4.3 when the log rate assumption is used.

Further results of the examination of the residuals are presented in Table 4.6. The tabulated P values confirm that the residuals are severely nonnormally distributed for all models when the response variable is untransformed and the lag time λ is modeled directly. The reason for this is that when untransformed, the large lag times that occur at temperatures of 8, 10, 12, and 14°C are not only poorly fitted by all four models, but also result in residuals that are often an order of magnitude larger than those at the higher end of the temperature scale. The results show that model 4.3 produces normally distributed residuals with the log rate transformation and the square-root transformation. What is surprising is the nonnormal distribution of the residuals for model 4.2 with the log rate stochastic assumption, which was the clear winner in terms of goodness-of-fit, well ahead of the other three models with that assumption. This result is due to a single data point, the lag time value of

TABLE 4.6
Probability Values Associated with the Test of Normality of the Residuals[a]

Model	Assumption		
	Untransformed	Log Rate	Square-Root Rate
(4.1) Hyperbola	<0.0001	0.489	0.020
(4.2) Extended hyperbola	<0.0001	0.0002	<0.0001
(4.3) Linear Arrhenius (T in Kelvin)	<0.0001	0.697	0.268
(4.4) Simple square root	<0.0001	0.010	<0.0001

[a] Data of Table A4.1.

TABLE 4.7
Probability Values Associated with the Test of Normality of the Residuals[a]

Model	Assumption		
	Untransformed	Log Rate	Square-Root Rate
(4.1) Hyperbola	<0.0001	0.450	0.027
(4.2) Extended hyperbola	<0.0001	0.128	0.031
(4.3) Linear Arrhenius (T in Kelvin)	<0.0001	0.471	0.357
(4.4) Simple square root	<0.0001	0.858	0.240

[a] Data of Table A4.1, with the data point for breast meat at 48°C eliminated.

1.6 h on breast meat at 48°C. Compared with all other values of lag time for temperatures above 35°C, that value is clearly too high (see Table A4.1). Eliminating that data point and refitting the model results in a set of normally distributed residuals, not only for model 4.2 but also for model 4.4, as shown in Table 4.7. Similarly, the table shows that use of model 4.4 in combination with the square-root stochastic assumption results in a set of normally distributed residuals.

4.2.5.3.2 Measures of Influence
One of the statistics measuring influence is the so-called "hat matrix" H, identifying those observations that are influential due to the values of the explanatory variable(s). An analogy is to consider children sitting on a seesaw. The further they sit from the fulcrum, the greater the "leverage," the word often employed to describe h_i, the ith element of the diagonal of that matrix. Since $\hat{y}_i$ may be written as a linear combination of the n observed y values,

$$\hat{y}_i = h_1 y_1 + h_2 y_2 + \dots + h_n y_n$$

the larger the value of h_i, the greater the weight given to the ith observation. The average value of h is conveniently given by the ratio of the number of parameters in the model to the total number of points n.

Another useful statistic is the Studentized residual, which is the ratio of the ordinary residual r_i to its standard error, the latter incorporating the leverage measure h_i. High values of this statistic, greater than two (say) in absolute magnitude, indicate a significantly large residual. Other measures of influence include Cook's D (Cook, 1979) and the Dffits statistic (Belsley et al., 1980). These statistics produce a combined measure of influence by coupling the effect of high leverage with the measure of whether the observation is an outlier. Hence, a large value of D or Dffits usually results from both the leverage and the residual being large. D_i is customarily compared to critical values of the F distribution with numerator df equal to the number of parameters estimated and denominator df equal to the residual df. If D_i exceeds the 50th percentile of this F distribution, the observation is deemed to be influential (see Mendenhall and Sincich, 1996).

Measures such as Cook's D and Dffits are intended for use with the "straight-line" model (4.7) or with the multiple regression model (4.8), just in the same way that R^2 or adjusted R^2 are intended to assess goodness-of-fit for such models. We have seen in Section 4.2.5.1 that moving the "fulcrum" from the center of the line or the plane to the origin of the coordinates results in an incorrect interpretation of R^2 or adjusted R^2 if the standard definition of those measures is not modified. Similarly, influence has to do with the distance that a point is from the fulcrum, and whereas such a distance is unambiguous with models such as 4.7 and 4.8, various problems of interpretation arise when one is dealing with a curvilinear regression, such as the polynomial models to be discussed in Section 4.3.1 or nonlinear regression models, ones like 4.5 and 4.6, in which the parameters appear nonlinearly. In any event, many of the methods for examining residuals are graphically based (such as normal probability plots and graphs of residuals vs. fitted values), and tests of significance should be considered to be only approximate. This is especially true for nonlinear regression models because of bias in the predicted response values, although such bias is typically small (see Ratkowsky, 1983, for discussion of the effect of the so-called "intrinsic" nonlinearity).

Table 4.8 shows some results of applying measures of influence to the data sets of Table A4.2, using the square-root model (4.5) coupled with the square-root stochastic assumption. Similar to the results shown in Table 4.3 and Table 4.4, little difference was observed between the parameter estimates obtained from the three stochastic assumptions, as well as between models 4.5 and 4.6.

For the CLD38 data set, the average value of the leverage h_i is $4/18 = 0.222$, since there are four parameters in the model and a total of 18 data points. Since $2(0.222) = 0.444$, only the last data point, the one corresponding to $t = 29.9°C$, exceeds this value and appears to be influential. Although the Studentized residual is far below 2.0, Cook's D of 2.43 exceeds the critical value of 1.52 for the F distribution with 4 and 14 df for the 50th percentile. Thus, the last data point should be considered to be significantly influential without being an outlier. There are indications that two of the interior points, the sixth and the ninth observations, have high residuals. This is confirmed by looking at Figure 4.2a, which shows a very

TABLE 4.8
Results of the Examination of Residuals from Use of Model 4.5 with the Square-Root Stochastic Assumption for the Data of Table A4.2[a]

Obs	Temp	Square Root of Rate	Predicted Square Root of Rate	Residual	Leverage h_i	Studentized Residual	Cook's D
				CLD 38			
1	1.3	0.07727	0.07535	0.00192	0.333	0.965	0.1164
2	2.8	0.08972	0.09030	−0.00058	0.229	−0.272	0.0055
3	4.0	0.10178	0.10224	−0.00046	0.170	−0.207	0.0022
4	5.2	0.11265	0.11415	−0.00150	0.131	−0.662	0.0165
5	7.2	0.13198	0.13391	−0.00192	0.104	−0.835	0.0202
6	7.9	0.14632	0.14079	0.00553	0.104	<u>2.402</u>	0.1671
7	9.2	0.15215	0.15349	−0.00134	0.113	−0.586	0.0109
8	11.4	0.17461	0.17472	−0.00011	0.145	−0.049	0.0001
9	12.4	0.17961	0.18422	−0.00461	0.161	<u>−2.066</u>	0.2047
10	17.2	0.22989	0.22729	0.00260	0.176	1.178	0.0740
11	18.3	0.23782	0.23621	0.00161	0.163	0.723	0.0254
12	19.5	0.24413	0.24531	−0.00117	0.148	−0.523	0.0119
13	21.7	0.25994	0.25957	0.00037	0.148	0.164	0.0012
14	23.0	0.26556	0.26598	−0.00043	0.179	−0.193	0.0020
15	24.3	0.27137	0.27031	0.00106	0.232	0.498	0.0188
16	25.6	0.27216	0.27191	0.00025	0.287	0.122	0.0015
17	28.1	0.26171	0.26406	−0.00235	0.315	−1.168	0.1566
18	29.9	0.24648	0.24534	0.00114	<u>0.861</u>	1.253	<u>2.4251</u>
				16L16			
1	0.0	0.07668	0.07469	0.00199	0.231	0.769	0.0444
2	1.5	0.09241	0.09078	0.00163	0.184	0.611	0.0210
3	2.4	0.10407	0.10043	0.00363	0.160	1.340	0.0852
4	4.1	0.11730	0.11867	−0.00137	0.122	−0.494	0.0085
5	5.2	0.12783	0.13047	−0.00264	0.103	−0.943	0.0256
6	7.1	0.14761	0.15085	−0.00323	0.081	−1.140	0.0287
7	7.9	0.16331	0.15943	0.00388	0.075	1.366	0.0381
8	9.3	0.17018	0.17443	−0.00426	0.071	−1.493	0.0426
9	11.3	0.19519	0.19586	−0.00067	0.075	−0.234	0.0011
10	13.3	0.21394	0.21725	−0.00331	0.089	−1.172	0.0337
11	14.3	0.22709	0.22792	−0.00083	0.099	−0.295	0.0024
12	17.4	0.26315	0.26080	0.00236	0.129	0.853	0.0270
13	18.5	0.27421	0.27233	0.00187	0.136	0.682	0.0182
14	19.5	0.27842	0.28272	−0.00429	0.138	−1.563	0.0978
15	22.0	0.31190	0.30791	0.00398	0.130	1.444	0.0782
16	23.1	0.31859	0.31843	0.00016	0.129	0.059	0.0001
17	24.5	0.33332	0.33093	0.00238	0.145	0.872	0.0322
18	25.6	0.33985	0.33973	0.00012	0.188	0.046	0.0001
19	28.1	0.35440	0.35372	0.00068	<u>0.381</u>	0.293	0.0133
20	29.9	0.35071	0.35427	−0.00356	<u>0.405</u>	−1.560	0.4142

TABLE 4.8 (Continued)
Results of the Examination of Residuals from Use of Model 4.5 with the Square-Root Stochastic Assumption for the Data of Table A4.2[a]

Obs	Temp	Square Root of Rate	Predicted Square Root of Rate	Residual	Leverage h_i	Studentized Residual	Cook's D
21	31.6	0.34220	0.34075	0.00145	0.928	1.821	10.6762
			M68				
1	19.0	0.04508	0.04554	−0.00046	0.327	−0.520	0.0329
2	19.8	0.04817	0.04736	0.00081	0.270	0.875	0.0708
3	21.5	0.05158	0.05121	0.00036	0.176	0.371	0.0074
4	23.5	0.05407	0.05573	−0.00166	0.111	−1.623	0.0826
5	24.7	0.05812	0.05843	−0.00031	0.094	−0.303	0.0024
6	25.6	0.06131	0.06045	0.00086	0.091	0.833	0.0173
7	26.8	0.06325	0.06312	0.00013	0.096	0.123	0.0004
8	28.2	0.06537	0.06619	−0.00082	0.112	−0.805	0.0204
9	29.3	0.06868	0.06855	0.00013	0.127	0.126	0.0006
10	30.3	0.07198	0.07064	0.00134	0.140	1.332	0.0721
11	31.6	0.07392	0.07323	0.00069	0.149	0.692	0.0210
12	33.0	0.07647	0.07576	0.00072	0.148	0.716	0.0223
13	34.5	0.07692	0.07797	−0.00105	0.143	−1.046	0.0457
14	36.0	0.07809	0.07931	−0.00122	0.166	−1.237	0.0763
15	37.4	0.07809	0.07923	−0.00114	0.242	−1.206	0.1164
16	38.8	0.07906	0.07697	0.00209	0.337	2.365	0.7107
17	40.0	0.07198	0.07224	−0.00026	0.344	−0.300	0.0119
18	41.5	0.06019	0.06039	−0.00020	0.925	−0.660	1.3528

[a] Significant statistics are underlined.

good fit of the square-root model (and the cardinal temperature model as well), with these two points being further from the fitted curve than any others.

For the 16L16 data set, the average value of h_i is 4/21 = 0.190, so the last three data points, especially the last one, have significant leverage. This is reflected in the highly significant value of Cook's D for the last point, but there do not appear to be any outliers. Figure 4.2b confirms this by displaying a very close fit between the square-root model and the data.

For the M68 data set, the average value of h_i is 0.222, so that the last data point exerts considerable leverage. Nevertheless, Cook's D of 1.35 is below the critical value of 1.52, so this point is not unduly influential. Observation 16 is seen to have a Studentized residual in excess of 2.0, which is confirmed by looking at Figure 4.2c.

We mention once again that the examination of residuals is an important tool, but that one should rely on graphical interpretation more than on significance testing, since the models are nonlinear regression models. Measures such as Cook's D are not strictly applicable, and like R^2 or adjusted R^2, they may be inappropriate or misleading for nonlinear regression models.

4.2.5.4 Measures of Nonlinear Behavior

Fitting of nonlinear regression models has become a relatively straightforward task with the use of modern statistical packages. However, regression modeling should not be viewed simply as a curve-fitting exercise, but one that requires thought and subsequent evaluation and testing. The examination of residuals, for example, which was the subject of the previous section, is part of the process of evaluation that logically follows the routine fitting of a mathematical model to a data set. A further step in that process is to ask whether there are other features of the model that may or may not be deemed desirable in a mathematical model. When the fitted model is a nonlinear regression model, one should ask whether the model is "close to linear" or not.

The concept of a "close to linear" nonlinear regression model was advanced in an earlier book (Ratkowsky, 1983). It was recognized then that some nonlinear regression models could have severely biased parameter estimates, have a probability distribution that was vastly different from that of a normal (Gaussian) distribution, typically being skewed with a long right-hand or left-hand tail, and have excess variance. This contrasts with linear regression models such as 4.7 and 4.8, which, when the stochastic assumption of a normally distributed error term is valid, have unbiased, normally distributed, minimum variance estimators. Although all nonlinear regression models have biased parameter estimators, the various models differ in the extent of the bias. The models that have only a very small bias in their estimates were called "close to linear" by Ratkowsky (1983), whereas those that exhibited severe bias were said to be "far from linear." In many models, parameter bias may be reduced by reparameterization, i.e., changing the form in which the parameters of the models appear. Other models can only be reparameterized at the price of producing an awkward-appearing model. (See Ratkowsky [1983, 1990] for a detailed discussion of these issues and for many examples of reparameterization.)

Several measures of nonlinear behavior have been advanced over the years, some of which have not withstood the test of time. A very reliable indicator of nonlinear behavior for an individual parameter estimator is based on Hougaard's measure of skewness, described in Ratkowsky (1990, pp. 27–28), which exploits the close connection between a nonlinear regression model's behavior and its expression in biased, skewed parameter estimators. This measure is available in recent releases of SAS© statistical software, PROC NLIN, using the option "Hougaard." Experience with this measure suggests that if the Hougaard skewness measure is less than 0.1 in absolute value, the estimator of the parameter is very close to linear, but that if its absolute value exceeds 0.25, the skewness is quite apparent (as may be seen, for example, by carrying out a simulation study), and if it exceeds 1.0, considerable nonlinear behavior of the estimator is present. Since skewness and bias (the difference between the mean value of a parameter's estimator and its true population value) are closely correlated, a high skewness measure can be taken to mean a high bias in the estimator of that parameter, and conversely, a low skewness measure equates to a low bias.

Table 4.9 presents results for Hougaard's skewness measure for the parameters of models 4.1 to 4.4, in combination with the data of Table A4.1 for the various

TABLE 4.9
Hougaard Skewness Measures for Models 4.1 to 4.4 in Their Original and Transformed Forms, Pooled Data of Table A4.1 for the Different Stochastic Assumptions

Model	Parameter	Assumption		
		Untransformed	Log Rate	Square-Root Rate
(4.1) Hyperbola	p	0.063	0.207	0.219
	q	−0.077	−0.360	−0.096
(4.2) Extended hyperbola	p	0.384	0.216	0.207
	q	−0.390	−0.540	−1.178
	m	0.346	0.279	0.632
(4.3) Linear Arrhenius	A	0.169	0	−0.055
	B	−0.192	0	0.053
	C	0.218	0	−0.051
(4.4) Simple square-root	b	0.119	0.017	0
	T_{min}	−0.089	−0.217	−0.214

stochastic assumptions. Although the extended hyperbola model (4.2) containing an exponent m fits the data much better than the simple hyperbola model (4.1), the parameter estimators for p and q are much more biased than they were for 4.1. For example, q substantially underestimates that parameter. The Linear Arrhenius model (4.3) has zero bias when the log assumption is used, reflecting the fact that that model is a linear regression model, and it has a small, nonperceptible bias for the square-root stochastic assumption. The simple square-root model also has parameters with low bias (the zero value for b with the square-root stochastic assumption reflecting the fact that the model is linear).

Table 4.10 presents results for Hougaard's skewness measure for the parameters of models 4.5 and 4.6, in combination with the three sets of data of Table A4.2 for the various stochastic assumptions. The four-parameter square-root model (4.5) and the cardinal temperature model (4.6) both contain the notional parameters T_{min} and T_{max} and the results for both models are in agreement in that the biases in T_{min} are both small, whereas the biases for T_{max} are quite large for all three data sets, particularly so for the 16L16 data. The bias for both models is positive, meaning that the estimates, on average, are larger than the true values.

4.2.6 BUILDING MATHEMATICAL MODELS

This section takes a look at the construction of models for predictive microbiology. Over the years, a variety of opinions have been expressed about the nature of models that might be used to describe the shelf life of food products and the rate at which food deteriorates. Many of these opinions are philosophical in nature. For example, some authors have been concerned with questions such as the differences between mechanistic and empirical models, among others (e.g., see Section 4.2.6.3). Herein, we will confine our attention to considerations that have led to the appearance of

TABLE 4.10
Hougaard Skewness Measures for Models 4.5 and 4.6 Fitted to the Data of Table A4.2 for the Different Stochastic Assumptions

Model	Parameter	Square-Root Rate	Rate	Log Rate
		Assumption		
CLD38 (Temperature Estimates in Kelvin)				
(4.5) 4-Parameter square root	T_{min}	0.001	0.002	0.019
	T_{max}	0.281	0.241	0.388
	b	0.274	0.284	0.284
	c	0.190	0.134	0.280
(4.6) Cardinal temperature	T_{min}	−0.038	−0.027	−0.026
	T_{max}	0.429	0.343	0.606
	T_{opt}	0.003	−0.006	0.021
	μ_{opt}	0.062	0.029	0.126
16L16 (Temperature Estimates in Kelvin)				
(4.5) 4-Parameter square root	T_{min}	0.012	0.030	0.023
	T_{max}	0.421	0.332	0.668
	b	0.197	0.219	0.195
	c	0.326	0.207	0.670
(4.6) Cardinal temperature	T_{min}	−0.006	0.007	0.024
	T_{max}	0.742	0.562	1.121
	T_{opt}	0.072	0.060	0.032
	μ_{opt}	0.137	0.069	0.314
M68 (Temperature Estimates in Kelvin)				
(4.5) 4-Parameter square root	T_{min}	−0.091	−0.098	−0.078
	T_{max}	0.342	0.363	0.330
	b	0.194	0.239	0.176
	c	0.221	0.213	0.245
(4.6) Cardinal temperature	T_{min}	−0.133	−0.138	−0.122
	T_{max}	0.434	0.451	0.431
	T_{opt}	0.014	0.006	0.025
	μ_{opt}	0.020	0.003	0.044

various classes of empirical models for practical use that have appeared in the predictive microbiology literature.

4.2.6.1 Why Polynomial Models Do Not Work

One class of models that is frequently encountered in the predictive food microbiology literature is "polynomial models." For example, if one were modeling specific growth rate constant (μ) as a function of temperature, modelers favoring polynomial models would use, instead of models such as 4.5 and 4.6, a model in which μ is expressed as a low-order polynomial in T, usually not exceeding the third order, i.e.,

$$\mu = a + bT + cT^2 + dT^3 \tag{4.11}$$

When there is more than a single environmental factor, the number of parameters of such models multiplies rapidly. For example, if temperature and salt concentration (NaCl) are the environmental factors, the model would become, if all terms of the polynomial up to third order are included,

$$\mu = a + bT + cT^2 + dT^3 + e\,(\text{NaCl}) + f(\text{NaCl})^2 + g(\text{NaCl})^3$$
$$+ hT(\text{NaCl}) + iT^2(\text{NaCl}) + jT^3(\text{NaCl}) + lT(\text{NaCl})^2 + mT^2(\text{NaCl})^2 \tag{4.12}$$
$$+ nT^3(\text{NaCl})^2 + oT(\text{NaCl})^3 + pT^2(\text{NaCl})^3 + qT^3(\text{NaCl})^3$$

A total of 16 parameters (coefficients) have to be evaluated for this complete third-order polynomial. This lack of parsimony often compels authors, for practical reasons, not to go beyond second-order terms, so that the model becomes

$$\mu = a + bT + cT^2 + d(\text{NaCl}) + e(\text{NaCl})^2 + fT(\text{NaCl}) + gT^2(\text{NaCl})$$
$$+ hT(\text{NaCl})^2 + iT^2(\text{NaCl})^2 \tag{4.13}$$

which has "only" nine parameters. Aside from its nonaesthetic appearance, there remains the practical question of whether such a model is capable of fitting real data.

As an example of the nonparsimonious nature of polynomials, let us look at one of the illustrative data sets of Table A4.1, that for CLD38. The graphs in Figure 4.3 display the fit of a quadratic, a cubic, and a quartic (i.e., fourth degree) polynomial, using the "rate" assumption (i.e., no transformation), and also show the observed data points. The quadratic fit, involving three parameters, is very poor, there being only four runs of like-signed residuals, with the predicted model underestimating the observed data at very low temperatures and at temperatures near the optimum, and

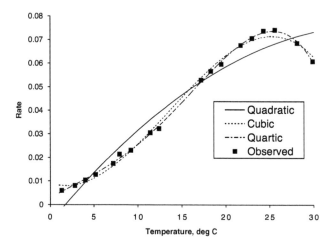

FIGURE 4.3 Observed and predicted rates (fitted quadratic, cubic, and quartic polynomials) vs. temperature. CLD38 data from Table A4.2.

overestimating at moderate temperatures and very high temperatures. The cubic fit, with four parameters, is a considerable improvement, but there are still only five runs of like-signed residuals, and the residual mean square is 3.67×10^{-6}, much higher than those for the square-root or cardinal temperature models. The quartic fit, with five parameters, does very well with 11 runs of like-signed residuals, with a residual mean square of 0.767×10^{-6}, which is almost as low as that of the square-root model and lower than that of the cardinal temperature model (see Table 4.4). To achieve this precision, however, one extra parameter beyond that required by the square-root or cardinal temperature models had to be employed, and none of the five parameters of the quartic polynomial is interpretable. In contrast, the square-root model has two interpretable parameters and the cardinal temperature model has four.

4.2.6.2 Comparison of Models Using *F* Tests

A frequently used test in the statistical literature for comparing models is the F test, employed extensively for formal testing of ANOVA models. In the food microbiology literature, it has been used, for example, to describe the combined effects of temperature, pH, and lactate on the growth of *Listeria innocua* (Houtsma et al., 1996), to quantify the interactions of spoilage microorganisms (Pin and Baranyi, 1998) and to determine if a simple, nested model was sufficient to describe the growth kinetics of a number of microorganisms (Delignette-Muller, 1998). Some limitations of this method have been noted (McMeekin et al., 1993), for example, (1) it cannot discriminate between models with the same number of parameters, or nonnested models; (2) the significance of the F test is only approximate for nonlinear regression models; (3) indiscriminate use of the F test may lead to overparameterized models, i.e., ones with more terms and parameters than are necessary. Some of these limitations are more serious than others; for example, the approximate nature of the F test in nonlinear regression models is not serious, as the bias in the F test depends upon the component of nonlinearity referred to as the "intrinsic" nonlinearity, and this bias is typically small in most nonlinear regression models, except for very small sample sizes (see Ratkowsky, 1983).

Models that typically are overparameterized are the polynomial models, criticized in Section 4.2.6.1. Some authors have tried to reduce the number of parameters by eliminating nonsignificant terms (e.g., Houtsma et al., 1996), but it is not clear what purpose is really served by that procedure. First, for correctness, the eliminated terms must be *jointly* nonsignificant, a conclusion that cannot be reached by applying stepwise procedures such as forward or backward elimination. One should compare the reduced model with the full model, which may be done using the F test, to see whether inclusion of the extra terms significantly improves the fit. In any event, the final model after terms have been eliminated is really no "better" than it was before the unnecessary terms were deleted; that is, although there may appear to be less unexplained variation in the response variable due to a smaller residual mean square, because there are now more df for error than before, the effect is mainly cosmetic. There is no substitute, when the goal is to produce accurate rate models or growth/no growth interface models for predicting food product shelf life, for trying to build the best possible mechanistic model for the process, or a close approximation to it.

4.2.6.3 Models with Several Environmental Factors

Although the earliest successful models in food microbiology involved only temperature, including the Bĕlehrádek-type models of which the square-root model is a special case (Ratkowsky et al., 1982, 1983) and the Arrhenius-type relationships (Schoolfield et al., 1981; Sharpe and DeMichele, 1977), it soon became apparent that other growth-limiting environmental factors had to be taken into account. Models containing water activity (a_w) in addition to temperature followed (e.g., Davey, 1989; McMeekin et al., 1987), and with time, the effect of hydrogen ion concentration in the form of pH (e.g., Adams et al., 1991) and the addition of weak acids such as acetic acid and lactic acid were being considered (e.g., Presser et al., 1997).

A parsimonious model involving the combined effect of temperature, water activity (or salt concentration), and the addition of a weak acid, such as lactic acid, cannot be successfully achieved using polynomials such as 4.11 to 4.13. Baranyi et al. (1996) promoted the desirability of models embodying known or assumed features of the phenomenon under consideration. Van Impe et al. (2001) considered models to be divided into three classes, following Ljung (1999), as white box models, black box models, and gray box models. Deductive white box models require full knowledge of the underlying physical mechanisms and a deep understanding of the physical, chemical, and biochemical laws driving the process, a situation that is rarely available at this moment in time. Black box models lie at the opposite end of the scale. They take the experimental data as input information and produce output variables with or without necessarily producing an equation or series of equations. This inductive approach includes polynomial modeling and the use of artificial neural networks, but models so produced cannot reflect physical considerations. A gray box model is a compromise between the two extremes and is probably the standard to which modelers in predictive food microbiology can realistically hope to achieve at this point of time. Another alternative, suggested by Geeraerd et al. (2002), is to retain the black box approach while incorporating *a priori* microbiological knowledge into the modeling process so that overfitting of the data and unrealistic parameter estimates are prevented from occurring.

The approach taken by Presser et al. (1997) was an attempt to incorporate some reasonable assumptions based upon physical chemistry into the modeling process. They used the observation reported by Cole et al. (1990) that the growth rate of a microorganism is directly proportional to the hydrogen ion concentration, and this led directly to an expression for the effect of pH. Similarly, the well-known Henderson–Hasselbalch equation of physical chemistry was used, which related the ratio of the undissociated to the dissociated forms of a weak organic weak to the pH and pK_a, the latter being the pH at which the concentrations of the two forms are equal. The resulting growth rate model (see Presser et al., 1997) for *E. coli* as a function of temperature, pH, and added lactic acid concentration contained only six parameters to be estimated. This is in sharp contrast to a polynomial model, which would have had to contain dozens of parameters to achieve the same level of prediction. If the model fit exhibits shortcomings, then it can be amended to improve its predictive ability, but the basic model form is a good foundation upon which to base further fine-tuning.

4.2.7 REPLICATED DATA SETS

We now examine the data in Table A4.3, which, unlike the data in Table A4.1, can be seen to be a group of data sets with genuine replication. The data of Table A4.1 served as a surrogate for replication, since the results for breast meat were indistinguishable from those on thigh meat, making it possible to consider the two sets of data to be replicates. The growth rate data for *L. monocytogenes* of Table A4.3 were obtained from five separate runs using a temperature gradient incubator, on samples of what was ostensibly the same material. Each data set is independent of the others.

The four-parameter square-root model (4.5) appears to be well suited to fit each of the individual data sets, as may be seen from Figure 4.4. Parameter estimates and their standard errors are given in Table 4.11. It is seen that there is a fair amount of variability in the estimates of the two cardinal temperatures, with the T_{min} estimates varying between –1.0 and 2.3°C and the T_{max} estimates varying between 45.5 and 48.3°C. Similarly, the measures of goodness-of-fit show considerable variation, with the residual mean squares ranging by a factor of 3 between 1.0×10^{-5} and 3.0×10^{-5} and the number of runs of like-signed residuals varying between only six runs for Data Set 4 to a rather substantial 20 runs for Data Set 3. Nevertheless, there is no correlation between number of runs of residuals and the normality of the set of residuals, with Data Set 4 having a close-to-normal set of residuals and Data Sets 2 and 3 being marginally nonnormal. As can be seen from Figure 4.5, which shows the pooled Data Sets 1 to 5 on a single graph with the square-root model (4.5) fitted to the pooled data, there is a group of five data points in Data Set 1 in the suboptimal temperature range of 27.5 to 32.8°C with much lower rates than those predicted by the overall fitted model.

Pooling the residual sum of squares from the individual data sets leads to 0.000661 + 0.000493 + 0.000333 + 0.000478 + 0.000250 = 0.002215, with 22 + 23 + 23 + 23 + 24 = 115 df. The residual sum of squares for the pooled data set of 135 points is 0.00677 (see Table 4.11) with 131 df. The difference between these two sums of squares is 0.00677 – 0.002215 = 0.00455 with 131 – 115 = 16 df. This leads to the following variance ratio test:

$$F_{16,115} = \frac{0.00455 / 16}{0.002215 / 115} = 14.76.$$

This is clearly a highly significant F value ($P < 0.001$) and indicates model inadequacy. The question is whether this is due to a poor model or poor data. The information in Figure 4.5 suggests that the model is adequate but that there are a number of aberrant data points. This is further borne out by Figure 4.6, which is a plot of the residuals vs. the fitted values from fitting 4.5 to the pooled set of 135 data points. There are seven data points that stand out as potential outliers, with residuals exceeding 0.015 in absolute magnitude. These are indicated using a larger font size.

Removing the data points with the seven biggest residuals and refitting model 4.5 to the remaining 128 data points results in a residual sum of squares of 0.00249 (see Table 4.11) with 124 df. The pooled residual sum of squares, recalculated from

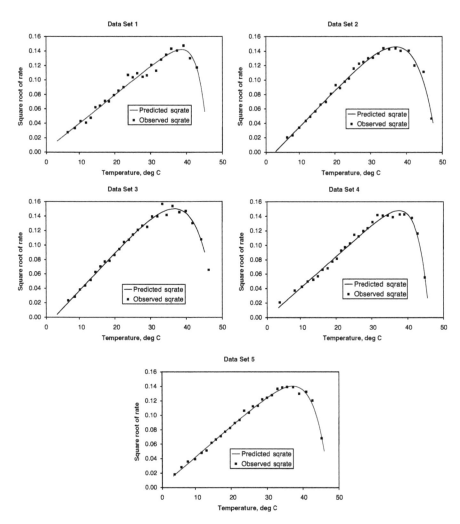

FIGURE 4.4 Five individual data sets and fitted square-root models. (Data from Nichols et al., *Appl. Environ. Microbiol.*, 68, 2809–2813, 2002.)

the five data sets with the outliers deleted, is 0.00134 with 108 df. The difference between these sums of squares is 0.00249 − 0.00134 = 0.00115 with 124 − 108 = 16 df. This leads to the following variance ratio test:

$$F_{16,108} = \frac{0.00115\,/\,16}{0.00134\,/\,108} = 5.79,$$

which is still highly significant ($P < 0.001$). From Table 4.11, we see that the P value for the normality test of the residuals is 0.445, indicating that the new set of residuals obtained by data elimination is close to having a normal distribution. Therefore, the significant variance ratio above cannot be attributed to outliers, but

TABLE 4.11
Results of Fitting Five Replicated Data Sets of Table A4.3

Data Set	n	Residual Sum of Squares	Residual Mean Square	P Value Test of Normality	No. of Runs of Residuals	Highest Hougaard Skewness	$b \pm$ SE	$T_{min} \pm$ SE	$c \pm$ SE	$T_{max} \pm$ SE
1	26	0.000661	0.000030	0.4837	15	0.956 (T_{max})	0.00388 ± 0.00016	−0.97 ± 0.91	0.3506 ± 0.0990	46.09 ± 0.99
2	27	0.000493	0.000021	0.0409	14	0.333 (T_{max})	0.00507 ± 0.00019	2.11 ± 0.59	0.1576 ± 0.0145	48.34 ± 0.25
3	27	0.000333	0.000014	0.0525	20	0.289 (T_{max})	0.00518 ± 0.00016	2.32 ± 0.48	0.1569 ± 0.0127	48.32 ± 0.23
4	27	0.000478	0.000021	0.8385	6	0.365 (T_{max})	0.00447 ± 0.00013	0.45 ± 0.60	0.2886 ± 0.0256	45.47 ± 0.17
5	28	0.000250	0.000010	0.1100	16	0.293 (T_{max})	0.00433 ± 0.00010	−0.23 ± 0.43	0.2192 ± 0.0163	46.32 ± 0.20
1–5	135	0.00677	0.000052	<0.0001	75	0.287 (T_{max})	0.00472 ± 0.00012	1.18 ± 0.43	0.1718 ± 0.0119	47.88 ± 0.22
1–5 (outliers deleted)	128	0.00249	0.000020	0.4447	74	0.200 (T_{max})	0.00474 ± 0.00008	1.22 ± 0.27	0.1678 ± 0.0077	48.20 ± 0.16

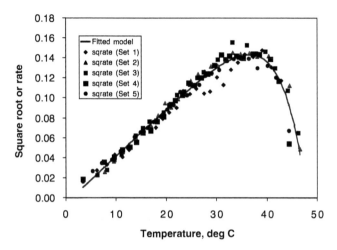

FIGURE 4.5 Pooled data and fitted overall square-root model (Data from Nichols et al., *Appl. Environ. Microbiol.*, 68, 2809–2813, 2002.)

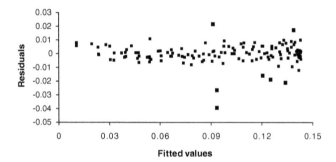

FIGURE 4.6 Residuals vs. fitted values from overall model ($n = 135$). The seven largest residuals are indicated using a larger font size. (Data from Nichols et al., *Appl. Environ. Microbiol.*, 68, 2809–2813, 2002.)

must be interpreted as indicating that there is significant variability among the five data sets.

Looking more closely at the five sets of data, we see from the dates given in Table A4.3 that Data Set 1 was obtained on 6 September 2000, Data Sets 2 and 3 were obtained on 25 September 2000, and Data Sets 4 and 5 were obtained on 14 December 2000. Data Sets 2 and 3 are, in fact, close replicates, as the "bar" run from which Data Set 2 was obtained used tubes on one side of the temperature gradient incubator, and the run from which Data Set 3 was obtained used tubes on the other side of the incubator. Different stock solutions were employed for the two runs, so that variation in the results would be a consequence of variation in the inocula and not in the ambient conditions of the room in which the incubator was housed. Similarly, Data Sets 4 and 5 are close replicates for the same reason.

We now compare Data Sets 2 and 3 using the same test we used above. From the results in Table 4.11, the pooled residual sum of squares is 0.000493 + 0.000333 = 0.000826 with 23 + 23 = 46 df. Combining the two data sets leads to a residual sum of squares of 0.000847 with 50 df. The difference between these sums of squares is 0.000847 – 0.000826 = 0.000021, with 50 – 46 = 4 df. This leads to the following variance ratio test:

$$F_{4,46} = \frac{0.000021/4}{0.000826/46} = 0.292$$

a nonsignificant F value that confirms that Data Sets 2 and 3 are close replicates.

Comparing Data Sets 4 and 5 in a similar way leads to a pooled residual sum of squares of 0.000728 with 47 df, while combining the two data sets leads to a residual sum of squares of 0.00103 with 51 df. The difference between these sums of squares is 0.00103 – 0.000728 = 0.000302 with 51 – 47 = 4 df, leading to the following variance ratio test:

$$F_{4,47} = \frac{0.000302/4}{0.000728/47} = 4.90$$

This variance ratio is significant ($P < 0.005$), so the two data sets, although obtained on the same bar run, cannot be considered to be close replicates. Removing the data point responsible for the largest residual ($T = 38.1°C$; sqrate = 0.1297; Data Set 5) still results in a significant variance ratio ($F_{4,46} = 3.49$), so the lack of closeness of agreement of these two data sets cannot be attributed to a single outlying data point. From Figure 4.5, one can visually observe that there are differences between Data Set 4 and Data Sets 2 and 3. A formal test leads to $F_{8,69} = 17.26$, which is highly significant. Hence, one must conclude that the data sets obtained on 14 December are different from those obtained on 25 September. All are different from Data Set 1, obtained on 6 September, which exhibited a series of low rates in the mid-temperature range.

Identifying deviant data points using analysis of residuals is desirable as a means of directing attention to the possibility of experimental errors. However, one should avoid deleting data points with large residuals unless there are good, objective reasons for doing so, because of belief that errors were committed during the experiment, resulting in erroneous readings. The indiscriminate use of data elimination may have the undesirable effect of leading to biased estimates of the parameters, and may produce a belief that the data set, modified by having its most deviant points deleted, is of a better quality than is really justified.

Because five replicate data sets for the growth of *L. monocytogenes* were available, it was possible to see that, although each data set looked to be good in isolation, the aberrant nature of Data Set 1 became apparent when all data were pooled and plotted on a single graph (Figure 4.5). To the author's knowledge, very few replicate data sets are available for turbidity measurements obtained using a temperature gradient incubator. Replication is desirable as it makes it possible to examine whether the variation in experimental data is significant or not.

What explanations might be offered to explain the variation among the five data sets discussed here? An obvious one is operator error in using the temperature gradient incubator, but this is unlikely in the present case because the experimenters were very experienced. Another explanation has to do with their use of the modified Gompertz function as a primary model to estimate the lag time and the maximum specific growth rate at each temperature. That model may not have been an appropriate one for determining these parameters, especially as the cultures were harvested when the individual incubation tubes reached a transmittance value of only 27 to 30%. A second explanation may be that the secondary model, the four-parameter square-root model (4.5), was inadequate or inappropriate. This is unlikely, as the information in Figure 4.4 and in Table 4.11 indicates that 4.5 is a good model for these data. Another explanation may lie with the equipment itself, or with the room in which the incubator was housed, as its ambient temperature is probably not adequately controlled. Although it is not possible to be certain about the true explanation for the variation observed among the data sets in question, it is always important for investigators to consider the various possibilities, so that one may improve the experimental procedure, or the modeling process that follows collection of the data, or both.

4.3 UNCERTAINTY IN LAG TIMES AND GENERATION TIMES, AND ITS CONSEQUENCES

In this section, we look at uncertainty in some of the basic parameters that are measured or derived from experimentation and data analysis in predictive food microbiology, and its consequences for food production and safety. Uncertainty is an ever-present phenomenon, one which may be reduced by careful quality control, but which may never be totally eliminated.

A fundamental issue of the discipline of predictive food microbiology (perhaps its most important one) is the accurate prediction of the shelf life of a food product so that the product remains edible throughout the whole of that period. We have seen in some of the earlier sections of this chapter (e.g., 4.2.4.2) that modeling of some of the important derived parameters such as the maximum specific growth rate μ occurs in a transformed rate domain; that is, μ, which has units of reciprocal time, is transformed by taking its square root (e.g., Equation 4.5 or Equation 4.6a) or its logarithm (e.g., Equation 4.5b or Equation 4.6b) when it is to be used in a modeling exercise or various other statistical calculations. The reason for applying one of these transformations is that the untransformed growth rate has undesirable statistical properties. Generally, the probability distribution of μ is nonnormal (i.e., not a Gaussian distribution), which means, among other things, that the variance of the distribution is not independent of its mean. By applying a transformation such as the square root or logarithm to the growth rate, the transformed variable comes close to having a normal distribution. The random variable will then be symmetrically distributed around the mean, with a variance that does not depend upon the mean. Since the mean and variance are the two parameters of the normal distribution, having a good estimate of these parameters from a sample of size n (say) enables the user to have confidence about the probability of obtaining an observation from the distribution of the transformed variable that falls outside of any specified bound. For example, the probability of an observation being more than two standard deviations from the mean can be calculated using a table of the t distribution with $n - 1$ df. If the random variable is not normally distributed, one may be able to make a similar calculation if its distribution is known, but usually with more difficulty than if the random variable is normal.

The situation is even more complicated when one considers time-based variables or derived parameters such as lag time, generation time, and shelf life. While square roots of rate or logarithms of rate may be normally distributed, lag times, generation times, and shelf lives never are, tending instead to have long-tailed distributions. In the following subsection, we review the distributional features of these time-domain quantities.

4.3.1 THE DISTRIBUTION OF LAG TIME AND GENERATION TIME

The distribution of lag time and generation time is a nonnormal distribution of which the variance usually falls somewhere between being proportional to the square of the mean or to the cube of the mean; that is, if we denote the mean (lag or generation) time by θ, then the variance, if denoted by V, is usually given by

$$V = c\theta^2 \tag{4.14}$$

or

$$V = c\theta^3 \tag{4.15}$$

or by an exponent of θ that is a noninteger value lying somewhere between 2 and 3, with c being a proportionality constant (see McMeekin et al., 1993; Ratkowsky et al., 1991; Schaffner, 1998). There are some well-known probability distributions with variances having the properties defined above. For example, the gamma and Weibull distributions have the variance proportional to the square of the mean, as given by 4.14, and the inverse-Gaussian distribution has its variance proportional to the cube of the mean, as given by 4.15. There are other lesser-known distributions with the same properties.

Data to test whether the exponent of θ is 2, 3, or some other value are hard to find, as they require many replicate runs carried out at each of a sequence of temperatures. Although one might expect to obtain a reasonable prediction of the population mean from a relatively small sample size, say 10 to 15 values, sample variances are more variable than sample means and many dozens of trials are needed to obtain a good prediction of the variance. Using the data of Neumeyer given in Appendix 4A.4 of McMeekin et al. (1993) on generation times for the growth of *Staphylococcus aureus* 3b from 12.5 to 35C, it was established that 4.15 was a reasonable model for the relationship between the variance and the mean (see Table 4.8 and associated text of McMeekin et al. [1993] for the methodology used).

Nevertheless, sample sizes were rather small, ranging from $n = 1$ to 18. Only two temperatures had $n > 10$.

A far larger data set ($n = 125$) for examining the question of the correct model between a probability distributions variance and mean is that of Macario, reported in Ratkowsky et al. (1996), for the generation time of *Pseudomonas fluorescens* in the temperature range of 2.4 to 16.3°C, obtained using nutrient broth in a temperature gradient incubator. Grouping the temperature data into 1°C intervals produced 15 intervals with sample sizes ranging between 2 and 17. The estimates of the means and variances were plotted against temperature as the ratios V/θ, V/θ^2, and V/θ^3. The regression line of V/θ^2 against temperature was the one with the least correlation, suggesting that 4.14 was the best model for those data. This suggested to Ratkowsky et al. (1996) that the Macario data could be modeled by 4.14 with the scale parameter $c = 0.006676$ estimated by assuming that a gamma distribution was a suitable probability distribution for the data (see Ratkowsky et al., 1996, for a detailed description of the methodology employed). It should have been realized that the gamma distribution, which contains only two parameters, is only one possible probability distribution having the property given by 4.14. Although the assumption of a gamma distribution led to predicted variances that were in the right "ball park" when compared with the experimental variances, the predicted standardized skewness coefficient of 0.163 indicated only a small amount of skewness, as the histograms in Figure 2 of Ratkowsky et al. (1996) were scarcely distinguishable from those of a normal distribution. Later, Hutchinson (1998) pointed out that the error made by Ratkowsky et al. (1996) was to believe that V/θ^2 being a constant told one something about the shape of the distribution, e.g., whether it was skewed.

Hutchinson (1998) pointed out that to determine the real shape of the distribution, one would have to look at the data for each temperature separately, and that would require dozens of observations at each temperature. Clearly, the Macario data set, averaging ca. 8 points per temperature, was too small for this. He further showed that if one could assume that the "shifted gamma distribution" were a suitable probability distribution for those data, having a third parameter that measured the extent to which the mean was shifted upwards from zero, one could obtain a fitted distribution with a considerable skewness. However, until such time as microbiologists can readily produce many replicated data sets at each temperature, the question of the true amount of skewness in data that obey models 4.14 and 4.15 will not be answered. In the following section, we explore how knowledge of the true distribution for lag time and generation time may be used to obtain accurate predictions about the likely shelf life of a food product.

4.3.2 THE PREDICTION OF SHELF LIFE

The shelf life (keeping time of a food product), subject to spoilage owing to a spoilage organism, can be predicted from lag time and generation time if there are appropriate models for these parameters expressed in terms of temperature T, water activity a_w, hydrogen ion concentration (pH), concentration of undissociated weak acid, and any other factors influencing them.

Denoting mean lag time by t_L and mean generation time by t_G, the shelf life may be estimated by the expression

$$\text{Shelf life} = t_L + t_G \ln(C_f/C_0)/\ln 2 \qquad (4.16)$$

where C_0 is the initial bacterial concentration and C_f is the maximum permissible concentration (e.g., in cfu/ml), where permissible means the product is deemed to be spoiled. The first term of the expression is the lag time before growth effectively begins, and the longer the value of t_L, the longer the shelf life. The second term of the expression calculates the number of generations through which the microorganism develops from its initial concentration to its final concentration. Note that the above assumes that the generation time is independent of the numbers of bacteria present. If that assumption is too naïve, or too crude an approximation, one may predict C_f from a model (such as the modified Gompertz curve — see chapter on Primary Modeling).

An example of the use of 4.16 will now be presented. Consider the data on the growth of *Aeromonas hydrophila* (aerobic atmosphere), from Palumbo et al. (1991). For $T = 19°C$, added salt = 3.5%, pH = 6.3 and added sodium nitrite = 50, and assuming $C_0 = 10$ and $C_f = 1.0 \times 10^7$ cfu/ml, the lag time t_L was determined to be 60 h and the generation time t_G was determined to be 2.6 h, using a square-root model. Using 4.16, the keeping time is calculated as

$$\text{Shelf life} = 60.0 + 2.6 \ln(10^7/10)/\ln 2 = 112 \text{ h}$$

The second approach, which does not assume a constant generation time, can be made using a computer package such as the pathogen modeling program (PMP), developed by the USDA/ARS/ERRC Microbial Food Safety Unit. From PMP, the lag phase duration t_L is 48.2 and the generation time t_G is 2.7 h, corresponding to the above environmental factors. The calculated time to reach the level of concern 10^7 cfu/ml is 105 h, which is similar to the 112 h calculated using 4.16.

The important thing about the above calculation is that in either approach only mean values of lag or generation times are used. The variability of these parameters, and their probability distributions, as discussed in Section 4.3.1, has not entered into the calculation. We will now take a brief look at how variability may affect the result obtained, and lead to a larger or smaller shelf life.

We reconsider the data of Macario, discussed in Section 4.3.1. It was found that it was reasonable to assume that the ratio of the variance to the square of the mean was constant, which is equivalent to assuming that the coefficient of variation is constant. Ratkowsky et al. (1996) further assumed, naively as Hutchinson (1998) later demonstrated, that this implied that one could use the gamma distribution, which has two parameters, for these data. Estimating the scale parameter to be $c = 0.006676$, we calculated predicted mean generation times for any temperature and the distribution of generation time under the assumption that the gamma distribution was an appropriate one for those data. For example, at $T = 2.4C$, the mean generation time was determined to be $\theta = 615.2$ min (see Table 2 of Ratkowsky et al., 1996) and a series of predicted probabilities, denoted by θ_0, were calculated at that temperature. Thus, for $P = 0.000001$, $\theta_0 = 405.1$ min, for $P = 0.001$, $\theta_0 = 471.5$ min, for $P = 0.999$, $\theta_0 = 782.3$ min, for $P = 0.999999$, $\theta_0 = 885.8$ min; and so on (see Table 4 of Ratkowsky et al., 1996). What do these numbers signify? For example, they tell us that although the *mean* generation time may be 615.2, there is a one in a thousand chance that the generation time may be as low as 471.5 and a one in a million chance that it may be as low as 405.1. From the point of view of food acceptability, these values, if used in 4.16 in place of 615.2, would lead to a much-reduced shelf life. At the other end of the scale, there is a one in a thousand chance that the generation time may be as high as 782.3 and a one in a million chance that it may be as high as 885.8. These values would lead to a prolonged shelf life compared to the expected shelf life based upon the mean generation time. Note that 471.5 and 782.3 are not equally distant from 615.2, and that neither are 405.1 and 885.8. This is a consequence of the asymmetry inherent in a distribution such as the gamma distribution. However, as pointed out by Hutchinson (1998), this distribution is not particularly skewed, and to get a good estimate of the true skewness would require many replicates at each temperature.

4.4 EPILOGUE

In this epilogue, the author raises a few issues that he has considered over the years. The first two subsections deal with beliefs, held by at least some modelers, that the author feels are erroneous. Because they have appeared in the food microbiology literature, these issues need to be raised. In the final two subsections, some newer or less familiar modeling methods are discussed, which have found, or will find, their way into the predictive modeling literature in the near future.

4.4.1 USE OF THE EXPRESSIONS "FAIL SAFE" AND "FAIL DANGEROUS" FOR MODELS

There seems to be a widespread use of the expressions "fail safe" and "fail danger-ous" in the food microbiological literature when applied to models. Assuming that one has unbiased, carefully collected data, fail-safe refers to a model that overpredicts the rate at which a spoilage or pathogenic organism will grow, or predicts overly stringent conditions at the growth/no growth interface for growth to occur. Conversely, fail-dangerous refers to a model that under-predicts the actual growth rate or fails to predict conditions at which growth will actually occur. In either case, the model must be deemed to be inadequate. To use the expression fail-safe to exonerate, exculpate, or absolve the modeler, scientist, or regulator from doing a better job seems to be just taking the easy way out. There is really only one kind of model that should find a place in the food microbiology literature, and that is a *good* model. Good models are those that closely mimic the rate at which spoilage or pathogenic organisms grow or which closely predict the position of the growth/no growth interface. Bad models are all other kinds.

The issue is not an academic one, but a practical one. Fail-safe can be taken to a ludicrous extreme, by adopting a "model" that always predicts that a pathogenic microorganism will grow, even under conditions so stringent as to be virtually impossible. Using this model, all food products would be declared unsafe to consume. We live on a planet in which a very high proportion of its human inhabitants do not get enough food to eat. Death by starvation is still a global problem and adequate nutrition, to help people ward off debilitating or potentially fatal diseases, is a worldwide problem. An extreme fail-safe model might suggest that food that is in reality safe to eat should be avoided or destroyed. We should always remember that death and disease could be caused as much by the nonavailability of food as well as by its being contaminated or spoiled. In the more affluent world economies, a fail-safe mentality also overlooks the producer's point of view. Food that is safe to be sold and consumed may, as a result of overzealous regulations resulting from inadequate modeling, have to be discarded, leading not only to lower profitability but less availability and choice to the consumer. In addition, any resulting added costs tend to get passed on to the consumer.

4.4.2 CORRELATION BETWEEN PARAMETERS

There seems to be a strong belief on the part of certain researchers that a regression model with low correlation between its parameters is, in some sense, "better" than one with high parameter correlation. For example, Rosso et al. (1993) found that the three cardinal temperature parameters in 4.6 were linearly correlated, a condition that they felt was "unexpected" (see title of their paper). No rational reasons were advanced as to why uncorrelated parameters are deemed to be superior to correlated ones. It is the present author's experience that there is no connection between parameter correlation and the properties of the estimators of the parameters in nonlinear regression models (Ratkowsky, 1983, 1990). It simply has nothing to do with the more important question of whether the estimators of the parameters are unbiased, jointly normally distributed, and attain the so-called "minimum variance bound." It might be nice to have a nonlinear regression model that not only is "close to linear" (see Section 4.2.5.4), but also has uncorrelated parameters. The present writer has never found such a model.

4.4.3 ARTIFICIAL NEURAL NETWORKS AS AN ALTERNATIVE TO STATISTICAL MODELING

Various alternatives to statistical modeling are starting to appear in the predictive microbiology literature, some of which are likely to prove appealing to a new generation of modelers. Probably the most important of these involves the use of artificial neural networks (ANN) in one form or other (they have also been referred to as computational neural networks, artificial computational neural networks, or general regression neural networks). An early exposition in the predictive microbiology literature of the methodology of this approach, which would be classified as a black box approach using the system of Ljung (1999), is that given by Hajmeer et al. (1997), who described some of the computational details and gave an example of its use on microbial growth data. ANNs operate by analogy to the human nervous system, where input variables provide an incoming signal to a neuron, are modified in a "hidden" neuron layer, and finally converted to an output signal by an appropriate "transfer function." The application by Hajmeer et al. (1997) to anaerobic growth data obtained by Zaika et al. (1994) on *Shigella flexneri* as a function of pH, NaCl, and $NaNO_2$ concentrations, used a hidden layer consisting of 20 nodes (neurons), because their "training cycles" indicated that their goodness-of-fit measures became stable when 20 nodes were used. Some compromises need to be made, since by increasing the number of training cycles and the number of hidden nodes indefinitely, one can fit "training sets" perfectly, but at the risk of poorly predicting subsequent validation (testing) sets. This is, of course, directly analogous to regression modeling, where the use of too many terms in a regression model may result in overfitting the original data set and a poor fit to validation data sets.

Geeraerd et al. (1998) used a low-complexity ANN to convert the incoming signal from environmental factors such as temperature, pH and salt concentration to an output signal embodied in parameters such as the maximum growth rate, the lag time, and the initial population size. They referred to their model as a "hybrid gray box" model because the ANN models were used only to describe the effect of an influencing factor such as temperature on an output parameter such as μ_{max}, which was then used in a dynamic growth model to predict the growth of the microbial population with time.

Jeyamkondan et al. (2001) used commercially available neural network software, which enabled them to examine different network structures quickly with little effort, a feature that is appealing to users who know some basic principles of neural networks but are not experts in neural network programming. They chose a general regression neural network (GRNN) to predict response parameters such as generation time and lag phase duration from input data involving changes in temperature, NaCl, pH, etc. for three microorganisms. They compared use of the GRNN to that of more traditional statistical models and found that the GRNN predictions were far superior to predictions from statistical models for training data sets, but similar to, or slightly worse than statistical model predictions for test data sets. They concluded that neural networks were adequate for food safety tests and for new product development. To assess goodness-of-fit, they investigated the performance of various statistical indices and concluded that the use of the mean absolute relative residual, the mean relative percentage residual, and the root mean square residual, in conjunction with graphical plots such as the bias plot and the residual plot, were sufficient for comparing competing models.

More recently, Hajmeer and Basheer (2002, 2003) proposed a probabilistic neural network (PNN) for use on growth/no growth data and compared its classificatory performance to that of a linear logistic regression model, a nonlinear logistic regression model of the kind proposed by Ratkowsky and Ross (1995), and a feedforward error backpropagation artificial neural network (FEBANN), and found that the optimal PNN gave the lowest misclassification rates. It should be pointed out, though, that the "nonlinear" logistic regression model that they derived using their training data subset was not a true nonlinear logistic model, since the cardinal parameters $a_{w\,min}$, T_{min}, and T_{max} were assumed to be fixed constants, taken from the paper of Salter et al. (2000), and not estimated as free parameters. Because of this, the resulting model was really a linear logistic model and not capable of doing as well as a true nonlinear logistic regression model.

At this early stage in the use of ANNs, it is difficult to forecast how valuable they might be in predictive microbiology to predict the conditions under which food products should be stored to guarantee their safety and quality. One question that arises is whether ANN models are "portable," i.e., whether other workers can readily use them in the way that they can use statistical models that have expressions in the form of mathematical equations. In this regard, hybrid models may prove to be attractive, where the ANN may serve to produce a primary model, in a similar way in which the modified Gompertz model or the model of Baranyi et al. (1993) is used, followed by a more traditional secondary model where the outputs from the ANN are then expressed as functions of the environmental factors.

4.4.4 PRINCIPAL COMPONENTS REGRESSION AND PARTIAL LEAST-SQUARES REGRESSION

There are other alternatives to ANN that might be used by a new generation of modelers. Jeyamkondan et al. (2001) mentioned principal component regression (PCR) and partial least-squares regression (PLS) as two statistical techniques of multivariate analysis that might be employed when the underlying relationships are not known. PCR has been known for some time, and is available in many standard statistical packages. Given a set of n experimental units on which measurements have been made of p explanatory variables, a principal component analysis can reduce those p variables to a smaller set (say 2 or 3) of "canonical" variates, upon which regression analysis of various response variables may then be performed. The canonical variates (i.e., the principal components) are linear combinations of the p explanatory variables and have the property that they are orthogonal (i.e., uncorrelated) to each other, unlike the original set of p variables, at least some of which are likely to be highly correlated. If the canonical variates are easily interpretable, the contributions of each variate to the explanation of the response variable can be quantified, because of their orthogonality. Hence, PCR has the potential to be a useful technique, provided that the canonical variates are subject to interpretation.

PLS is a much newer technique and at the moment is poorly understood by users, but is gaining increasing application. It can be contrasted with PCR and with another technique, called reduced rank regression (RRR), when the response variables form a multivariate set. Whereas PCR extracts successive linear combinations of the explanatory variables to explain as much *predictor* sample variation as possible, RRR extracts successive linear combinations of the set of response variables to explain as much *response* sample variation as possible. PLS tries to balance the two objectives by simultaneously explaining response variation and predictor variation. The same caveats that applied to ANN modeling apply here as well. Just as the use of too many nodes or too many training cycles can lead to over-fitting the training set of data and poor prediction in test data sets for ANN modeling, extracting too many factors in PLS can also lead to overfitting. Like ANN modeling, the use of validation, by splitting one's data set into a training set and a test set, is an integral part of the modeling process.

Time can only tell how useful such techniques might be to predictive microbiology, but one has to anticipate that a new generation of modelers is certain to come forward with applications employing one or more of these procedures. One must retain an open mind to their use, but at the same time avoid uncritical acceptance of them. In addition, the restriction inherent in all these techniques, which involve *linear* combinations of variables, may limit the general applicability of the methodology. After all, the world we live in is not a linear one, and it is a rare circumstance in mathematical modeling when a linear model explains natural phenomena adequately.

APPENDIX

TABLE A4.1
Data Set on Lag Time

	Lag Time (hr)	
$T(°C)$	Breast	Thigh
8	43.8	46.8
10	19.6	21.6
12	14.9	10.3
14	11.3	9.1
16	6.5	5.7
18	5.3	4.5
20	3.7	3.9
22	3.8	3.2
24	3.3	2.8
26	2.5	2.4
28	2.2	2.2
30	2.1	2.3
32	1.6	1.5
34	1.4	1.6
36	1.4	1.4
38	1.3	1.4
40	1.3	1.2
42	1.0	1.5
44	1.1	1.1
46	0.9	1.0
48	1.6	1.0

Source: From Oscar, T.P., *Int. J. Food Microbiol.*, 76, 177–190, 2002. With permission.

TABLE A4.2
Specific Growth Rate Constant μ vs. Temperature for Three Data Sets

Alteromonas sp. (CLD38)		*Pseudomonas* sp. (16L16)		*Morganella morganii* (M68)	
T (°C)	μ	T (°C)	μ	T (°C)	μ
1.3	0.00597	0.0	0.00588	19.0	0.002032
2.8	0.00805	1.5	0.00854	19.8	0.002320
4.0	0.01036	2.4	0.01083	21.5	0.002660
5.2	0.01269	4.1	0.01376	23.5	0.002924
7.2	0.01742	5.2	0.01634	24.7	0.003378
7.9	0.02141	7.1	0.02179	25.6	0.003759
9.2	0.02315	7.9	0.02667	26.8	0.004000
11.4	0.03049	9.3	0.02896	28.2	0.004273
12.4	0.03226	11.3	0.03810	29.3	0.004717
17.2	0.05285	13.3	0.04577	30.3	0.005181
18.3	0.05656	14.3	0.05157	31.6	0.005464
19.5	0.05960	17.4	0.06925	33.0	0.005848
21.7	0.06757	18.5	0.07519	34.5	0.005917
23.0	0.07052	19.5	0.07752	36.0	0.006098
24.3	0.07364	22.0	0.09728	37.4	0.006098
25.6	0.07407	23.1	0.1015	38.8	0.006250
28.1	0.06849	24.5	0.1111	40.0	0.005181
29.9	0.06075	25.6	0.1155	41.5	0.003623
		28.1	0.1256		
		29.9	0.1230		
		31.6	0.1171		

TABLE A4.3
Growth Rate Data of *Listeria monocytogenes*, Presented as Square Root of Rate vs. Temperature, Five Replicates

Data Set 1 (6 Sept 2000)		Data Set 2 (25 Sept 2000)		Data Set 3 (25 Sept 2000)		Data Set 4 (14 Dec 2000)		Data Set 5 (14 Dec 2000)	
T (°C)	$\sqrt{\mu}$	T (°C)	$\sqrt{\mu}$	T (°C)	$\sqrt{\mu}$	T (°C)	$\sqrt{\mu}$	T (°C)	$\sqrt{\mu}$
1.8	0	6.2	0.0232	6.2	0.0229	0.7	0	0.8	0
3.8	0	7.7	0.0262	8.1	0.0280	3.4	0.0185	3.4	0.0163
6.1	0.0277	9.5	0.0365	9.6	0.0378	7.6	0.0356	5.3	0.0266
8.2	0.0335	11.3	0.0459	11.1	0.0434	9.6	0.0407	7.1	0.0345
9.7	0.0432	12.6	0.0516	12.6	0.0509	11.2	0.0480	9.1	0.0379
11.3	0.0411	14.1	0.0591	13.9	0.0622	12.8	0.0506	10.9	0.0465
12.6	0.0478	15.5	0.0681	15.4	0.0697	14.1	0.0558	12.3	0.0500
14.0	0.0623	17.0	0.0719	16.7	0.0765	15.4	0.0650	13.7	0.0607
15.3	0.0646	18.4	0.0827	18.0	0.0776	16.9	0.0672	14.9	0.0654
16.7	0.0712	19.7	0.0948	19.5	0.0856	18.2	0.0766	16.3	0.0700
17.9	0.0703	21.0	0.0909	20.8	0.0938	19.5	0.0811	17.5	0.0761
19.3	0.0792	22.3	0.0994	22.1	0.1032	20.7	0.0927	18.9	0.0814
20.6	0.0854	23.5	0.1036	23.5	0.1065	21.8	0.0971	20.2	0.0884
22.0	0.0902	24.8	0.1173	24.7	0.1137	23.2	0.1024	21.5	0.0926
23.3	0.1071	26.3	0.1240	26.1	0.1201	24.4	0.1150	22.8	0.1059
24.7	0.1039	27.6	0.1262	27.4	0.1257	25.6	0.1125	24.1	0.1027
26.0	0.1092	28.9	0.1314	28.8	0.1243	27.0	0.1201	25.2	0.1121
27.5	0.1045	30.2	0.1320	30.2	0.1384	28.2	0.1240	26.6	0.1128
28.7	0.1064	31.8	0.1375	31.7	0.1388	29.6	0.1326	27.8	0.1218
30.0	0.1207	33.2	0.1455	33.1	0.1559	31.0	0.1429	29.1	0.1242
31.3	0.1129	34.7	0.1437	34.3	0.1407	32.5	0.1419	30.5	0.1278
32.8	0.1284	36.5	0.1449	36.0	0.1530	33.9	0.1421	32.1	0.1368
34.2	0.1353	38.0	0.1414	37.9	0.1445	35.3	0.1398	33.4	0.1386
35.6	0.1435	40.3	0.1419	39.7	0.1462	37.2	0.1437	34.7	0.1393
37.2	0.1410	41.9	0.1214	41.6	0.1296	38.5	0.1436	36.4	0.1391
39.1	0.1478	44.4	0.1128	44.0	0.1071	40.7	0.1389	38.1	0.1297
40.9	0.1304	46.6	0.0490	46.2	0.0650	42.3	0.1166	40.1	0.1323
42.8	0.1175					44.3	0.0540	41.8	0.1201
45.0	0					46.5	0	44.3	0.0671
47.1	0							46.3	0

REFERENCES

Adams, M.R., Little, C.L., and Easter, M.C. (1991). Modelling the effect of pH, acidulant and temperature on the growth rate of *Yersinia enterocolitica*. *J. Appl. Bacteriol.* 71: 65–71.

Baranyi, J., Roberts, T.A., and McClure, P.J. (1993). A non-autonomous differential equation to model bacterial growth. *Food Microbiol.* 10: 43–59.

Baranyi, J., Ross, T., McMeekin, T.A., and Roberts, T.A. (1996). Effects of parameterization on the performance of empirical models used in predictive microbiology. *Food Microbiol.* 13: 83–91.

Belsley, D.A., Kuh, E., and Welsch, R.E. (1980). *Regression Diagnostics: Identifying Influential Data and Sources of Collinearity.* New York: Wiley.

Cole, M.B., Jones, M.V., and Holyoak, C. (1990). The effect of pH, salt concentration and temperature on the survival and growth of *Listeria monocytogenes*. *J. Appl. Bacteriol.* 69: 63–72.

Cook, R.D. (1979). Influential observations in linear regression. *J. Am. Statist. Assoc.* 74:169–174.

Dantigny, P. and Molin, P. (2000). Influence of the modeling approach on the estimation of the minimum temperature for growth in Bělehrádek-type models. *Food Microbiol.* 15, 185–189.

Davey, K.R. (1989). A predictive model for combined temperature and water activity on microbial growth during the growth phase. *J. Appl. Bact.* 67: 483–488.

Davey, K.R. (1991). Applicability of the Davey (linear Arrhenius) predictive model to the lag phase of microbial growth. *J. Appl. Bact.* 70: 253–257.

Davey, K.R. and Amos, S.A. (2002). Letter to the editor. *J. Appl. Microbiol.* 92:583–584.

Delignette-Muller, M.L. (1998). Relation between the generation time and the lag time of bacterial growth kinetics. *Int. J. Food Microbiol.* 43: 97–104.

Fox, J. (1991). *Regression Diagnostics* (Sage University Paper series on quantitative applications in the social sciences, Series No. 07-079). Newbury Park, CA: Sage.

Geeraerd, A.H., Herremans, C.H., Cenens, C., and Van Impe, J.F. (1998). Application of artificial neural networks as a non-linear modular modeling technique to describe bacterial growth in chilled food products. *Int. J. Food Microbiol.* 44: 49–68.

Geeraerd, A.H., Valdramidis, V.P., Devlieghere, F., Bernaerts, H., Debevere, J., and Van Impe, J.F. (2002). Development of a novel modelling methodology by incorporating *a priori* microbiological knowledge in a black box approach. In *Proceedings and Abstracts of the 18th International ICFMH Symposium, Food Micro 2002*, Lillehammer, Norway, 18–23 August 2002, Axelsson, L., Tronrud, E.S., and Merok, K.J. (Eds.), pp. 135–138.

Hajmeer, M. and Basheer, I. (2002). A probabilistic neural network approach for modeling and classification of bacterial growth/no-growth data. *J. Microbiol. Methods* 51: 217–226.

Hajmeer, M. and Basheer, I. (2003). Comparison of logistic regression and neural network-based classifiers for bacterial growth. *Food Microbiol.* 20: 43–55.

Hajmeer, M.N., Basheer, I.A., and Najjar, Y.M. (1997). Computational neural networks for predictive microbiology II. Application to microbial growth. *Int. J. Food Microbiol.* 34: 51–66.

Houtsma, P.C., Kant-Muermans, M.L., Rombouts, F.M., and Zwietering, M.H. (1996). Model for the combined effects of temperature, pH and sodium lactate on growth rates of *Listeria innocua* in broth and Bologna-type sausages. *Appl. Environ. Microbiol.* 62: 1616–1622.

Hutchinson, T.P. (1998). Note on probability distributions for generation time. Letter to the editor. *J. Appl. Microbiol.* 85: 192–193.

Jeyamkondan, S., Jayas, D.S., and Holley, R.A. (2001). Microbial growth modelling with artificial neural networks. *Int. J. Food Microbiol.* 64: 343–354.

Ljung, L. (1999). *System Identification Theory for the User*, 2nd ed. Upper Saddle River, NJ: Prentice Hall.

Lowry, R.K. and Ratkowsky, D.A. (1983). A note on models for poikilotherm development. *J. Theor. Biol.* 105: 453–459.

McMeekin, T.A., Chandler, R.E., Doe, P.E., Garland, C.D., Olley, J., Putro, S., and Ratkowsky, D.A. (1987). Model for the combined effect of temperature and water activity on the growth rate of *Staphylococcus xylosus*. *J. Appl. Bacteriol.* 62: 543–550.

McMeekin, T.A., Olley, J.N., Ross, T., and Ratkowsky, D.A. (1993). *Predictive Microbiology: Theory and Application*. Taunton, Somerset: Research Studies Press.

Mendenhall, W. and Sincich, T. (1996). *A Second Course in Statistics: Regression Analysis*, 5th ed. Upper Saddle River, NJ: Prentice-Hall.

Nagelkerke, N.J.D. (1991). A note on a general definition of the coefficient of determination. *Biometrika* 78: 691–692.

Nichols, D.S., Presser, K.A., Olley, J., Ross, T., and McMeekin, T.A. (2002). Variation of branched-chain fatty acids marks the normal physiological range for growth in *Listeria monocytogenes*. *Appl. Environ. Microbiol.* 68: 2809–2813.

Oscar, T.P. (2002). Development and validation of a tertiary simulation model for predicting the potential growth of *Salmonella typhimurium* on cooked chicken. *Int. J. Food Microbiol.* 76: 177–190.

Palumbo, S.A., Williams, A.C., Buchanan, R.L., and Phillips, J.G. (1991). Model for the aerobic growth of *Aeromonas hydrophila* K144. *J. Food Protection* 54: 429–435.

Pin, C. and Baranyi, J. (1998). Predictive models as a means to quantify the interactions of spoilage organisms. *Int. J. Food Microbiol.* 41: 59–72.

Presser, K.A., Ratkowsky, D.A., and Ross, T. (1997). Modelling the growth rate of *Escherichia coli* as a function of pH and lactic acid concentration. *Appl. Environ. Microbiol.* 63: 2355–2360.

Ratkowsky, D.A. (1983). *Nonlinear Regression Modeling: A Unified Practical Approach*. New York: Marcel Dekker.

Ratkowsky, D.A. (1990). *Handbook of Nonlinear Regression Models*. New York: Marcel Dekker.

Ratkowsky, D.A. and Ross, T. (1995). Modelling the bacterial growth/no growth interface. *Lett. Appl. Microbiol.* 20: 29–33.

Ratkowsky, D.A., Olley, J., McMeekin, T.A., and Ball, A. (1982). Relationship between temperature and growth rate of bacterial cultures. *J. Bacteriol.* 149: 1–5.

Ratkowsky, D.A., Lowry, R.K., McMeekin, T.A., Stokes, A.N., and Chandler, R.E. (1983). Model for bacterial growth rate throughout the entire biokinetic temperature range. *J. Bacteriol.* 154: 1222–1226.

Ratkowsky, D.A., Ross, T., McMeekin, T.A., and Olley, J.N. (1991). Comparison of Arrhenius-type and Bĕlehrádek-type models for the prediction of bacterial growth in foods. *J. Appl. Bacteriol.* 71: 452–459.

Ratkowsky, D.A., Ross, T., Macario, N., Dommett, T.W., and Kamperman, L. (1996). Choosing probability distributions for modeling generation time variability. *J. Appl. Bacteriol.* 80: 131–137.

Rosso, L., Lobry, J.R., and Flandrois, J.P. (1993). An unexpected correlation between cardinal temperatures of microbial growth highlighted by a new model. *J. Theor. Biol.* 162: 447–463.

Salter, M.A., Ratkowsky, D.A., Ross, T., and McMeekin, T.A. (2000). Modelling the combined temperature and salt (NaCl) limits for growth of a pathogenic *Escherichia coli* strain using nonlinear logistic regression. *Int. J. Food Microbiol.* 61: 159–167.

SAS Institute Inc. (1990). *SAS/STAT User's Guide*, Version 6, 4th ed. Cary, NC: SAS Institute Inc. (The SAS/STAT User's Guide for more recent releases of SAS such as Version 8 are available as on-line documentation by connecting to the SAS website.)

Schaffner, D.W. (1998). Predictive food microbiology Gedanken experiment: why do microbial growth data require a transformation? *Food Microbiol.* 15: 185–189.

Schoolfield, R.M., Sharpe, P.J.H., and Magnuson, C.E. (1981). Non-linear regression of biological temperature-dependent rate models based on absolute reaction-rate theory. *J. Theor. Biol.* 88:719–731.

Sen, A. and Srivastava, M. (1990). *Regression Analysis: Theory, Methods, and Applications.* New York: Springer-Verlag.

Sharpe, P.J.H. and DeMichele, D.W. (1977). Reaction kinetics and poikilotherm development. *J. Theor. Biol.* 64:649–670.

Van Impe, J.F., Bernaerts, K., Geeraerd, A.H., Poschet, F., and Versyck, K.J. (2001). Modelling and prediction in an uncertain environment. In *Food Process Modelling*, Tijskens, L.M.M., Hertog, M.L.A.T.M., and Nicolai, B.M. (Eds.), Woodhead Publishing, Cambridge, U.K., pp.156–179.

Zaika, L.L., Moulden, E., Weimer, L., Phillips, J.G., and Buchanan, R.L. (1994). Model for the combined effects of temperature, initial pH, sodium chloride and sodium nitrite concentrations on anaerobic growth of *Shigella flexneri. Int. J. Food Microbiol.* 23: 345–358.

5 Challenge of Food and the Environment

Tim Brocklehurst

CONTENTS

5.1 ROLE OF FOOD HETEROGENEITY

Foods are typically not homogeneous. The structure of the food creates local chemical or physical environments that affect the spatial distribution of microorganisms

0-8493-1237-X/04/$0.00+$1.50
© 2004

TABLE 5.1
Examples of the Heterogeneity of Foods

Structure of the Food	Examples of Food	Model Experimental Systems Used to Mimic This Food Structure
Liquid	Soups, juices (with some suspended material)	Broth culture medium
Gel	Pate, jellies, skimmed milk cheeses, such as cottage cheese	Cells immobilized in agar or gelatin (including in a specifically designed Gel Cassette System)
Oil-in-water emulsion	Dairy cream, milk, salad cream, mayonnaise	Alkane:culture medium emulsions
Water-in-oil emulsion	Butter, margarine, low fat spread	Culture medium:alkane emulsions
Gelled emulsion	Whole milk cheese	Alkane:culture medium emulsions, where the aqueous phase is gelled with agarose
Surface	Vegetable tissues, meat tissues	Agar or gelatin (including a modified version of the Gel Cassette System)

as well as their survival and growth.[197] Microorganisms occupy the aqueous phase of foods, and structural features of this phase (Table 5.1) relevant to the length scale of microorganisms can influence their growth. The effects of these structural features on microbial growth include constraints on the mechanical distribution of water,[77,78] the redistribution of organic acids, including those used as food preservatives,[31,32] and constraints on the mobility of microorganisms.[30,60,61,109,110,139,152,153,201]

Many foods will contain a number of microstructural features, and the behavior of microorganisms is influenced differently in each. For example, Parker et al.[140] described the effect of microstructure on the distribution and growth of microorganisms in Serra cheese. Some growth occurred in liquid regions, while other microorganisms formed colonies on surfaces and within the protein gel of the curd (Figure 5.1). Predictions based on data obtained from broth systems can be applied successfully to organisms growing in structured foods. However, where the structure of the food results in a different behavior, this is described below, together with model experimental systems for its study. In many cases growth is "fail-safe," in that organisms grow more slowly in structured systems than in broths. Wilson et al.[197] suggested that this may explain the differences that food manufacturers sometimes observe, where challenge testing of real foods indicates growth at a slower rate than suggested from predictive models. Additionally, the complexity of food structure has been identified as a major contribution to the "overall error" included in microbiological modeling predictions.[144]

5.1.1 AQUEOUS PHASE

Growth in a liquid aqueous phase is typically planktonic, with motility allowing taxis to preferred regions of the food. Diffusive transport of nutrients to

a)

b)

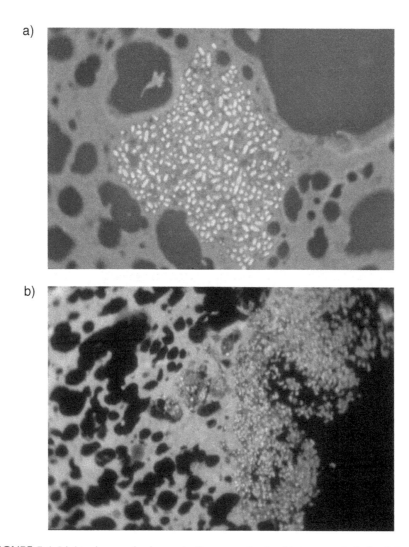

FIGURE 5.1 Light micrographs demonstrating some structural heterogeneity in hard cheese, and showing (a) a colony embedded within the gelled protein of the cheese curd and (b) a colony growing on the surface. The black irregular shapes are embedded globules of milk fat.

microorganisms and of their metabolites away can result in a locally stable equilibrium environment until accumulation of microbial biomass and metabolites cause bulk chemical changes. This is typically manifested by changes in pH or in gaseous composition. When broth culture medium is used in microbiological experiments it is this environment that is mimicked, and, with few exceptions, models for bacterial growth and death have been developed in such simple broth systems. The complexity of foods has been recognized for many years, and it has been suggested that the development of detailed models to account for all aspects of microbial growth in foods may be too costly, and will not yield useful intermediate models.[14] Simplifying assumptions can be made, and models derived in this way have proved useful.[14]

However, with improving knowledge and the advent of mechanistic modeling approaches it is possible to make predictions of the behavior of microorganisms in structured foods.

5.1.2 GELLED AQUEOUS PHASE

In gelled regions microorganisms are immobilized. This can occur as single isolated cells, or when these multiply, they are constrained to grow as colonies.[60,61,84,88,140,201]

Model experimental systems for studying colonial growth include agar[17,124] and gelatin in a specifically designed Gel Cassette System.[28] Immobilized growth as colonies results in local depletion of oxygen[200,201] and local accumulation of end-products of metabolism, which results in a local decrease in pH within and around the colonies.[104,192,201] Immobilized bacteria also differ from planktonic cultures in their susceptibility to antimicrobial compounds, their energy metabolism, and their metabolic end-products.[165,193] Accordingly, in gelled regions of foods, the growth of microorganisms will result in local changes in the concentration of their growth requirements and metabolites. This results in growth at a slower rate and to a lower yield than planktonic, or free-living cells.[30,152] A unifying theory of microbial growth, which includes proposed equations for a structured-cell mathematical model, influences of local environmental conditions on growth, influences of the microorganisms themselves on the environment, transport of solutes between phases, and physical expansion of colonies,[152] has been developed to attempt explanation of these growth characteristics.[79] Experimental data demonstrate both a decrease in growth rate and shrinkage of habitat domain in the case of *Listeria monocytogenes*, *Listeria innocua*, and *Bacillus cereus*. In all of these cases, the use of a predictive model based on data from the broth experiments would lead to a "fail-safe" prediction in the gelled system. However, Wilson et al.[197] described the growth of *Staphylococcus aureus* as a function of sucrose concentration. In the absence of sucrose, growth was slower than in the broth cultures when the cells were immobilized in gel. However, as the concentration of sucrose was increased, the growth rate in broth decreased, but remained unaffected in gel. Hence, these authors identified conditions of a concentration of sucrose above ca. 15% (w/v) at pH 6 where growth was faster in the case of cells immobilized in gel than for cells in broth (i.e., "fail-dangerous" if a model prediction was based on data from broth cultures).

Growth of cells immobilized in gelatin has been examined under nonisothermal conditions.[28] This study showed that immobilized cells differ from planktonic bacteria during temperature cycling when stressed by high salt or low pH. A finite-difference scheme has been used to combine thermal inactivation modeling with thermal conduction modeling to simulate inactivation of bacteria immobilized within agar blocks.[17]

The local accumulation of metabolic end-products within and around colonies can result in interaction between them. Such competition resulting from close spatial distribution has been termed propinquity, and occurs up to a separation distance of between 1400 and 2000 μm.[177,201] The authors of these works go on to emphasize that a gap exists between model systems and food, and that to bridge it requires the combined efforts of food microbiologists and microbial physiologists.[201]

5.1.3 OIL-IN-WATER EMULSIONS

Here, structure is affected by the concentration and form of the oil phase. The concentration of oil in food varies considerably,[32] and in milk is typically between 3 and 5% (v/v), but in mayonnaise may be between 26 and 85% (v/v). The oil phase exists as polydispersed droplets with a mean diameter that is typically between 0.15 and 8 μm. In concentrated emulsions, the space of the interstices between the droplets is of the same order of size, which is also the same order of size as many bacteria.

In model experimental systems a relationship exists between the concentration of oil and the form of growth of microorganisms.[139] Where the concentration of lipid phase was low (30% v/v) the growth of bacteria was as free-living (or planktonic) cells. An increase in the concentration of the oil phase had no effect on the form of growth of bacteria until it was increased to 83% (v/v). Here the bacteria became immobilized between the close-packed oil droplets. This entrapment resulted in growth not as planktonic cells, but as discrete colonies. The droplets within emulsions confer opacity, and hence visualization of microorganisms is difficult. A mixture of chloroform and methanol was used to selectively remove the oil phase and allow the examination of colonies *in situ*.[30,139] The investigators showed that the colonies are formed from a single bacterium, and as they expanded they displaced the emulsion droplets. Immobilization of bacteria by the lipid component and subsequent growth as colonies resulted in a decreased rate of growth and a shrinkage of the habitat domain compared with growth as planktonic cells — essentially, similar results to the consequences of colonial growth in gels.

5.1.4 WATER-IN-OIL EMULSIONS

These consist of an internal aqueous phase dispersed as discrete spherical or irregularly shaped droplets within an outer oil phase, which may contain a mixture of fluid and crystalline fats. In the case of margarines the droplets of aqueous phase are typically irregular in shape, and can range between 0.3 and 30 μ in diameter.[186]

Droplets can be contaminated with microorganisms at the point of emulsion manufacture.[186] The proportion of droplets occupied by microorganisms is small, and a model to predict microbiological contamination based on a function of the initial contamination, and the numbers of droplets exceeding the minimum size for occupancy, has been developed.[186]

Classical theories to describe microbial growth rely on the maintenance of discrete compartmentalized droplets that restrict the availability of water, space, and nutrients for growth. On the basis of these assumptions, Verrips and Zaalberg[186] and Verrips et al.[187] used a mechanistic approach to predict the growth of bacteria within discrete droplets related to the dimensions of the occupied droplets. This was expanded further by modeling the energy demands of the contained bacteria.[175] Models are useful here to predict states that are difficult to measure, and predictions confirm that bacteria in the droplets can grow well, but that their numbers remain small when expressed per unit volume of emulsion (although their local number density within a droplet is extremely high). Additionally, microorganisms cease to grow when the concentration of metabolic end-products (typically organic acids) becomes toxic or if a requirement for growth, such as oxygen or a carbon source, is exhausted. Models confirm that bacterial growth is restricted when the food structure remains intact (i.e., when coalescence of the droplets does not occur). This was observed in model experimental systems where an increase in numbers of bacteria in water-in-oil emulsions was always accompanied by coalescence of the droplets of aqueous phase.[31]

5.1.5 GELLED EMULSIONS

Many food emulsions are gelled. This can occur by the deliberate addition of gums or thickeners to increase the bulk viscosity (such as in sausages) or the denaturation of protein to form protein micelles (such as in cheese). Microorganisms are immobilized and constrained to form colonies much as in gelled systems described above.[60,61,140]

5.1.6 SURFACES

The simplest form of food structure is the surface. Growth of bacteria on the surface of food has been measured on Canadian wieners,[118] pâté,[69] and vegetable tissues.[27] Model experimental systems are numerous and include agar gels,[53,115,168,179,199,202] agar film,[115] two-dimensional gradient plates,[178–180,203,204] and a modification of the Gel Cassette mentioned above.[29]

Nicolai et al.[132] modeled surface growth with the assumption that it was in a surface film of liquid. However, growth on a surface is typically colonial. Hence, constraints on growth are similar to those described in the case of gels. Some key differences are important in modeling. Crucially, diffusion limitations are greater at a surface than within an enveloping gel. This was confirmed by Wimpenny and Coombs,[200] Peters et al.,[141] and Robinson et al.[155] who measured the depletion of oxygen and accumulation of protons immediately beneath the colony and extending into the substratum. Colonial growth on surfaces results in decreased growth rates, and comparisons of the growth rates of *Salmonella typhimurium* affected by increasing salt or sucrose followed the order: broth > immersed colonies > surface colonies.[29] This suggests that the rate of growth on surfaces may not be well predicted by models derived from broth systems.[29] Spatial distribution on a surface leads to interactions between colonies.[176] Spatial and temporal variations have a major influence on the potential of surfaces to support bacterial growth. In foods, it is particularly the availability of water.[50] Drying of a food may be deliberate to inhibit growth, and desiccation of microorganisms has been reviewed.[146] A solid surface model system was developed to study the effect of gas atmosphere on growth of several psychrotrophic pathogens.[21] This system demonstrated that increased CO_2 markedly inhibited the growth of all pathogens. The model system can be applied to examination of the growth of pathogens on minimally processed produce under modified atmospheres. Radial growth of colonies of *B. cereus* on a solid agar surface was dependent on interaction between agar concentration and water activity.[168]

5.2 MODELING THE FOOD ENVIRONMENT

In order to predict the growth of microorganisms in foods reliably, it is vital to use the correct initial chemical conditions. The structural heterogeneity of foods results in a chemical heterogeneity, which is often complicated by dynamics within the food that create a "new" chemical environment. Models of varying complexity exist that can predict the true initial chemical state of foods. Microorganisms occupy the aqueous phase of foods,[30,184] and hence, it is the chemical composition of this phase that requires accurate prediction.

Many foods rely for their preservation on the concentration of organic acids (e.g., acetic, lactic, benzoic, or sorbic acid). In addition, the concentration of sugars or salts can contribute to preservation. It is, therefore, no surprise that many predictive models use combinations of pH and water activity (although often expressed as concentration of NaCl) together with temperature as the three major determinants of growth. What follows is a summary of available models that can predict the initial environmental conditions within foods.

5.2.1 ORGANIC ACIDS

Acetic, lactic, benzoic, and sorbic acid (and their salts) are added as preservatives in many foods, although acetic and lactic acids are also produced in fermented foods as end-products of microbial metabolism. Their preservative action is by virtue of a combination of their effect on the pH of the food and the antimicrobial properties of the undissociated form of the molecule. Accordingly, their antimicrobial effect is influenced by the fundamental thermodynamic characteristics of dissociation and partition. It is these that must be modeled to predict the potential of foods to inhibit the growth of microorganisms.

5.2.2 DISSOCIATION

Weak organic acids dissociate (or separate) into their component parts. In the case of acetic acid, this occurs as:

$$CH_3COOH \quad \Leftrightarrow \quad CH_3COO^- \quad and \quad H^+$$

acetic acid	acetate	hydrogen ion
(undissociated)	(dissociated)	(proton)

This dissociation is key to prediction of the concentration of the undissociated form of the acid, which has the predominant antimicrobial effect in foods.[11,65,166]

The Henderson–Hasselbalch equation relates the pH of the food to the pK_a and the relative proportions of dissociated and undissociated acid in foods have been predicted[198] as follows:

$$pH = pK + \log_{10} \frac{[acid]_{dissociated}}{[acid]_{undissociated}} \qquad (5.1)$$

Rearrangement gives the concentration of weak acid in its undissociated (i.e., micro-biologically active) form, $[HA]_{aq}$, given the pH, pK_a, and total concentration of weak acid, $[HA]_T$, as follows:

$$[HA]_{aq} = \frac{[HA]_T}{1 + 10^{(pH - pK_a)}} \qquad (5.2)$$

where $[HA]_{aq}$ is the concentration of undissociated organic acid in the aqueous phase and $[HA]_T$ is the total concentration of organic acid. pK_a is the negative logarithm of the dissociation constant K_a, which is a thermodynamic constant controlling the dissociation equilibrium shown above:

$$pK_a = -\log(K_a) \qquad (5.3)$$

K_a is typically a small number, and published values are available.[205] pK_a varies slightly with temperature, and an empirical equation that predicts this variation has been published[154]:

$$pK_a = \left(\frac{A}{T}\right) - B + (CT) \qquad (5.4)$$

where T is the temperature in Kelvin (K), and A, B, and C are shown in Table 5.2. Such predictions are important preliminaries in dealing with the challenge of food and the environment. Without such knowledge it is quite simple to apply an incorrect initial environmental condition when using predictive microbiology tools, and this can easily result in erroneous predictions.

Predictions must also be reiterative. For example, once dissolved, the organic acid will dissociate depending upon local pH, but will then perturb this pH. The dissociation is also dependent on local buffering capacity of the food, and this is

TABLE 5.2
Values of A, B, and C to Be Inserted into Equation 5.4 for Calculation of the Effect of Temperature on pK_a[154]

Acid	A	B	C
Acetic	1170.48	3.1649	0.013399
Lactic	1286.49	4.8607	0.014776
Benzoic	1590.2	6.394	0.01765

extremely difficult to predict. However, Wilson et al.[198] developed a method for performing calculations describing the reiterative dissociation of organic acids, and hence predicting the true chemical composition of foods. This not only allows microbial growth models to predict growth, but also allows the changes in pH caused by microbial metabolism to be predicted. These authors used a theory describing the behavior of weakly dissociating systems, and knowledge of dissociation constants and concentrations. They make the point that food is too complex for solutions to be achieved through complex calculation. Hence, the authors characterized the buffering behavior of food by a titration with a strong (i.e., completely dissociating) acid, and then used knowledge of the dissociation constants of weak acid preservatives to predict the behavior of these in the food. Their calculation scheme may also be applied to a mixture of weak acids including polyacid species such as the tricarboxylic acids (e.g., citric acid).[198]

5.2.3 Partitioning into Oil Phases

In biphasic foods, which contain aqueous and lipid phases, the antimicrobial undissociated acids partition between the aqueous and lipid components.[32] This decreases the concentration of undissociated acid in the aqueous phase. Partition coefficients of acetic, lactic, and sorbic acids between sunflower oil and water have been reported as 0.02, 0.033, and 2.15, respectively,[32] demonstrating the potential for, particularly, the undissociated form of sorbic acid to decrease in the oil phase of biphasic foods.

As a complication, the pH of foods preserved using organic acids is typically in a region where weak organic acids are present in both the undissociated and the dissociated forms. Calculation of the residual concentration of the undissociated form following partition is thus difficult because the concentration is subject to the effects of partition, and to the dissociation equilibrium based on the new pH of the system and the new residual concentration of undissociated acid.

A modified form of the Henderson–Hasselbalch equation has been developed,[198] which takes these effects into account and gives the proportion of the total weak acid in a two-phase system that is present in its undissociated form in the aqueous phase, given the pH, the volume fraction of oil, and the partition coefficient for the undissociated weak acid. It was cast as:

$$\frac{[HA]_{aq}}{[HA]_T} = \frac{1}{1 + K_p\left(\dfrac{\phi}{1-\phi}\right) + 10^{(pH-pK_a)}} \tag{5.5}$$

where K_p is the partition coefficient and ϕ is the fraction volume of the oil phase. Predictions have been validated in aqueous and biphasic foods.[198]

5.2.4 Water Activity

Water activity (a_w) is a measure of the concentration of available water in a food and can be defined as the tendency of water to escape from a solution relative to its ability to escape from pure water at a specific temperature. Water activity is equal to the equilibrium relative humidity divided by 100. Pure water has an a_w of 1.000, and an environment where water is absent has an a_w of 0.000.[49,182] Most microorganisms require a high a_w for growth, and a_w is included in many predictive microbiology models. The a_w of foods can be adjusted by the addition of solutes (humectants), such as sodium chloride, sucrose, or glycerol. In some cases, the solute itself may have toxic effects, and the inhibition of growth of microorganisms when sodium chloride is used to adjust a_w can be greater than when glycerol is used, due to the toxicity of high concentrations of sodium chloride.[12,75,182] Care must be taken, therefore, to use only those predictive models that use the same humectant as the food of interest. Prediction of the initial a_w of the food can be achieved from first principles using a variety of equations, such as Raoult's law,[49,93] which was derived by Christian[49] as:

$$\mathrm{Log}_e\, a_w = \frac{-vm\phi}{55.51} \qquad (5.6)$$

where m is the molal concentration of the solute, v is the number of ions generated by each molecule of the solute, and ϕ is the molal osmotic coefficient. Commercial software to predict water activity from a list of food ingredients in a recipe is available (e.g., ERH CALC™).

5.3 HURDLE CONCEPT

Hurdle technology involves the use of combinations of physical or physicochemical preservation techniques at subinhibitory levels to control the growth of food-borne microorganisms.[98] This has the effect of conferring microbial safety and stability while maintaining acceptable nutritional and sensorial attributes,[160] an approach that is important for minimally processed extended shelf life foods.[108] With the development of new food products that depend on multiple barriers to ensure safety, it becomes necessary to develop the means to apply predictive microbiology to hurdle technology.[43,97] Careful definition of the conditions defining the boundaries of growth or survival will allow industry to design foods with the appropriate level of safety;[149,160] however, there have been few attempts to provide a quantitative assessment of hurdles.[160]

Examples of interactions include CO_2, pH, and NaCl on *L. monocytogenes*;[71] temperature, pH, citric acid, and NaCl in reduced calorie mayonnaise on *Salmonella* spp.;[121] pH, acid, and salt on *Staphylococcus aureus*;[64] salt, pH, and nitrite on *Escherichia coli* O157:H7 in pepperoni;[151] temperature and pH on *E. coli* O157:H7 in Lebanon bologna;[66] and nisin and leucosin on *L. monocytogenes*.[138]

While it is clear that combinations of hurdles can influence food-borne microorganisms, it is not clear to what extent these factors interact. When the square root model is used to describe the effect of several hurdles such as temperature, pH, and a_w, these factors are usually considered to act independently, with no interactions.[119] Ratkowsky and Ross[149] described a combined probability/kinetic model for *Shigella flexneri* in which temperature, pH, a_w, and nitrite were shown to act individually. It would be expected, however, that interactions must occur between certain hurdles. For example, interactions between organic acids and pH would be expected (due to the influence of pH on the extent of dissociation as described in Section 5.2) and have been observed.[38,121,147] Effects on heat resistance of *E. coli* due to the interactions between combinations of temperature, pH, NaCl, and sodium pyrophosphate have been modeled.[86,87]

Polynomial models can be used to describe interactions between a wide variety of hurdles. This is because the regression methods used facilitate the search for quadratic or interactive effects. Combination effects have also been modeled using Belehradek and Arrhenius models.[90,91] The growth of *L. monocytogenes* at 9°C as influenced by sodium nitrite, pH, sodium chloride, sodium lactate, and sodium acetate has been modeled,[130] and predictions compared with the growth of organisms in real sausage and predictions from Food MicroModel. Food MicroModel is a software package developed in the U.K. that contains secondary models of the effects of environmental factors (mainly pH, concentration of NaCl, and temperature) on the survival, growth, and thermal death of major food-borne pathogenic bacteria in broth. Predictions were on average within 20% of the Food MicroModel predictions based on 10 experiments although predictions of growth in sausage were, on average, 16% below the observed values based on inoculation of four sausages. This is perhaps related to the effects of structure as described in Section 5.1. The effect of previous growth temperature, previous cell concentration, and previous pH on the lag time and specific growth rate of *Salmonella typhimurium* has been investigated using response surface models.[135–137] In all cases the previous growth history did not influence the predictions of the model.

Some authors contend that predictive models of the combined effects of temperature and water activity and the combined effects of temperature and pH suggest that the effect of the combinations on growth rate is independent.[120] However, these authors go on to state that the factors are interactive at the no-growth interface (i.e., the point where growth ceases). Such interface models quantify the probability of growth and define conditions at which the growth rate is zero or the lag time is infinite. Such new growth interface or habitat domain models have been published.[116,181] Square root models and response surface models were developed to look at the effects of interactions between dissolved carbon dioxide and water activity on the growth and lag time of *Lactobacillus sarcae*.[58] The response surface models showed the best correlation although at low water activities, predictions were illogical. Both models, however, proved to be useful in the prediction of the shelf life of meat products, and were validated by comparison with an existing model.[196] Similarly, a quadratic response surface model was built to predict the combined effects of temperature and modified gaseous atmosphere on the growth of *Yersinia enterocolitica*.[143]

Predictive models have been used to predict the response of *Listeria monocytogenes* exposed to acid, alkaline, or osmotic shock at the time of inoculation on the subsequent effects of temperature, concentration of NaCl, and pH.[47] The authors found that predictive models were unreliable, highlighting potential problems of variable conditions, but failing to consider the implications of adaptation of the organisms to osmotic or pH effects. An important development is the use of the gamma concept, which assumes that the effects of controlling variables can be multiplied and that the cardinal parameters of temperature, pH, and water activity are not a function of other variables.[196] Accordingly, these authors developed a model based on the prediction of growth rate as a function of temperature and water activity and another where growth rate was predicted as a function of temperature and pH. The two models were multiplied to produce one overall model, which was validated against new experiments. Additive interaction between inhibitors has been observed.[24] These authors used a response surface methodology to model the response of *L. monocytogenes* to a bacteriocin (curvaticine) and sodium chloride: the model showed that the combination of the two inhibitors was greater than the effect of each individually. Interactions between inhibitory compounds were also investigated[8] by using a series of secondary models[7] describing independently the effects of environmental factors.[8] The authors of the latter work then went on to show that, by taking into account interactions between environmental factors, the model decreased the frequency of fail-safe growth predictions from 13.5 to 12.1%, while the frequency of fail-dangerous no-growth predictions decreased from 16.1 to 7.1%. These findings suggest that interactions are occurring within the system, and that the models were taking them into account.[8] However, even with multiplicative models the predictions are less accurate to describe lag time and growth rate near the limits of growth of microorganisms,[7] and lag time models were particularly vulnerable to error.

Inactivation modeling is less common in response to a combination of hurdles. Death kinetics as a function of pH, storage temperature, and concentration of essential oil have been described using a quadratic function, and used to predict successfully the death of *Salmonella* in home-made salads.[89] A regression model describing the heat inactivation of *L. monocytogenes* was based on the Gompertz Equation.[48] The equation enabled separate characterization of the parameters of the shoulder, the maximum slope, and the tail. Interactive effects were then derived from the regression model. This showed that the shoulder region of the survival curve was affected by pH, and the maximum slope by temperature, fat content, and interaction of temperature and milk fat. Model validation was successful for temperatures only above 62°C, however. The combined effects of pH and ethanol on the heat inactivation of *B. cereus*, *S. typhimurium*, and *Lactobacillus delbrueckii* were modeled using a series of second-order polynomial equations to describe variations in D values resulting from changes in pH or added ethanol.[45] The heat inactivation of *B. cereus* spores was modeled using a new concept of z-value modeling using a z(pH) value,[96] where z(pH) was defined as the difference in pH from a reference pH value required to effect a 10-fold reduction in the D value. A linear relationship between the calculated z(pH) value and the lowest of the pK values of organic acids used to effect heat resistance was found. The heat resistance of *Listeria monocytogenes* in logarithmic phase cells that had been heat shocked at 42°C for 1 h and subcultures of cells that were resistant to prolonged heating has been modeled.[9] A better fit for the survivor curves was found using sigmoidal equations compared with the classical log-linear models. Comparisons between models showed that an increase of thermal tolerance was induced by sublethal heat shock or by the selection of the heat-resistant population. Both isothermal and nonisothermal heat inactivation effects on the germination and heat resistance of *B. cereus* spores have been modeled.[70] An inactivation model was developed for *Salmonella enteritidis*.[96] It modeled the response of the organism to a range of concentrations of oregano essential oil and temperatures at two pH values. Quadratic functions were then used to predict the growth of this organism in home-made salads. The inactivation kinetics of *E. coli* O157:H7 were modeled using the Baranyi model (based on a set of nonautonomous differential equations)[13] as a function of time to estimate the kinetic parameters.[164] Quadratic models were then developed with natural logarithms taken of the shoulder and death rate as a function of temperature, pH, and concentration of oregano essential oil. The predicted values from the model were validated using viable count measurements made within real salads.

Modeling spore responses (other than inactivation) is unusual. The germination kinetics of spores of proteolytic *Clostridium botulinum* 56A as a function of temperature, pH, and concentration of sodium chloride have been modeled.[46] The germination kinetics were collected and expressed as the accumulated fraction of germinated spores with time and each environmental condition, and this accumulated fraction was then described by an exponential distribution. Quadratic polynomial models were developed by regression analysis of the exponential parameter and the extent of germination as a function of the variables under study. Validation experiments confirmed that the predictions were acceptable, and in most cases were fail-safe.

5.4 COMPETITION WITH OTHER MICROORGANISMS

Existing published models include a wide range of environmental, physical, or chemical factors; however, the competitive influence of microorganisms has not yet been incorporated into them. Competition may not be an issue in many foods, since interactions would not be expected until cell numbers had reached a potential hazard or caused spoilage.[160] On the other hand, growth of *L. monocytogenes* in dairy products is influenced by the natural microflora, and interactions may be difficult to model.[33] Therefore, it has been suggested that competition must be considered in the development of predictive models.[163]

Competition between microorganisms in a solid matrix such as food depends to a large extent on proximity of colonies to each other.[201] Cells growing on surfaces generate gradients of redox potential, pH, oxygen concentration, and nutrients, which can influence the growth of neighboring colonies. This phenomenon can be observed in foods, for example, where "nests" of lactic acid bacteria in fermented sausage influence the survival of food-borne pathogens,[201] and also in dairy products where interactions between natural microflora and *L. monocytogenes* are influenced by the nature of the food matrix.[33]

A related concept is the idea of "maximum carrying capacity" of a food product,[160] in which inhibition of pathogens by other microorganisms takes place when the competing flora have reached numbers at which the environment can support no further growth. This was observed with cocultures of *L. monocytogenes* and *Carnobacterium piscicola*.[35] In this study, the maximum population density of *L. monocytogenes* was reduced by the competing lactic acid bacteria, and this was attributed to nutrient depletion. It is by no means clear to what extent competition is related to depletion of nutrients. The thermal tolerance of *S. typhimurium* was enhanced by the presence of competing microflora, and it was suggested that the presence of competitors may have influenced the pathogen to induce stationary-phase gene expression.[62]

The interaction of spoilage microorganisms has recently been quantitated by Pin and Baranyi.[142] Polynomial models were developed for a number of microorganisms, and the growth of groups of strains was compared individually and in the total mixture. This approach allowed the identification of the dominant group on the basis of its growth rate and lag time. These authors also showed that reduced growth rate could be attributed to microbial interactions. Competition from naturally occurring microflora has been documented.[94] Here, predictions of the growth of *Pseudomonas* and *Listeria* in meat were made. Predictive models worked well in predicting the growth of both organisms in decontaminated meat and in decontaminated meat inoculated with each organism, together or individually. However, the presence of naturally occurring microflora in non-decontaminated meat prevented the initiation of growth of *Listeria* and the predictive models failed.

A related aspect of interaction is that of the potential for quorum sensing between microorganisms.[101] At low inoculation concentrations, modifications to modeling approaches were necessary to take into account inoculum size variation. Modeling the effects of inoculum size stochastics, however, confirmed that the growth rate was independent of inoculum concentration but that variability occurred as the inoculum concentration decreased.[209,210]

5.4.1 INTERACTIONS BASED ON THE END-PRODUCTS OF METABOLISM OF ONE SPECIES

This is a complex modeling task, but stoichiometric modeling can be used to relate the end-products of metabolism to the inhibition of the same or an accompanying organism. It assumes a "reaction scheme," and seeks to choose the simplest representation of a system that embodies the behavior of interest.

Thus, a stoichiometric model can predict the local changes in weak acid concentration resulting from microbial growth. This must then be used to predict changes in local pH. This can be done by an empirical characterization, merely by using a titration of the growth environment with the acid of interest, and fitting a curve to these data. Alternatively a quasi-mechanistically based approach may be taken,[132] or use made of a Buffering theory[198] described in Section 5.2. An advantage of the latter is that the model may be easily applied to systems of differing buffering capacity, and can combine the effects of mixtures of weak acids. Diffusion is an integral part of such modeling, and a standard model of Fickian diffusion using published diffusion coefficients in aqueous solution is usually appropriate.

For growth in liquid systems, a cardinal growth model has been combined with cardinal pH data.[99] Cardinal models use the cardinal values (minimum, optimum, and maximum values) of the environmental factors that constrain growth. Instantaneous growth rates from this model were used in a modified Baranyi growth model,[13] together with stoichiometric parameters determined from bioreactor experiments.[197]

The change in pH from production of lactic acid was determined by use of a Buffering theory.[198] Very close agreement was found between the model and the data.

5.4.2 MIXED CULTURE

Application of stoichiometric approaches to mixed cultures also works well. Wilson et al.[197] showed the growth of a mixed culture of *Lactococcus lactis* and *Listeria innocua* in a bioreactor at pH 4.5. Predictions used cardinal model parameters,[99] and stoichiometric parameters from bioreactor experiments.[197] A Buffering theory[198] was used to predict changes in pH. Such an approach provided good prediction of both the rate and extent of growth of the two organisms. Of interest in these approaches is that a stationary phase was not incorporated into the primary growth model, but emerged from the prediction in response to the accumulation of metabolites.

Interactions resulting from the production of antimicrobial bacteriocins by lactic acid bacteria in conjunction with the inhibition resulting from production of lactic acid have been modeled.[44] These authors used a modification to logistical equations that described the combined (although not additive) effects of two or more inhibitory compounds. They then applied their findings to the inhibition of *Leuconostoc mesenteroides*. The inhibition of growth of *Enterobacter cloacae* by *Lactobacillus curvatus* resulted from the production of lactic acid by the latter, and the concomitant decrease in pH,[105] which was also inhibitory to *L. curvatus*. This interaction has been modeled using a set of first-order differential equations describing growth, consumption, and production rates for both microorganisms.[107] Parameters were obtained from pure culture studies and from the literature, and the equations were solved using a combination of analytical and numerical methods. Predictions of growth of mixed cultures used parameters from pure culture experiments, which were close to the experimental data. The models also showed that interactions occurred when the antagonistic bacterium, in this case *L. curvatus*, reached 10^8 cfu/ml.

5.5 ADAPTATION AND INJURY

5.5.1 EFFECTS OF ENVIRONMENT ON ADAPTATION

Predictive microbiology should deal with bacterial stress within populations.[6] An example is the extension of the lag time of *Listeria monocytogenes* under suboptimal conditions when the inoculum was stressed.[6] More important, considerable interest has arisen recently in the problems of adaptive responses of bacteria and in the cross-resistance that this can confer. For example, adaptation of bacteria to methods of preservation can result in survival or growth that is better than predicted if the adaptive response is ignored. Accordingly, adaptation of bacteria can lead to unsafe or spoiled food.[34] The implications of adaptation can be demonstrated by reference to the acid tolerance response (ATR). The ATR in *L. monocytogenes* has been attributed to the *de novo* synthesis of proteins (sometimes referred to as acid shock proteins) when exposed to a decrease in extracellular pH.[134] Such biochemical changes confer acid resistance on the organisms, but O'Driscoll et al. also noted that *L. monocytogenes* that had been induced to show the ATR also had an increased resistance to thermal, osmotic, and cold stresses.[134] ATR has been defined as the resistance of cells to low pH when they have been grown at moderately low pH or when exposed to a low pH for some time,[59] and is typically demonstrated in broth culture, where a pH of 4.8 to 5.0 is reported to give an optimum ATR.[56] Many foods fall into this region of pH, and, more important, many microorganisms can experience this pH transiently during food production or sanitation protocols. Adapted populations could then result.

Additionally, it is clear from the above sections that one of the key effects of food structure is the immobilization of microorganisms and their resultant growth as colonies. This results in local changes in the concentration of substrates[201] and, particularly, a local accumulation of acidic metabolic end-products leading to a decline in pH within and around the colony[104,192] with a pH gradient extending into the surrounding menstrum.[192,201] In the case of *S. typhimurium*, the pH gradient extended from the original pH 7.0 in the surrounding medium to pH 4.3 inside the colony.[192] Such a local decline in pH within the colony is greater than the change required to stimulate an ATR in *Salmonella* and other Gram-negative enteric bacteria[95] and in *L. monocytogenes*.[92] It is conceivable, therefore, that cells of food-borne pathogenic bacteria immobilized as colonies embedded in a food matrix may undergo a self-induced ATR stimulated by a localized pH that has declined by virtue of the colony's own metabolic processes. It is known that acid shock proteins are synthesized and exported from cells experiencing adaptation in broths. Should this also be the case in colonies, it would result in cells within the colony becoming acid tolerant.

Despite the importance of adaptation in food microbiology, attempts to model it are rare. Authors have acknowledged that organisms behaved differently when exposed to changes in pH or sodium chloride concentration, and that exposure to these agents during exponential phase had a more dramatic effect than during the lag phase when adaptation was possibly induced.[47] However, no attempt to incorporate adaptive responses into models was made. A cross-resistance between high hydrostatic pressure and mild heat, acidity, oxidants, and osmotic stresses was demonstrated for *E. coli* O157.[20] Differences were most dramatic in stationary-phase cells; the only exception being acid resistance where differences were also apparent in the exponential phase, although, again, no attempt to incorporate these into a model was made. In one attempt to model adaptation, a model to describe the influence of temperature and the duration of preincubation on the lag time of *L. monocytogenes* was developed.[10]

5.5.2 Effects of Sublethal Injury

Subjection of bacteria to inimical processes can result in the cumulative injury of the bacteria, resulting in death. Sublethal injury is the reversible damage inflicted on bacteria that is insufficient to cause a loss of viability, and from which the bacteria can recover.[5,80,150] It is an important phenomenon to recognize when collecting data for modeling, because bacteria can often fail to form colonies on conventional selective microbiological culture medium used for their enumeration.[2,127] They can also fail to respond positively to viability stains.[26] However, the cells can remain viable and the injury can be repaired in foods, where the bacteria can then increase in numbers.[82,206] The severity of treatment that results in sublethal injury differs between species, although serotypes of *Salmonella* have been found to respond similarly to one another.[128]

5.5.2.1 Enumeration of Sublethally Injured Bacteria

A range of methods have been used to determine the extent of injury of microorganisms. These include differential plate counts on selective and nonselective agars[15,150] or on minimal and more complex media,[102] extension of the lag phase,[4,102] and changes in bioluminescence.[67] Such methods can be used to optimize both the recovery medium and the time and temperature of incubation. For example, it has been shown that cells of *L. monocytogenes* that were subjected to sublethal injury by heat exhibited a broad optimum temperature for recovery, with an optimum between 20 and 25°C, but that incubation at 2 or 5°C failed to allow repair.[103] The time taken for repair of injury to complete can be determined by measuring the time before equivalent counts are found on a selective medium (which will not support the growth of sublethally injured bacteria) and a nonselective culture medium (which will allow the growth of sublethally injured bacteria).[103] Some modeling of resuscitation has been published.[117] Predictions of response might be possible: for example, a relationship was found between the concentration of sodium chloride in the heating menstrum and its concentration in the growth medium used for the resuscitation and subsequent enumeration of *S. typhimurium*.[106]

5.6 VALIDATION IN FOODS

One of the most important aspects of model development is ensuring that predictions made by the model are applicable to real situations. This is the validation process. It should involve comparisons of the predictions of the model with observed measurements, which should be different data to those used to construct the original model. Although some predictive models have been constructed in real foods (see later in this chapter), the vast majority of models have been constructed from experiments performed in laboratory culture media (typically broth). In all cases the validation process should, ideally, include comparisons with the behavior of microorganisms in real foods or during real food processes. However, due often to cost but also other factors, validation can be done in model systems, or using previously published data. A validated model should be consistently "fail-safe," that is, predictions should fail on the side of safety (i.e., predicted growth rate and lag time should be faster and shorter, respectively, than experimental values). Predictive models can be crucial aspects of HACCP protocols. Imaginary scenarios depicting the way in which predictive models can be incorporated into HACCP concepts have been published,[122] as has a useful review of the application of predictive food microbiology in the meat industry.[113] Similarly, predictive microbiology is an important element of Quantitative Microbial Risk Assessment (QMRA). Models are useful decision support tools, but it should be remembered that models are, at best, only a simplified representation of reality. The application of model predictions should be tempered with previous experience and with knowledge of other microbial ecology principles that may be experienced in the food by the organism.[158] Sources of data and models relevant to the growth of *L. monocytogenes* in seafood and that could be part of a QMRA have been published.[158]

5.6.1 BIAS AND ACCURACY

Some criticism of the term "validation" revolves around the difficulty in quantifying just how well models perform their predictive role. Error occurs implicitly in the use of data for modeling and the use of those models for the prediction of growth of microorganisms. There are a number of potential sources of error: the homogeneity of foods; the completeness of the environmental factors used to collect the data; conversion of empirical results to a mathematical function; and fitting the models to the data.[159] For example, the overall errors in the application of growth models to the growth of *Pseudomonas* species in food and in laboratory media have been quantified.[144] The authors made the point that the error was small in the case of culture medium but great in the case of food, and went on to quantify the influence of food structure and composition on the overall error. Sutherland et al.[172] found that much of the published work on *E. coli* O157:H7 was done under conditions outside of the experimental values used to develop their growth model. These workers also reported that validation with data from cheeses and meats was difficult because the original authors often did not report experimental conditions such as NaCl content or pH. In these cases, poor predictions were often made. Similar observations were made when a growth model for *B. cereus* was being validated.[170]

It is clear in the above cases that some quantification of the deviation of the predictions from the observed values would be useful. Many measures of such quantification of error in the validation process have been made.[157] Additionally, however, Ross[157] has proposed using simple indices of the performance of models as a step towards an objective definition of the term "validated model." These indices give an indication of the confidence with which those models can be used (accuracy factor), and whether the model displays bias towards fail-dangerous predictions (bias factor). The accuracy factor is defined as:

$$\text{Accuracy factor} = 10^{\left(\Sigma\left|\log(GT_{predicted}/GT_{observed})\right|/n\right)} \tag{5.7}$$

where $GT_{predicted}$ is the predicted generation time and $GT_{observed}$ is the observed generation time, and n is the number of observations. The less accurate the predictions the larger the accuracy factor.

The bias factor is defined as:

$$\text{Bias factor} = 10^{\left(\Sigma\log(GT_{predicted}/GT_{observed})/n\right)} \tag{5.8}$$

If no disagreement between predicted and observed values occurs then the bias factor is equal to 1. However, a value of the bias factor greater than 1 indicates a fail-dangerous model because it will predict generation times longer than actually observed. It should be noted, however, that when rate values are used to compute the bias factor, a fail-dangerous model will have a bias factor of less than 1.

As mathematical techniques advance, so does the process of comparing models. The use of artificial neural networks has been identified as a useful alternative technique for modeling microbial growth. Neural networks also lend themselves to quantifying comparisons between models and suitable indices have been suggested.[85]

5.6.2 VALIDATION USING LITERATURE VALUES

The most common method of validation is the use of literature data. This is based on the assumption that if the published experiments were performed under well-defined conditions that do not differ markedly from those used to develop the model, then the model predictions should be reasonably reflected in the published data. A large number of models have been validated using published information including models for *Y. enterocolitica*,[22,100,171] *Aeromonas hydrophila*,[112] *Clostridium botulinum*,[76] *S. enteritidis*[23] and *E. coli* O157:H7,[23,173] *L. monocytogenes*,[71,111] and a number of other microorganisms.[55]

There are, however, some potentially serious limitations to the use of literature data for validation of predictive models. Additional food components are frequently responsible for deviations between predicted and observed values in validation experiments. For example, Tienungoon et al.[181] predicted the growth limits of *L. monocytogenes* as a function of temperature, pH, NaCl, and lactic acid. The authors used two strains of *L. monocytogenes*, Scott A (a pathogenic strain) and L5 (a wild-type strain isolated from cold-smoked salmon). Experiments were carried out in broth culture at a wide range of environmental conditions. Aliquots of the inoculated media were observed for a period of 90 days to determine whether the conditions supported growth. Data from the experimental program were modeled using a probability model for growth. Figure 5.2 shows the growth boundary predicted by the model for the case of no added lactic acid, and a water activity of 0.992 (representing 0.5% NaCl in a typical culture medium) as a function of temperature and pH. This boundary is plotted alongside the data from the literature (Table 5.3). Generally, the model predicted values that were in good agreement with literature values. However, where deviation from the observed measurements occurred, this was usually explained by additional identifiable preservative factors in the system, and these are described in Table 5.3.

A similar issue arises using the growth boundary model of McKellar and Lu,[116] which predicts the growth limits of *E. coli* O157:H7 as a function of temperature, pH, NaCl, sucrose, and acetic acid. These authors used five strains of *E. coli* O157:H7 growing in broth culture for a period of 72 h to determine whether the conditions supported growth. Data from the experimental program were modeled using a probability model for growth.

This boundary is plotted alongside the data from the literature (Table 5.4) in Figure 5.3. As above, the model predicted values that were in good agreement with literature values. Again, however, deviation from the observed measurements occurred, due to additional identifiable preservative factors, which are described in Table 5.4.

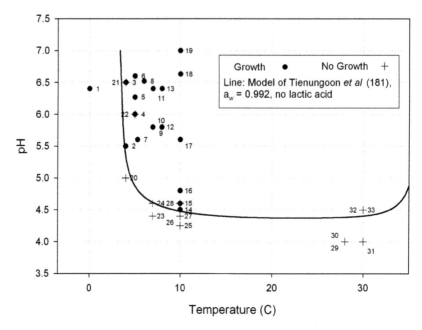

FIGURE 5.2 The growth boundary predicted by the model of Tienungoon et al.[181] for the case of no added lactic acid, and a water activity of 0.992 (representing 0.5% NaCl in a typical culture medium) as a function of temperature and pH. This boundary is plotted alongside the data from the literature described in Table 5.3.

Other sources of error associated with the use of literature data include lack of information on preincubation conditions that might result in the development of acid tolerance; use of selective media for enumerating microorganisms; lack of estimates on variability; and presence of factors in foods that are not taken into account in models (e.g., preservatives).[39] It appears that the most appropriate method for validation might be to use data derived under well-controlled conditions, so that the model's performance will not be unfairly biased.[157] Unsafe predictions and lack of published information on error also limit the usefulness of literature data, and emphasize the need to validate against new data.[57]

5.6.3 VALIDATION IN FOODS

The most common method for validating models using new data is to carry out experiments directly in the food product of concern. Thus, several models have been validated directly in food products including survival of *L. monocytogenes* in uncooked-fermented meat,[195] fishery products,[158] or pâté;[69] survival of *Campylobacter jejuni* in a variety of foods;[54] growth of *L. monocytogenes* in dairy products;[129] growth of *L. innocua* in Bologna-type sausage;[81] growth of *Staphylococcus aureus* in sterile foods;[190] growth of *E. coli* O157:H7 on raw ground beef;[188] growth of *L. monocytogenes* in sterile foods;[189] growth of *Shigella flexneri* in sterile foods;[207] growth of *E. coli* on raw displayed pork;[73] growth of *Y. enterocolitica* in seafood;[143] and growth of *Listeria* in a range of foods.[174]

TABLE 5.3
Literature Values Used in the Validation of the Growth Boundary Model Shown in Figure 5.2

Data Ref.	Temp. (°C)	pH	Other Hurdles[a]	Matrix[b]	Ref.	Obs. Time[c]
			Listeria monocytogenes — Growth Data			
1	0	6.4		Chicken broth	191	
2	4	5.5	0.5% NaCl	TSBYG	71	
3	4	6.5	4% NaCl	TSBYG	71	
4	5	6	4.5% NaCl	Tryptose phosphate broth	42	
5	5	6.27	0.05% NaCl	Minced beef	111	
6	5	6.6	0.05% NaCl	UHT milk	111	
7	5.3	5.6	0.05% NaCl	Vacuum packed lean beef	111	
8	6	6.52	0.05% NaCl	Chicken legs	111	
9	7	5.8	0.05 [0.004]% acetic acid[d]	Tryptose broth	3	
10	7	5.8	0.05% citric acid	Tryptose broth	3	
11	7	6.4	0.05% NaCl	Nonfat milk	111	
12	8	5.8	0.05% NaCl	Minced beef	111	
13	8	6.4	0.05% NaCl	Skimmed milk	111	
14	10	4.5		TSBYE	181	
15	10	4.6	Poised with citric acid	TSB	167	
16	10	4.8	Poised with lactic acid	TSB	167	
17	10	5.6		Tryptic meat broth	16	
18	10	6.63	0.277% NaCl + 170 ppm nitrite	Vacuum packed ham	111	
19	10	7	$a_w = 0.96$	Tryptic meat broth	16	
			Listeria monocytogenes — No Growth Data			
20	4	5		TSBYG	72	28 d
21	4	6.5	8% NaCl	TSBYG	71	70 d
22	5	6	4.5% NaCl + nitrite	Tryptose phosphate broth	42	NS
23	7	4.4	0.2% citric acid	Tryptose broth	3	400 h
24	7	4.6		TSBYG	72	28 d
25	10	4.25		TSBYE	181	NS
26	10	4.4		TSBYG	72	28 d
27	10	4.4	Poised with citric acid	TSB	167	28 d
28	10	4.6	Poised with lactic acid	TSB	167	28 d
29	28	4	6 [5.12]% acetic acid	BHI	40	62 d
30	28	4	9 [3.78]% lactic acid	BHI	40	62 d
31	30	4	0.029% citric acid	TSBYE	51	42 d
32	30	4.5	0.068 [0.043]% acetic acid	TSBYE	51	42 d
33	30	4.5	0.043 [0.008]% lactic acid	TSBYE	51	42 d

Note: NS = Not stated.

[a] These are responsible for the deviation of the data points from the growth boundary predicted by the model.

[b] The following matrices refer to commonly used microbiological growth media: TSBYG; Tryptose-phosphate broth; Tryptose broth; TSBYE; TSB; Tryptic meat broth; BHI.

[c] Time for which no growth was observed.

[d] Concentration of acetic and lactic acids expressed as total, with undissociated in square brackets.

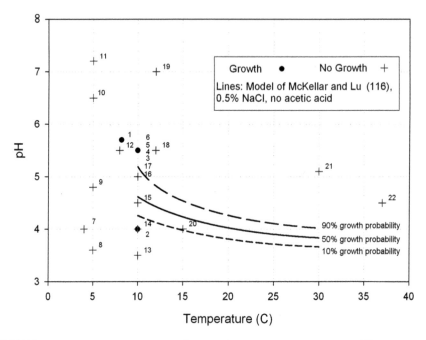

FIGURE 5.3 The growth boundary predicted by the model of McKellar and Lu[116] that predicts the growth limits of *Escherichia coli* O157:H7 as a function of temperature, pH, NaCl, sucrose, and acetic acid. This boundary is plotted alongside the data from the literature described in Table 5.4.

Dynamic modeling has also been validated,[25] where predictions from FoodMicroModel have been applied to the growth of *L. monocytogenes* and *Salmonella* in a range of foods incubated under constant as well as fluctuating temperatures. The authors found that generally the accuracy of prediction under the fluctuating temperatures was similar to the isothermal conditions, although inhibition by natural microflora did decrease the expected growth of *L. monocytogenes* in milk. Significant deviation of predictions of the growth of bacteria growing as colonies when immobilized in gel occurred when predictions were made from isothermal growth in broth.[125,126]

Validation of combined growth of the spoilage bacteria *Pseudomonas*, *Shewanella putrefaciens*, *Brochothrix thermosphacta*, and lactic acid bacteria was made in modified atmosphere packaged fish as a function of temperature and concentration of carbon dioxide.[91] Combined models based on polynomial, Belehradek, and Arrhenius equations were developed and validated by comparison with experimental growth rates of these bacteria obtained on three Mediterranean fish species. Predictions of the models based on the Belehradek and Arrhenius equations were judged satisfactory overall. This approach has been modified[90] to determine a procedure for modeling the shelf life of fish. Similarly, a quadratic response surface model has been used to describe the maximum specific growth rate of *Y. enterocolitica*. The model predicted growth rates as a function of refrigeration temperature and

TABLE 5.4
Literature Values Used in the Validation of the Growth Boundary Model Shown in Figure 5.3

Data Ref.	Temp. (°C)	pH	Other Hurdles[a]	Matrix[b]	Ref.	Obs. Time[c]
			E. coli — Growth Data			
1	8.2	5.7		Ground mutton	83	
2	10	4	0.5% NaCl	TSB	116	
3	10	5.5	Poised with lactic acid	TSBYE	52	
4	10	5.5	Poised with citric acid	TSBYE	52	
5	10	5.5	0.5% NaCl	BHI	37	
6	10	5.5	5% NaCl	BHI	41	
			E. coli — No Growth Data			
7	4	4		TSBYE	52	21 d
8	5	3.6		"Condiments"	183	7 d
9	5	4.8	$a_w = 0.99$	TSB	156	48 h
10	5	6.5	5% NaCl	BHI	41	12 d
11	5	≈7.2		Cucumber slices	1	10 d
12	8	5.5	0.5% NaCl	BHI	37	10 d
13	10	3.5	0.5% NaCl	TSB	116	72 h
14	10	4	Poised with acetic acid	TSBYE	52	21 d
15	10	4.5	5% NaCl	BHI	41	12 d
16	10	5	Poised with lactic acid	TSBYE	52	21 d
17	10	5	Poised with citric acid	TSBYE	52	21 d
18	12	5.5	30% sucrose, $a_w = 0.972$	BHI	36	24 h
19	12	≈7		Shredded carrot	1	10 d
20	15	4	0.5% NaCl	TSB	116	72 h
21	30	5.1	0.1 [0.03]% acetic acid as vinegar[d]	Nutrient agar	68	4 d
22	37	4.5	Poised with lactic acid	TSBYE	74	14 d

Note: NS = Not stated.

[a] These are responsible for the deviation of the data points from the growth boundary predicted by the model.

[b] The following matrices refer to commonly used microbiological growth media: TSBYE; TSB; BHI; Nutrient agar.

[c] Time for which no growth was observed.

[d] Concentration of acetic acid expressed as total, with undissociated in square brackets.

modified atmosphere and comparisons of the model predictions were made with growth rates obtained in seafood deliberately inoculated with *Y. enterocolitica*.[17]

Validations of the growth of *L. monocytogenes* in tryptose phosphate broth and in chicken and in beef have been made as a function of changing the pH and sodium chloride concentration.[133] Predictions of the growth of *L. monocytogenes* were then made using either a square root model[148] or a response surface polynomial model.[42] The square root model predicted growth rates at between 0 and 25°C with a coefficient of determination of between 98.36 and 99.63%. The response surface polynomial model, however, predicted generation times at 5 to 25°C with between 0 and 17.4% difference between the observed and expected generation times in broth. Of greater significance in terms of validation in food here are the large differences observed in the generation time at pH 5.6 and 8°C (25.5 h) and the generation time predicted by the Pathogen Modeling Program (PMP) in these conditions in tryptose phosphate broth (5.3 h). The PMP is a web-based package developed in the U.S. that contains secondary models of the effects of environmental factors (mainly pH, concentration of NaCl, and temperature) on the survival, growth, and inactivation of major food-borne pathogenic bacteria in broth. A divergence from predicted values was also shown at temperatures between 0 and 3.5°C in the square root model.

Predictions of the growth of *Bacillus cereus* from PMP were validated for its growth from spores in boiled rice.[114] An analysis of variance showed that there was no statistically significant difference between the observed and measured growth rates in boiled rice and predictions made from PMP. Modeled predictions were failsafe for generation time and exponential growth rate at all temperatures. Although the model was fail-safe for lag phase duration at 20 and 30°C, it was not at 15°C.

Modeling the growth of filamentous fungi is rare. The growth of three strains of heat-resistant fungi, as influenced by water activity adjusted using sucrose was modeled using the Baranyi model[13] to fit the changing colony diameter.[185] Modeling the growth of filamentous fungi has also been done using a model derived from the cardinal model family. The model was successfully fitted on data sets from a range of filamentous fungi whose growth was affected by a range of humectants including sodium chloride, glucose/fructose as a mixture, and glycerol and at different pH values. Further cardinal values were extracted from the literature and the model was used to predict the evolution of the radial growth of *Penicillium rocqueforti* and *Paecilomyces variotii*.[162]

In spite of the effort expended to develop and validate models, it is rare to find a model developed in broth that accurately predicts behavior in food systems. Models tend to fail-safe, and provide somewhat conservative predictions.[23,54,73,81,100,114,129,170,172] Indeed, the use of faster-growing strains has been suggested to provide a margin of safety.[123,131,194] Although many validations of models show that there is a fail-safe tendency and hence a margin of safety in growth prediction, some manufacturers of foods find that the error is unacceptable and the margin of safety provided by such models may well be more conservative than is desirable for many food applications. There are, however, examples of situations where the model makes what are clearly unsafe predictions, and these usually involve an overestimation of the extent of lag time.[69,188–190]

An alternative approach is to develop models directly in food products. This is not possible in many cases, due to the requirement of appropriate facilities for incorporating pathogens into the process under carefully controlled conditions. In spite of this limitation, models have been developed for growth of *L. monocytogenes* on vacuum-packed cooked meats[63] and liver pâté;[69] inactivation of *Salmonella typhimurium* in reduced calorie mayonnaise;[121] inactivation of *Enterobacteriaceae* and clostridia[18] and growth of *Lactobacillus* spp. in dry fermented sausage;[19] growth of

Clostridium botulinum in processed cheese;[169] thermal inactivation of *L. monocytogenes*[145] and *Enterococcus faecium*[161] during high-speed short-time pasteurization; and the thermal inactivation of *E. faecium* during cooking of Bologna sausage.[208] These models generally provide good estimates of the behavior of foodborne pathogens in food processes. However, it is questionable if effort should be expended developing models specific for all food processes. Improved validation techniques for models derived in broth or other model systems would appear to have more general applicability.

It has been suggested that models should only be regarded as first estimates of the behavior of pathogens, and that additional studies with products giving poor predictions should be undertaken.[195] Inclusion of additional data into models will often improve their predictive ability[207]; however, it is important that users of these models take great care in their use, and ensure that predictions are carefully validated in any product of concern.

REFERENCES

1. Abdulraouf, U.M., L.R. Beuchat, and M.S. Ammar. 1993. Survival and growth of *Escherichia coli* O157-H7 on salad vegetables. *Appl. Environ. Microbiol.* 59:1999–2006.
2. Abiss, J.S. 1983. Injury and resuscitation of microbes with reference to food microbiology. *Irish J. Food Sci. Technol.* 7:69–81.
3. Ahamad, N. and E.H. Marth. 1989. Behavior of *Listeria monocytogenes* at 7, 13, 21, and 35°C in tryptose broth acidified with acetic, citric, or lactic-acid. *J. Food Prot.* 52:688–695.
4. Alexandrou, O., C.D.W. Blackburn, and M.R. Adams. 1995. Capacitance measurement to assess acid-induced injury to *S. enteritidis* PT4. *Int. J. Food Microbiol.* 27(1):27–36.
5. Andrews, W.H. 1986. Resuscitation of injured *Salmonella* spp. and coliforms from foods. *J. Food Prot.* 49:62–75.
6. Augustin, J.C., A. Brouillaud-Delattre, L. Rosso, and V. Carlier. 2000. Significance of inoculum size in the lag time of *Listeria monocytogenes*. *Appl. Environ. Microbiol.* 66:1706–1710.
7. Augustin, J.C. and V. Carlier. 2000. Mathematical modelling of the growth rate and lag time for *Listeria monocytogenes*. *Int. J. Food Microbiol.* 56:29–51.
8. Augustin, J.C. and V. Carlier. 2000. Modelling the growth rate of *Listeria monocytogenes* with a multiplicative type model including interactions between environmental factors. *Int. J. Food Microbiol.* 56:53–70.
9. Augustin, J.C., V. Carlier, and J. Rozier. 1998. Mathematical modelling of the heat resistance of *Listeria monocytogenes*. *J. Appl. Microbiol.* 84:185–191.
10. Augustin, J.C., L. Rosso, and V. Carlier. 2000. A model describing the effect of temperature history on lag time for *Listeria monocytogenes*. *Int. J. Food Microbiol.* 57:169–181.
11. Baird-Parker, A.C. 1980. Organic acids. In *Microbial Ecology of Foods, Vol. 1: Factors Affecting Life and Death of Micro-Organisms*. Academic Press, New York, pp. 126–135. International Commission on Microbiological Specifications for Foods.

12. Baird-Parker, A.C. and B. Freame. 1967. Combined effect of water activity, pH and temperature on the growth of *Clostridium botulinum* from spore and vegetative cell inocula. *J. Appl. Bacteriol.* 30:420–429.

13. Baranyi, J. and T.A. Roberts. 1994. A dynamic approach to predicting bacterial growth in food. *Int. J. Food Microbiol.*, 23:277–294.

14. Baranyi, J. and T.A. Roberts. 1995. Mathematics of predictive food microbiology. *Int. J. Food Microbiol.* 26:199–218.

15. Barrell, R.A.E. 1988. The survival and recovery of *Salmonella typhimurium* phage type U285 in frozen meats and tryptone soya yeast extract broth. *Int. J. Food Microbiol.* 6:309–316.

16. Begot, C., I. Lebert, and A. Lebert. 1997. Variability of the response of 66 *Listeria monocytogenes* and *Listeria innocua* strains to different growth conditions. *Food Microbiol.* 14:403–412.

17. Bellara, S.R., P.J. Fryer, C.M. McFarlane, C.R. Thomas, P.M. Hocking, and B.M. Mackey. 1999. Visualization and modelling of the thermal inactivation of bacteria in a model food. *Appl. Environ. Microbiol.* 65:3095–3099.

18. Bello, J. and M.A. Sanchezfuertes. 1995. Application of a mathematical model for the inhibition of Enterobacteriaceae and clostridia during a sausage curing process. *J. Food Prot.* 58:1345–1350.

19. Bello, J. and M.A. Sanchezfuertes. 1995. Application of a mathematical model to describe the behaviour of the *Lactobacillus* spp. during the ripening of a Spanish dry fermented sausage (chorizo). *Int. J. Food Microbiol.* 27:215–227.

20. Benito, A., G. Ventoura, M. Casadei, T. Robinson, and B. Mackey. 1999. Variation in resistance of natural isolates of *Escherichia coli* O157 to high hydrostatic pressure, mild heat, and other stresses. *Appl. Environ. Microbiol.* 65:1564–1569.

21. Bennik, M.H.J., E.J. Smid, F.M. Rombouts, and L.G.M. Gorris. 1995. Growth of psychrotrophic foodborne pathogens in a solid surface model system under the influence of carbon dioxide and oxygen. *Food Microbiol.* 12:509–519.

22. Bhaduri, S., C.O. Turnerjones, R.L. Buchanan, and J.G. Phillips. 1994. Response surface model of the effect of pH, sodium chloride and sodium nitrite on growth of *Yersinia enterocolitica* at low temperatures. *Int. J. Food Microbiol.* 23:333–343.

23. Blackburn, C.D., L.M. Curtis, L. Humpheson, C. Billon, and P.J. McClure. 1997. Development of thermal inactivation models for *Salmonella enteritidis* and *Escherichia coli* O157:H7 with temperature, pH and NaCl as controlling factors. *Int. J. Food Microbiol.* 38:31–44.

24. Bouttefroy, A., M. Linder, and J.B. Milliere. 2000. Predictive models of the combined effects of curvaticin 13, NaCl and on the behaviour of *Listeria monocytogenes* ATCC 15313 in broth. *J. Appl. Microbiol.* 88; 919–929.

25. Bovill, R., J. Bew, N. Cook, M. D'Agostino, N. Wilkinson, and J. Baranyi. 2000. Predictions of growth for *Listeria monocytogenes* and *Salmonella* during fluctuating temperature. *Int. J. Food Microbiol.* 59:157–165.

26. Bovill, R.A., J.A. Shallcross, and B.M. Mackey. 1994. Comparison of the fluorescent redox dye 5-cyano-2,3-ditolyltetrazolium chloride with *p*-iodonitrotetrazolium violet to detect metabolic activity in heat-stressed *Listeria monocytogenes* cells. *J. Appl. Bacteriol.* 77(4):353–358.

27. Brocklehurst, T.F. 1994. Delicatessen salads, and chilled prepared fruit and vegetable products. In *Evaluation of Shelf Life of Foods—Principles and Practice*, Man, D. and Jones, A. (Eds.). Chapman & Hall, Glasgow, pp. 87–126.

28. Brocklehurst, T.F., G.A. Mitchell, Y.P. Ridge, R. Seale, and A.C. Smith. 1995. The effect of transient temperatures on the growth of *Salmonella typhimurium* LT2 in gelatin gel. *Int. J. Food Microbiol.* 27:45–60.

29. Brocklehurst, T.F., G.A. Mitchell, and A.C. Smith. 1997. A model experimental gel-surface for the growth of bacteria on foods. *Food Microbiol.* 14:303–311.

30. Brocklehurst, T.F., M.L. Parker, P.A. Gunning, H.P. Coleman, and M.M. Robins. 1995. Growth of food-borne pathogenic bacteria in oil-in-water emulsions. II. Effect of emulsion structure on growth parameters and form of growth. *J. Appl. Bacteriol.* 78:609–615.

31. Brocklehurst, T.F., M.L. Parker, P.A. Gunning, and M.M. Robins. 1993. Microbiology of emulsions: physicochemical aspects. *Lipid Technol.* July/August: 83–88.

32. Brocklehurst, T.F. and P.D.G. Wilson. 2000. The role of lipids in controlling microbial growth. *Grasas y Aceitas* 51:66–73.

33. Brouillaud-Delatre, A., M. Maire, C. Collette, C. Mattei, and C. Lahellec. 1997. Predictive microbiology of dairy products: influence of biological factors affecting growth of *Listeria monocytogenes*. *J. AOAC Int.* 80:913–919.

34. Brul, S. and P. Coote. 1999. Preservative agents in foods: mode of action and microbial resistance mechanisms. *Int. J. Food Microbiol.* 50:1–17.

35. Buchanan, R.L. and L.K. Bagi. 1997. Microbial competition: effect of culture conditions on the suppression of *Listeria monocytogenes* Scott A by *Carnobacterium piscicola*. *J. Food Prot.* 60:254–261.

36. Buchanan, R.L. and L.K. Bagi. 1997. Effect of water activity and humectant identity on the growth kinetics of *Escherichia coli* O157:H7. *Food Microbiol.* 14:413–423.

37. Buchanan, R.L., L.K. Bagi, R.V. Goins, and J.G. Phillips. 1993. Response-surface models for the growth-kinetics of *Escherichia coli* O157H7. *Food Microbiol.* 10:303–315.

38. Buchanan, R.L. and M.H. Golden. 1998. Interactions between pH and malic acid concentration on the inactivation of *Listeria monocytogenes*. *J. Food Safety* 18:37–48.

39. Buchanan, R.L., M.H. Golden, and J.G. Phillips. 1997. Expanded models for the non-thermal inactivation of *Listeria monocytogenes*. *J. Appl. Microbiol.* 82:567–577.

40. Buchanan, R.L., M.H. Golden, and R.C. Whiting. 1993. Differentiation of the effects of pH and lactic or acetic acid concentration on the kinetics of *Listeria monocytogenes* inactivation. *J. Food Prot.* 56:474–478 and 484.

41. Buchanan, R.L. and L.A. Klawitter. 1992. The effect of incubation-temperature, initial pH, and sodium chloride on the growth-kinetics of *Escherichia coli* O157-H7. *Food Microbiol.* 9:185–196.

42. Buchanan, R.L., H.G. Stahl, and R.C. Whiting. 1989. Effects and interactions of temperature, pH, atmosphere, sodium chloride and sodium nitrite on the growth of *Listeria monocytogenes*. *J. Food Prot.* 52:844–851.

43. Buchanan, R.L. and R.C. Whiting. 1996. Risk assessment and predictive microbiology. *J. Food Prot.* 31–36.

44. Cabo, M.L., M.A. Murado, M.P. Gonzalez, and L. Pastoriza. 2000. Dose–response relationships. A model for describing interactions, and its application to the combined effect of nisin and lactic acid on *Leuconostoc mesenteroides*. *J. Appl. Microbiol.* 88:756–763.

45. Casadei, M.A., R. Ingram, E. Hitchings, J. Archer, and J.E. Gaze. 2001. Heat resistance of *Bacillus cereus, Salmonella typhimurium* and *Lactobacillus delbrueckii* in relation to pH and ethanol. *Int. J. Food Microbiol.* 63:125–134.

46. Chea, F.P., Y.H. Chen, T.J. Montville, and D.W. Schaffner. 2000. Modeling the germination kinetics of *Clostridium botulinum* 56A spores as affected by temperature, pH, and sodium chloride. *J. Food Prot.* 63:1071–1079.
47. Cheroutre-Vialette, M. and A. Lebert. 2000. Growth of *Listeria monocytogenes* as a function of dynamic environment at 10°C and accuracy of growth predictions with available models. *Food Microbiol.* 17:83–92.
48. Chhabra, A.T., W.H. Carter, R.H. Linton, and M.A. Cousin. 1999. A predictive model to determine the effects of pH, milk fat, and temperature on thermal inactivation of *Listeria monocytogenes*. *J. Food Prot.* 62:1143–1149.
49. Christian, J.H.B. 1980. Reduced water activity. In *Microbial Ecology of Foods, Vol. 1: Factors Affecting Life and Death of Micro-Organisms*. Academic Press, New York, pp. 70–91.
50. Clayson, D.H.F. and R.M. Blood. 1957. Food perishability: the determination of the vulnerability of food surfaces to bacterial infection. *J. Sci. Food Agri.* 8:404–414.
51. Conner, D.E., V.N. Scott, and D.T. Bernard. 1990. Growth, inhibition, and survival of *Listeria monocytogenes* as affected by acidic conditions. *J. Food Prot.* 53:652–655.
52. Conner, D.E. and J.S. Kotrola. 1995. Growth and survival of *Escherichia coli* O157-H7 under acidic conditions. *Appl. Environ. Microbiol.* 61:382–385.
53. Cooper, A.L., A.C.R. Dean, and C. Hinshelwood. 1968. Factors affecting the growth of bacterial colonies on agar plates. *Proc. Royal Soc. B.* 171:175–199.
54. Curtis, L.M., M. Patrick, and C.D. Blackburn. 1995. Survival of *Campylobacter jejuni* in foods and comparison with a predictive model. *Lett. Appl. Microbiol.* 21:194–197.
55. Daughtry, G.J., K.R. Davey, and K.D. King. 1997. Temperature dependence of growth kinetics of food bacteria. *Food Microbiol.* 14:21–30.
56. Davis, M. J., P.J. Coote, and C.P. O'Byrne. 1996. Acid tolerance in *Listeria monocytogenes*: the adaptive acid tolerance response (ATR) and growth-phase-dependent acid resistance. *Microbiology (UK)* 142:2975–2982.
57. Delignettemuller, M.L., L. Rosso, and J.P. Flandrois. 1995. Accuracy of microbial growth predictions with square root and polynomial models. *Int. J. Food Microbiol.* 27:139–146.
58. Devlieghere, F., B. VanBelle, and J. Debevere. 1999. Shelf life of modified atmosphere packed cooked meat products: a predictive model. *Int. J. Food Microbiol.* 46:57–70.
59. Dilworth, M. J., A.R. Glenn, W.N. Konings, I.R. Booth, R.K. Poole, T.A. Krulwich, R.J.Rowbury, J.B. Stock, J.L. Slonczewski, G.M. Cook, E. Padan, H. Kobayashi, G.N. Bennett, A. Matin, and V. Skulachev. 1999. Problems of adverse pH and bacterial strategies to combat it. *Bacterial Response to pH*, Novartis Foundation Symposium 221. John Wiley & Sons, New York. 221:4–18.
60. Dodd, C. 1990. Detection of microbial growth in food by cryosectioning and light microscopy. *Food Sci. Tech. Today* 4(3):180–182.
61. Dodd, C. and W.M. Waites. 1991. The use of toluidine blue for *in situ* detection of micro-organisms in foods. *Lett. Appl. Microbiol.* 13:220–223.
62. Duffy, G., A. Ellison, W. Anderson, M.B. Cole, and G.S.A.B. Stewart. 1995. Use of bioluminescence to model the thermal inactivation of *Salmonella typhimurium* in the presence of a competitive microflora. *Appl. Environ. Microbiol.* 61:3463–3465.
63. Duffy, L.L., P.B. Vanderlinde, and F.H. Grau. 1994. Growth of *Listeria monocytogenes* on vacuum-packed cooked meats: effects of pH, a_w, nitrite and ascorbate. *Int. J. Food Microbiol.* 23:377–390.
64. Eifert, J.D., C.R. Hackney, M.D. Pierson, S.E. Duncan, and W.N. Eigel. 1997. Acetic, lactic, and hydrochloric acid effects on *Staphylococcus aureus* 196E growth based on a predictive model. *J. Food Sci.* 62:174–178.

65. Eklund, T. 1983. The antimicrobial effects of dissociated and undissociated sorbic acid at different pH levels. *J. Appl. Bacteriol.* 54:383–389.

66. Ellajosyula, K.R., S. Doores, E.W. Mills, R.A. Wilson, R.C. Anantheswaran, and S.J. Knabel. 1998. Destruction of *Escherichia coli* O157:H7 and *Salmonella typhimurium* in Lebanon bologna by interaction of fermentation pH, heating temperature, and time. *J. Food Prot.* 61:152–157.

67. Ellison, A., S.F. Perry, and G.S.A.B. Stewart. 1991. Bioluminescence as a real-time monitor of injury and recovery in *Salmonella typhimurium*. *Int. J. Food Microbiol.* 12:323–332.

68. Entani, E., M. Asai, S. Tsujihata, Y. Tsukamoto, and M. Ohta. 1998. Antibacterial action of vinegar against food-borne pathogenic bacteria including *Escherichia coli* O157:H7. *J. Food Prot.* 61:953–959.

69. Farber, J.M., R.C. McKellar, and W.H. Ross. 1995. Modelling the effects of various parameters on the growth of *Listeria monocytogenes* on liver pâté. *Food Microbiol.* 12:447–453.

70. Fernandez, A., M.J. Ocio, P.S. Fernandez, and A. Martinez. 2001. Effect of heat activation and inactivation conditions on germination and thermal resistance parameters of *Bacillus cereus* spores. *Int. J. Food Microbiol.* 63:257–264.

71. Fernandez, P.S., S.M. George, C.C. Sills, and M.W. Peck. 1997. Predictive model of the effect of CO_2, pH, temperature and NaCl on the growth of *Listeria monocytogenes*. *Int. J. Food Microbiol.* 37:37–45.

72. George, S.M., B.M. Lund, and T.F. Brocklehurst. 1988. The effect of pH and temperature on initiation of growth of *Listeria monocytogenes*. *Lett. Appl. Microbiol.* 6:153–156.

73. Gill, C.O., G.G. Greer, and B.D. Dilts. 1998. Predicting the growth of *Escherichia coli* on displayed pork. *Food Microbiol.* 15:235–242.

74. Glass, K.A., J.M. Loeffelholz, J.P. Ford, and M.P. Doyle. 1992. Fate of *Escherichia coli* O157/H7 as affected by pH or sodium chloride and in fermented, dry sausage. *Appl. Environ. Microbiol.* 58:2513–2516.

75. Gomez, R.F. and Herrero, A.A. 1983. Chemical preservation of foods. In *Food Microbiology,* Rose, A.H. (Ed.). Academic Press, London, pp. 78–116.

76. Graham, A.F., D.R. Mason, and M.W. Peck. 1996. A predictive model of the effect of temperature, pH and sodium chloride on growth from spores of non-proteolytic *Clostridium botulinum*. *Int. J. Food Microbiol.* 31:69–85.

77. Hills, B.P., C.E. Manning, Y.P. Ridge, and T.F. Brocklehurst. 1996. NMR water relaxation, water activity and bacterial survival in porous media *J. Sci. Food Agric.* 71:185–194.

78. Hills, B.P., C.E. Manning, Y.P. Ridge, and T.F Brocklehurst. 1997. Water availability and the survival of *Salmonella typhimurium* in porous systems. *Int. J. Food Microbiol.* 36:187–198.

79. Hills, B.P. and K.M. Wright. 1994. A new model for bacterial growth in heterogeneous systems. *J. Theor. Biol.* 168:31–41.

80. Hoffmans, C.M., D.Y.C. Fung, and C.L. Kastner. 1997. Methods and resuscitation environments for the recovery of heat-injured *Listeria monocytogenes*: a review. *J. Rapid Meth. Automat. Microbiol.* 5(4):249–268.

81. Houtsma, P.C., M.L. Kant-Muermans, F.M. Rombouts, and M.H. Zwietering. 1996. Model for the combined effects of temperature, pH, and sodium lactate on growth rates of *Listeria innocua* in broth and Bologna-type sausages. *Appl. Environ. Microbiol.* 62:1616–1622.

82. Hurst, A. 1977. Bacterial injury: a review. *Can. J. Microbiol.* 23:935–944.
83. International Commission on Microbiological Specifications for Foods. 1996. *Microorganisms in Foods, Vol. 5: Microbiological Specifications of Food Pathogens.* Blackie Academic and Professional, London.
84. Johnson, S., E. Parsons, S. Stringer, T.F. Brocklehurst, C. Dodd, M. Morgan, and W. Waites. 1996. Multixenic growth of micro-organisms in food. *Food Sci. Tech. Today* 12(1):53–56.
85. Jeyamkondan, S., D.S. Jayas, and R.A. Holley. 2001. Microbial growth modeling with artificial neural networks. *Int. J. Food Microbiol.* 64:343–354.
86. Juneja, V.K. and B.S. Eblen. 1999. Predictive thermal inactivation model for *Listeria monocytogenes* with temperature, pH, NaCl, and sodium pyrophosphate as controlling factors. *J. Food Prot.* 62:986–993.
87. Juneja, V.K., B.S. Marmer, and B.S. Eblen. 1999. Predictive model for the combined effect of temperature, pH, sodium chloride, and sodium pyrophosphate on the heat resistance of *Escherichia coli* O157:H7. *J. Food Safety* 19:147–160.
88. Katsaras, K. and L. Leistner. 1991. Distribution and development of bacterial colonies in fermented sausages. *Biofouling* 5:1145–124.
89. Koutsoumanis, K., K. Lambropoulou, and G.J.E. Nychas. 1999. A predictive model for the non-thermal inactivation of *Salmonella enteritidis* in a food model system supplemented with a natural antimicrobial. *Int. J. Food Microbiol.* 49:63–74.
90. Koutsoumanis, K. and G.J.E. Nychas. 2000. Application of a systematic experimental procedure to develop a microbial model for rapid fish shelf life predictions. *Int. J. Food Microbiol.* 60:171–184.
91. Koutsoumanis, K.P., P.S. Taoukis, E.H. Drosinos, and G.J.E. Nychas. 2000. Applicability of an Arrhenius model for the combined effect of temperature and CO_2 packaging on the spoilage microflora of fish. *Appl. Environ. Microbiol.* 66:3528–3534.
92. Kroll, R.G. and R.A. Patchett. 1992. Induced acid tolerance in *Listeria monocytogenes. Lett. Appl. Microbiol.* 14:224–227.
93. Labuza, T.P. and Bell, L.N. 1984. *Moisture Sorption. Practical Isotherm Measurement and Use,* 2nd ed. American Association of Cereal Chemists, St. Paul, MN.
94. Lebert, I., V. Robles-Olvera, and A. Lebert. 2000. Application of polynomial models to predict growth of mixed cultures of *Pseudomonas* spp. and *Listeria* in meat. *Int. J. Food Microbiol.* 61:27–39.
95. Lee, I.S., J.L. Slonczewski, and J.W. Foster. 1994. A low-pH-inducible, stationary-phase acid tolerance response in *Salmonella typhimurium. J. Bacteriol.* 176:1422–1426.
96. Leguerinel, I. and P. Mafart. 2001. Modelling the influence of pH and organic acid types on thermal inactivation of *Bacillus cereus* spores. *Int. J. Food Microbiol.* 63:29–34.
97. Leistner, L. 1992. Food preservation by combined methods. *Food Res. Int.* 25:151–158.
98. Leistner, L. and L.G.M. Gorris. 1995. Food preservation by hurdle technology. *Trends Food Sci. Tech.* 6:41–46.
99. Le Marc, Y., V. Huchet, C.M. Bourgeois, J.P. Guyonnet, P. Mafart, and D. Thuault. 2002. Modelling the growth kinetics of *Listeria* as a function of temperature, pH and organic acid concentration. *Int. J. Food Microbiol.* 73:219–237.
100. Little, C.L., M.R. Adams, W.A. Anderson, and M.B. Cole. 1994. Application of a log-logistic model to describe the survival of *Yersinia enterocolitica* at sub-optimal pH and temperature. *Int. J. Food Microbiol.* 22:63–71.
101. Llaudes, M.K., L.H. Zhao, S. Duffy, and D.W. Schaffner. 2001. Simulation and modelling of the effect of small inoculum size on time to spoilage by *Bacillus stearothermophilus. Food Microbiol.* 18:395–405.

102. Mackey, B.M. and C.M. Derrick. 1982. The effects of sub-lethal injury by heating, freezing, drying and gamma-radiation on the duration of the lag phase of *Salmonella typhimurium. J. Appl. Bacteriol.* 53:243–251.

103. Mackey, B.M., E. Boogard, C.M. Hayes, and J. Baranyi. 1994. Recovery of heat-injured *Listeria monocytogenes. Int. J. Food Microbiol.* 22(4):227–237.

104. Malakar, P., T.F. Brocklehurst, A.R. Mackie, P.D.G. Wilson, M.H. Zwietering, and K. Van't Riet. 2000. Microgradients in bacterial colonies: use of fluorescent ratio imaging, a non-invasive technique. *Int. J. Food Microbiol.* 56:71–80.

105. Malakar, P.K., D.E. Martens, M.H. Zwietering, C. Beal, and K. Van't Riet. 1999. Modelling the interactions between *Lactobacillus curvatus* and *Enterobacter cloacae.* II. Mixed cultures and shelf life predictions. *Int. J. Food Microbiol.* 51:67–79.

106. Manas, P., R. Pagan, I. Leguerinel, S. Condon, P. Mafart, and F. Sala. 2001. Effect of sodium chloride concentration on the heat resistance and recovery of *Salmonella typhimurium. Int. J. Food Microbiol.* 63:209–216.

107. Martens, D.E., C. Beal, P.K. Malakar, M.H. Zwietering, and K. Van't Riet. 1999. Modelling the interactions between *Lactobacillus curvatus* and *Enterobacter cloacae.* I. Individual growth kinetics. *Int. J. Food Microbiol.* 51:53–65.

108. Marth, E.H. 1998. Extended shelf life refrigerated foods: microbiological quality and safety. *Food Tech.* 52:57–62.

109. Mattila, T. and A.J. Frost. 1988. The growth of potential food poisoning organisms on chicken and pork muscle surfaces. *J. Appl. Bacteriol.* 65:455–461.

110. Mattila, T. and A.J. Frost. 1988. Colonization of beef and chicken muscle surfaces by *Escherichia coli. Food Microbiol.* 5:219–230.

111. McClure, P.J., A.L. Beaumont, J.P. Sutherland, and T.A. Roberts. 1997. Predictive modelling of growth of *Listeria monocytogenes:* the effects on growth of NaCl, pH, storage temperature and NaNO2. *Int. J. Food Microbiol.* 34:221–232.

112. McClure, P.J., M.B. Cole, and K.W. Davies. 1994. An example of the stages in the development of a predictive mathematical model for microbial growth: the effects of NaCl, pH and temperature on the growth of *Aeromonas hydrophila. Int. J. Food Microbiol.* 23:359–375.

113. McDonald, K. and D.W. Sun. 1999. Predictive food microbiology for the meat industry: a review. *Int. J. Food Microbiol.* 52:1–27.

114. McElroy, D.M., L.A. Jaykus, and P.M. Foegeding. 2000. Validation and analysis of modeled predictions of growth of *Bacillus cereus* spores in boiled rice. *J. Food Prot.* 63:268–272.

115. McKay, A.L. and A.C. Peters. 1995. The effect of sodium chloride concentration and pH on the growth of *Salmonella typhimurium* colonies on solid medium. *J. Appl. Bacteriol.* 79:353–359.

116. McKellar, R.C. and X.W. Lu. 2001. A probability of growth model for *Escherichia coli* O157:H7 as a function of temperature, pH, acetic acid, and salt. *J. Food Prot.* 64:1922–1928.

117. McKellar, R.C., G. Butler, and K. Stanich. 1997. Modelling the influence of temperature on the recovery of *Listeria monocytogenes* from heat injury. *Food Microbiol.* 14:617–625.

118. McKellar, R.C., R. Moir, and M. Kaleb. 1994. Factors influencing the survival and growth of *Listeria monocytogenes* on the surface of Canadian retail wieners. *J. Food Prot.* 57:387–392.

119. McMeekin, T.A., J.N. Olley, T. Ross, and D.A. Ratkowsky. 1993. *Predictive Microbiology: Theory and Application.* Wiley, New York.

120. McMeekin, T.A., K. Presser, D. Ratkowsky, T. Ross, M. Salter, and S. Tienungoon. 2000. Quantifying the hurdle concept by modelling the bacterial growth/no growth interface. *Int. J. Food Microbiol.* 55:93–98.
121. Membre, J.M., V. Majchrzak, and I. Jolly. 1997. Effects of temperature, pH, glucose, and citric acid on the inactivation of *Salmonella typhimurium* in reduced calorie mayonnaise. *J. Food Prot.* 60:1497–1501.
122. Miles, D.W. and T. Ross. 1999. Identifying and quantifying risks in the food production chain. *Food Aust.* 51:298–303.
123. Miles, D.W., T. Ross, J. Olley, and T.A. McMeekin. 1997. Development and evaluation of a predictive model for the effect of temperature and water activity on the growth rate of *Vibrio parahaemolyticus*. *Int. J. Food Microbiol.* 38:133–142.
124. Mitchell, A.J. and J.W.T. Wimpenny. 1997. The effects of agar concentration on the growth and morphology of submerged colonies of motile and non–motile bacteria. *J. Appl. Bacteriol.* 83:76–84
125. Mitchell, G.A., T.F. Brocklehurst, R. Parker, and A.C. Smith. 1994. The effect of transient temperatures on the growth of *Salmonella typhimurium* LT2. I. Cycling within the growth region. *J. Appl. Bacteriol.* 77:113–119.
126. Mitchell, G.A., T.F. Brocklehurst, R. Parker, and A.C. Smith. 1995. The effect of transient temperatures on the growth of *Salmonella typhimurium* LT2. II. Excursions outside the growth region. *J. Appl. Bacteriol.* 79:128–134.
127. Mossel, D.A.A., J.E.L. Corry, C.B. Struijk, and R.M. Baird. 1995. *Essentials of the Microbiology of Foods: A Textbook for Advanced Studies.* Wiley, Chichester, UK.
128. Morinigo, M.A., R. Cornax, D. Castro, E. Martinez-Manzanares, and J.J. Borrego. 1990. Viability of *Salmonella* spp. and indicator micro-organisms in seawater using membrane diffusion chambers. *Antonie van Leeuwenhoek* 57(2):109–118.
129. Murphy, P.M., M.C. Rea, and D. Harrington. 1996. Development of a predictive model for growth of *Listeria monocytogenes* in a skim milk medium and validation studies in a range of dairy products. *J. Appl. Bacteriol.* 80:557–564.
130. Nerbrink, E., E. Borch, H. Blom, and T. Nesbakken. 1999. A model based on absorbance data on the growth rate of *Listeria monocytogenes* and including the effects of pH, NaCl, Na-lactate and Na-acetate. *Int. J. Food Microbiol.* 47:99–109.
131. Neumeyer, K., T. Ross, and T.A. McMeekin. 1997. Development of a predictive model to describe the effects of temperature and water activity on the growth of spoilage pseudomonads. *Int. J. Food Microbiol.* 38:45–54.
132. Nicolai, B.M., J.F. van Impe, B. Verlinden, T. Martens, J. Vandewalle, and J. De Baerdemaeker. 1993. Predictive modelling of surface growth of lactic acid bacteria in vacuum-packed meat. *Food Microbiol.* 10:229–238.
133. Nyati, H. 2000. Survival characteristics and the applicability of predictive mathematical modelling to *Listeria monocytogenes* growth in sous vide products. *Int. J. Food Microbiol.* 56:123–132.
134. O'Driscoll, B., C.G.M. Gahan, and C. Hill. 1996. Adaptive acid tolerance response in *Listeria monocytogenes*: isolation of an acid-tolerant mutant which demonstrates increased virulence. *Appl. Environ. Microbiol.* 62:1693–1698.
135. Oscar, T.P. 1999. Response surface models for effects of temperature and previous growth sodium chloride on growth kinetics of *Salmonella typhimurium* on cooked chicken breast. *J. Food Prot.* 62:1470–1474.
136. Oscar, T.P. 1999. Response surface models for effects of temperature and previous temperature on lag time and specific growth rate of *Salmonella typhimurium* on cooked ground chicken breast. *J. Food Prot.* 62:1111–1114.

137. Oscar, T.P. 1999. Response surface models for effects of temperature, pH and previous growth pH on growth kinetics of *Salmonella typhimurium* in brain heart infusion broth. *J. Food Prot.* 62:106–111.
138. Parente, E., M.A. Giglio, A. Ricciardi, and F. Clementi. 1998. The combined effect of nisin, leucocin F10, pH, NaCl and EDTA on the survival of *Listeria monocytogenes* in broth. *Int. J. Food Microbiol.* 40:65–75.
139. Parker, M.L., T.F. Brocklehurst, P.A. Gunning, H.P. Coleman, and M.M. Robins. 1995. Growth of food-borne pathogenic bacteria in oil-in-water emulsions. I. Methods for investigating the form of growth of bacteria in model oil-in-water emulsions and dairy cream. *J. Appl. Bacteriol.* 78:601–608.
140. Parker, M.L., P.A. Gunning, A.C. Macedo, F.X. Malcata, and T.F. Brocklehurst. 1998. The microstructure and distribution of micro-organisms within mature Serra cheese. *J. Appl. Microbiol.* 84:523–530.
141. Peters, A.C., J.W.T. Wimpenny, and J.P. Coombs. 1987. Oxygen profiles in, and in the agar beneath, colonies of *Bacillus cereus*, *Staphylococcus albus* and *Escherichia coli. J. Gen. Microbiol.* 133:1257–1263.
142. Pin, C. and J. Baranyi. 1998. Predictive models as means to quantify the interactions of spoilage organisms. *Int. J. Food Microbiol.* 41:59–72.
143. Pin, C., J. Baranyi, and G. de Fernando. 2000. Predictive model for the growth of *Yersinia enterocolitica* under modified atmospheres. *J. Appl. Microbiol.* 88:521–530.
144. Pin, C., J.P. Sutherland, and J. Baranyi. 1999. Validating predictive models of food spoilage organisms. *J. Appl. Microbiol.* 87:491–499.
145. Piyasena, P., S. Liou, and R.C. McKellar. 1998. Predictive modelling of inactivation of *Listeria* spp. in bovine milk during HTST pasteurization. *Int. J. Food Microbiol.* 39:167–173.
146. Potts, M. 1994. Desiccation tolerance of prokaryotes. *Microbiol. Rev.* 58:755–805.
147. Presser, K.A., T. Ross, and D.A. Ratkowsky. 1998. Modelling the growth limits (growth no growth interface) of *Escherichia coli* as a function of temperature, pH, lactic acid concentration, and water activity. *Appl. Environ. Microbiol.* 64:1773–1779.
148. Ratkowsky, D.A., R.K. Lowry, T.A. McMeekin, A.N. Stokes, and R.E. Chandle. 1983. Model for bacterial culture-growth rate throughout the entire biokinetic temperature-range. *J. Bacteriol.* 154(3):1222–1226.
149. Ratkowsky, D.A. and T. Ross. 1995. Modelling the bacterial growth/no growth inter-face. *Lett. Appl. Microbiol.* 20:29–33.
150. Ray, B. 1979. Methods to detect stressed micro-organisms. *J. Food Prot.* 42:346–355.
151. Riordan, D.C., G. Duffy, J.J. Sheridan, B.S. Eblen, R.C. Whiting, I.S. Blair, and D.A. Mcdowell. 1998. Survival of *Escherichia coli* O157:H7 during the manufacture of pepperoni. *J. Food Prot.* 61:146–151.
152. Robins, M.M. and P.D.G. Wilson. 1994. Food structure and microbial growth. *Trends Food Sci. Tech.* 5:289–293.
153. Robins, M.M., T.F. Brocklehurst, and P.D.G. Wilson. 1994. Food structure and the growth of pathogenic bacteria. *Food Technol. Int. (Eur.)* 31–36.
154. Robinson, R.A. and R.H. Stokes. 1959. *Electrolyte Solutions.* Butterworths, London.
155. Robinson, T.P., J.W.T. Wimpenny, and R.G. Earnshaw. 1991. pH gradients through colonies of *Bacillus cereus* and the surrounding agar. *J. Gen. Microbiol.* 137:2885–2889.
156. Rocelle, M., S. Clavero, and L.R. Beuchat. 1996. Survival of *Escherichia coli* O157:H7 in broth and processed salami as influenced by pH, water activity, and temperature and suitability of media for its recovery. *Appl. Environ. Microbiol.* 62:8–2740.

157. Ross, T. 1996. Indices for performance evaluation of predictive models in food microbiology. *J. Appl. Bacteriol.* 81:501–508.
158. Ross, T., P. Dalgaard, and S. Tienungoon. 2000. Predictive modelling of the growth and survival of *Listeria* in fishery products. *Int. J. Food Microbiol.* 62:231–245.
159. Ross, T. and J. Olley. 1997. Problems and solutions in the application of predictive microbiology. In *Seafood Safety, Processing, and Biotechnology,* Shahidi, F., Jones, Y., and Kitts, D.D. (Eds.). Technomic Publishing Company Inc., Lancaster, PA, pp. 101–118.
160. Ross, T., T.A. McMeekin, and J. Baranyi. 1999. Predictive microbiology and food safety. In *Encyclopedia of Food Microbiology,* Robinson, R.K., Batt, C.A., and Patel, P. (Eds.). Academic Press, New York, pp. 1699–1710.
161. Ross, W.H., H. Couture, A. Hughes, T. Gleeson, and R.C. McKellar. 1998. A non-linear mixed effects model for the destruction of *Enterococcus faecium* in a high-temperature short-time pasteurizer. *Food Microbiol.* 15:567–575.
162. Rosso, L. and T.P. Robinson. 2001. A cardinal model to describe the effect of water activity on the growth of moulds. *Int. J. Food Microbiol.* 63:265–273.
163. Schaffner, D.W. and T.P. Labuza. 1997. Predictive microbiology: where are we, and where are we going? *Food Tech.* 51:95–99.
164. Skandamis, P.N. and G.J.E. Nychas. 2000. Development and evaluation of a model predicting the survival of *Escherichia coli* O157:H7 NCTC 12900 in homemade eggplant salad at various temperatures, pHs, and oregano essential oil concentrations. *Appl. Environ. Microbiol.* 66:1646–1653.
165. Skandamis, P., E. Tsigarida, and G-J.E. Nychas. 2000. Ecophysiological attributes of *Salmonella typhimurium* in liquid culture and within a gelatin gel with or without the addition of oregano essential oil. *World J. Microbiol. Biotech.* 16(1):31–35.
166. Sofos, J.N. and F.F. Busta. 1981. Antimicrobial activity of sorbate. *J. Food Prot.* 44:614–622, 647.
167. Sorrells, K.M., D.C. Enigl, and J.R. Hatfield. 1989. Effect of pH, acidulant, time, and temperature on the growth and survival of *Listeria monocytogenes. J. Food Prot.* 52:571–573.
168. Stecchini, M.L., M. Del Torre, S. Donda, and E. Maltini. 2000. Growth of *Bacillus cereus* on solid media as affected by agar, sodium chloride, and potassium sorbate. *J. Food Prot.* 63:926–929.
169. Steeg, P.F.T. and H.G.A.M. Cuppers. 1995. Growth of proteolytic *Clostridium botulinum* in process cheese products. II. Predictive modeling. *J. Food Prot.* 58:1100–1108.
170. Sutherland, J.P., A. Aherne, and A.L. Beaumont. 1996. Preparation and validation of a growth model for *Bacillus cereus*: the effects of temperature, pH, sodium chloride, and carbon dioxide. *Int. J. Food Microbiol.* 30:359–372.
171. Sutherland, J.P. and A.J. Bayliss. 1994. Predictive modelling of growth of *Yersinia enterocolitica*: the effects of temperature, pH and sodium chloride. *Int. J. Food Microbiol.* 21:197–215.
172. Sutherland, J.P., A.J. Bayliss, and D.S. Braxton. 1995. Predictive modelling of growth of *Escherichia coli* O157:H7: the effects of temperature, pH and sodium chloride. *Int. J. Food Microbiol.* 25:29–49.
173. Sutherland, J.P., A.J. Bayliss, D.S. Braxton, and A.L. Beaumont. 1997. Predictive modelling of *Escherichia coli* O157:H7: inclusion of carbon dioxide as a fourth factor in a pre-existing model. *Int. J. Food Microbiol.* 37:113–120.
174. te Giffel, M.C. and M.H. Zwietering. 1999. Validation of predictive models describing the growth of *Listeria monocytogenes. Int. J. Food Microbiol.* 46:135–149.

175. ter Steeg, P.F., F.H. Pieterman, and J.C. Hellemons. 1995. Effects of air/nitrogen, temperature and pH on energy-dependent growth and survival of *Listeria innocua* in continuous culture and water-in-oil emulsions. *Food Microbiol.* 12:471–485.

176. Thomas, L.V. and J.W.T. Wimpenny. 1993. Method for investigation of competition between bacteria as a function of three environmental factors varied simultaneously. *Appl. Environ. Microbiol.* 59:1991–1997.

177. Thomas L.V., J.W.T. Wimpenny, and G.C. Barker. 1997. Spatial interactions between subsurface bacterial colonies in a model system: a territory model describing the inhibition of *Listeria monocytogenes* by a nisin-producing lactic acid bacterium. *Microbiology (UK)* 143(Part 8): 2575–2582.

178. Thomas, L.V., J.W.T. Wimpenny, and J.G. Davis. 1993. Effect of three preservatives on the growth of *Bacillus cereus*, verocytotoxigenic *Escherichia coli* and *Staphylococcus aureus* on plates with gradients of pH and sodium chloride concentration. *Int. J. Food Microbiol.* 17:289–301.

179. Thomas, L.V., J.W.T. Wimpenny, and A.C. Peters. 1991. An investigation of four variables on the growth of *Salmonella typhimurium* using two types of gradient gel plates. *Int. J. Food Microbiol.* 14:261–275.

180. Thomas, L.V., J.W.T. Wimpenny, and A.C. Peters. 1992. Testing multiple variables on the growth of a mixed inoculum of *Salmonella* strains using gradient plates. *Int. J. Food Microbiol.* 15:165–175.

181. Tienungoon, S., D.A. Ratkowsky, T.A. McMeekin, and T. Ross. 2000. Growth limits of *Listeria monocytogenes* as a function of temperature, pH, NaCl, and lactic acid. *Appl. Environ. Microbiol.* 66:4979–4987.

182. Troller, J.A. 1983. Effect of low moisture environments on the microbial stability of foods. In *Food Microbiology,* Rose, A.H. (Ed.). Academic Press, London, pp. 173–198.

183. Tsai, Y.W. and S.C. Ingham. 1997. Survival of *Escherichia coli* O157:H7 and *Salmonella* spp. in acidic condiments. *J. Food Prot.* 60:751–755.

184. Tuynenburg Muys, G. 1971. Microbial safety in emulsions. *Process Biochem.* 6(6):25–28.

185. Valik, L. and E. Pieckova. 2001. Growth modelling of heat-resistant fungi: the effect of water activity. *Int. J. Food Microbiol.* 63:11–17.

186. Verrips, C.T. and J. Zaalberg. 1980. The intrinsic microbial stability of water-in-oil emulsions. I. Theory. *Eur. J. Appl. Microbiol. Biotech.* 10:187–196.

187. Verrips, C.T., D. Smid, and A. Kerkhof. 1980. The intrinsic microbial stability of water-in-oil emulsions. II. Experimental. *Eur. J. Appl. Microbiol. Biotech.* 10:73–85.

188. Walls, I. and V.N. Scott. 1996. Validation of predictive mathematical models describing the growth of *Escherichia coli* O157:H7 in raw ground beef. *J. Food Prot.* 59:1331–1335.

189. Walls, I. and V.N. Scott. 1997. Validation of predictive mathematical models describing the growth *of Listeria monocytogenes. J. Food Prot.* 60:1142–1145.

190. Walls, I., V.N. Scott, and D.T. Bernard. 1996. Validation of predictive mathematical models describing growth of *Staphylococcus aureus*. *J. Food Prot.* 59:11–15.

191. Walker, S.J., P. Archer, and J.G. Banks. 1990. Growth of *Listeria monocytogenes* at refrigeration temperatures. *J. Appl. Bacteriol.* 68:157–162.

192. Walker, S.L., T.F. Brocklehurst, and J.W.T. Wimpenny. 1997. The effects of growth dynamics upon pH gradient formation within and around subsurface colonies of *Salmonella typhimurium*. *J. Appl. Microbiol.* 82:610–614.

193. Walker, S.L., T.F. Brocklehurst, and J.W.T. Wimpenny. 1998. Adenylates and adenylate-energy charge in submerged and planktonic cultures of *Salmonella enteritidis* and *Salmonella typhimurium*. *Int. J. Food Microbiol.* 44:107–113.

194. Whiting, R.C. 1997. Microbial database building: what have we learned? *Food Tech.* 51:82–109.
195. Whiting, R.C. and M.O. Masana. 1994. *Listeria monocytogenes* survival model validated in simulated uncooked-fermented meat products for effects of nitrite and pH. *J. Food Sci.* 59:760–762.
196. Wijtzes, T., F.M. Rombouts, M.L.T. Kant-Muermans, K. Van't Riet, and M.H. Zwietering. 2001. Development and validation of a combined temperature, water activity, pH model for bacterial growth rate of *Lactobacillus curvatus*. *Int. J. Food Microbiol.* 63:57–64.
197. Wilson, P.D.G., T.F. Brocklehurst, S. Arino, D. Thuault, M. Jakobsen, M. Lange, J. Farkas, J.W.T. Wimpenny, and J.F. Van Impe. 2002. Modelling microbial growth in structured foods: towards a unified approach. *Int. J. Food Microbiol.* 73:275–289.
198. Wilson, P.D.G., D.R. Wilson, and C.R. Waspe. 2000. Weak acids: disassociation in complex buffering systems and partitioning into oils. *J. Sci. Food Agric.* 80:471–476.
199. Wimpenny, J.W.T. 1979. The growth and form of bacterial colonies. *J. Gen. Microbiol.* 114:483–486.
200. Wimpenny, J.W.T. and J.P. Coombs. 1983. Penetration of oxygen into bacterial colonies. *J. Gen. Microbiol.* 129:1239–1242.
201. Wimpenny, J.W.T., L. Leistner, L.V. Thomas, A.J. Mitchell, K. Katsaras, and P. Peetz. 1995. Submerged bacterial colonies within food and model systems: their growth, distribution and interactions. *Int. J. Food Microbiol.* 28:299–315.
202. Wimpenny, J.W.T. and M.W.A. Lewis. 1977. The growth and respiration of bacterial colonies. *J. Gen. Microbiol.* 103:9–18.
203. Wimpenny, J.W.T. and P. Waters. 1984. Growth of microorganisms in gel-stabilized two-dimensional gradient systems. *J. Gen. Microbiol.* 130:2921–2926.
204. Wimpenny, J.W.T. and P. Waters. 1987. The use of gel-stabilized gradient plates to map the responses of microorganisms to three or four factors varied simultaneously. *FEMS Microbiol. Lett.* 40:263–267.
205. Windholz, M. 1983. *The Merck Index.* Merck and Co Inc., Rahway, NJ.
206. Yu, L.S. and D.Y.C. Fung. 1993. Five-tube most-probable-number method using the Fung-Yu tube for enumeration of *Listeria monocytogenes* in restructured meat products during refrigerated storage. *Int. J. Food Microbiol.* 18(2):97–106.
207. Zaika, L.L., J.G. Phillips, J.S. Fanelli, and O.J. Scullen. 1998. Revised model for aerobic growth of *Shigella flexneri* to extend the validity of predictions at temperatures between 10 and 19°C. *Int. J. Food Microbiol.* 41:9–19.
208. Zanoni, B., C. Peri, C. Garzaroli, and S. Pierucci. 1997. A dynamic mathematical model of the thermal inactivation of *Enterococcus faecium* during bologna sausage cooking. *Food Sci. Technol. Lebensm. Wiss.* 30:727–734.
209. Zhao, L., T.J. Montville, and D.W. Schaffner. 2000. Inoculum size of *Clostridium botulinum* 56A spores influences time-to-detection and percent growth-positive samples. *J. Food Sci.* 65:1369–1375.
210. Zhao, L., T.J. Montville, and D.W. Schaffner. 2002. Time to detection, percent-growth-positive and maximum growth rate models for *Clostridium botulinum* 56A at multiple temperatures. *Int. J. Food Microbiol.* 77:187–197.

6 Software Programs to Increase the Utility of Predictive Microbiology Information*

Mark Tamplin, József Baranyi, and Greg Paoli

CONTENTS

6.1 INTRODUCTION

The advent of computer technology and associated advances in computational power have made it possible to perform complex mathematical calculations that otherwise would be too time-consuming for useful applications in predictive microbiology. Computer software programs provide an interface between the underlying mathematics and the user, allowing model inputs to be entered and estimates to be observed through simplified graphical outputs. Examples of model software packages that

---

* Mention of brand or firm names does not constitute an endorsement by the U.S. Department of Agriculture over others of a similar nature not mentioned.

have gained wide use in the food industry and research communities include the Pathogen Modeling Program (PMP)* and the Food MicroModel (FMM).

Behind predictive software programs are the raw data upon which the models are built. Access to these data has become important for validating the robustness of models, for bringing transparency to microbial risk assessment, and for advancing modeling techniques. Recent initiatives, such as the relational database, *ComBase*, developed by the UK Institute of Food Research, Norwich, and US Department of Agriculture-Agricultural Research Service are compiling tens of thousands of predictive microbiology data sets to describe the growth, survival, and inactivation of microorganisms, and to accelerate model development and validation. Software databases and spreadsheets have made it possible to organize large quantities of data, and to search and retrieve specific items of information.

Along with advancements in databases and microbial modeling, a growing need has developed for decision-support tools for navigating across large quantities of data and retrieving specific information. Such information management systems have been used in other scientific fields, but have not been adequately developed and applied to predictive microbiology.

In this chapter, we present an overview of the role of commonly used software applications in the field of predictive microbiology, and we provide examples of software use in model interfaces, relational databases, and expert systems. Continued development and application of software programs in this field will improve the tools that are available to researchers and risk managers for enhancing the safety and quality of the food supply.

6.2 MODEL INTERFACES

Software programs have markedly enhanced the use of microbial models by the food industry, risk assessors, and food microbiologists. Well-designed interfaces with intuitive features allow users to define parameter inputs and then easily observe model outputs in graphic formats. This section describes more widely used predictive microbiology software with demonstrated applications in food safety and quality.

6.2.1 Pathogen Modeling Program

The PMP is a free software package of microbial models that describes growth, survival, inactivation, and toxin production under various conditions defined by the user (Buchanan, 1993). The current version, 6.1, contains 37 models for 10 bacterial pathogens that predict their growth and thermal and nonthermal inactivation. In addition, the PMP contains dynamic temperature models for the growth of *Clostridium perfringens* and *Clostridium botulinum*. Such dynamic models are increasingly sought by food industries that must meet performance standards for cooked–cooled meat products. Depending on the specific model, environmental variable inputs include atmosphere (aerobic or anaerobic), temperature (°F or °C), pH, water activity,

---

* U.S. Department of Agriculture-Agricultural Research Service *Pathogen Modeling Program* (www.arserrc.gov/mfs/pathogen.htm).

ionizing radiation, varying concentrations of lactic acid, sodium chloride, nitrite, and sodium pyrophosphate, or all of these. Model outputs for lag phase duration, generation time, and time to a user-defined level of interest are displayed in either hours or days. In addition, growth/inactivation curves are displayed in both graphical and tabular formats, along with associated upper and lower confidence limits. New features in the 6.1 version include links between specific models and associated publications, separate printing of output tables and figures, and input of time–temperature data sets in comma-separated-variable (csv) format. Future versions of the PMP will include direct links to the data sets underlying individual models, optional outputs for lag and no-lag scenarios, and operating the PMP on-line using Microsoft.Net database technology.

Similar to studies found in the scientific literature, the majority of the PMP 6.1 models were developed in defined microbiological media. Consequently, the PMP informs the user that there can be no guarantee that the predicted values will match those that would occur in a specific food system. To make the PMP models more useful to the food industry, as well as to food microbiologists and risk assessors, more food-specific models are under development. Examples of PMP 6.1 models include dynamic temperature for *C. perfringens* in cured beef and cured chicken (Juneja et al., 2001; Juneja and Marks, 2002), and growth models for *Salmonella typhimurium* on chicken breast meat (Oscar, 1999) with inputs for previous growth at variable temperatures and NaCl concentrations.

For the majority of the PMP models, bacterial growth and survival are represented by the Gompertz model. Although this approach is normally satisfactory for simple sigmoidal curves, the Gompertz model lacks the desired flexibility for modeling dynamic conditions that are relevant to commercial food production, particularly for thermally processed food. To meet this demand, future versions of the PMP will incorporate the dynamic model described by Baranyi and Roberts (1994).

6.2.2 SEAFOOD SPOILAGE PREDICTOR

The Seafood Spoilage Predictor* (SSP), produced by the Danish Institute of Fisheries Research, is a predictive microbiology software package for the microbial spoilage of fisheries products. It is more versatile than other similar programs by being able to predict food spoilage with both fixed and fluctuating temperatures. Also, the SSP has two model forms, relative rate of spoilage and microbial spoilage, which estimate spoilage as a function of organoleptic change and change in microbial levels, respectively.

6.2.3 INITIATIVES LEADING TO COMBASE

The UK Institute of Food Research has produced curve-fitting softwares** that are increasingly used by researchers to develop predictive models. The PC-based Microfit and DMFit programs both use the model described by Baranyi and Roberts (1994) to fit curves to time versus colony-forming unit bacterial growth data. Microfit

---

* Danish Institute of Fisheries Research (http://www.dfu.min.dk/micro/ssp/).
** Institute of Food Research (www.ifr.bbsrc.ac.uk).

allows the user to compare the specific growth rates of different bacterial growth curves and to measure statistical significance. The DMFit program is an Excel®-based add-in feature that provides different types of curve-fitting programs and it plots the fit and the data for each data set. The DMFit software is an advantageous tool when the user anticipates incorporating the data sets into the ComBase relational database (described in the next section) because of similar data input formats.

The Food MicroModel (FMM)* was a commercial package of models with a graphical user interface, similar to the PMP (McClure et al., 1994); however, it has been discontinued in its original form. The FMM received extensive use by the food industry and offered users simple formats for inputting parameter values and obtaining model output data. It contained microbiological broth and food-specific models for pathogens and spoilage bacteria. Because of the commercial nature of the FMM, the user had limited access to its underlying models and data sets. Recently, the data behind the software have been merged with those that were generated for the PMP as well as data from international collaborators and from the published literature. The combined database (ComBase) is freely available at http://wyndmoor.arserrc.gov/combase/. The intention is to combine data and predictions in a unified Web-based database and model package called ComBase-PMP.

6.3   DATABASES

Predictive microbiology software packages are based on two main pillars: databases and mathematical models, and the scientific study of these belongs to both Bioinformatics and Biomathematics. Most chapters of this book deal with mathematical models; however, the present chapter is primarily dedicated to the other two members of this triangle (Figure 6.1).

The mutual dependence between mathematical models and databases is also confirmed by the fact that the ultimate tests for predictions are comparisons with observations. This can be done quickly and efficiently on large amounts of data only if the data-recording format is strictly standardized and harmonized with the respective mathematical variables.

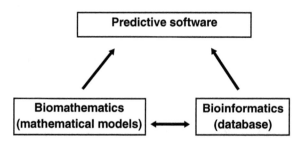

FIGURE 6.1 Relationship between predictive software, biomathematics, and bioinformatics.

---

* Leatherhead Food Research Association, UK (www.leatherheadfood.com/lfi/index.htm).

A database is a large collection of data organized especially for rapid search and retrieval. As an example, a typical, everyday-used database is a personal address book. Its computer form can be conceived as a table of rows and columns, where the rows and columns are termed *records* and *fields*, respectively. In an address book, typical fields are the name, address, and phone number. Fields belonging together (in our example, belonging to one person) form a record.

When defining the fields, one must define the "resolution" of the database according to the tasks for which it is used. A field, such as the address in our example, could be further divided into town and street, thus increasing the resolution of the record. However, this may cause problems when such an address as "Oak House, Kingston" is to be recorded, with no street name. Therefore, if there is no special reason for further resolution, it is unnecessary to subdivide the field. A valid reason for further refinement could be if it was a frequent task to list those persons who live on the same street.

Similar problems are encountered when building a database on microbial responses to the environment. For the benefit of a fast data search, those elements of information that are likely to be searched should be recorded in specified fields, in a specified format. Therefore, the first step when building such a database is to define the fields and their syntax, i.e., what values can be in a field. Typically, these values are either numeric (for quantities) or alphabetic (for categories). It is common to provide an interval for the values found in the numeric fields; similarly a list of values can be used to define the content of a category field.

Because of the complexity of the food environment and the tedious nature of food microbiology measurements, predictive food microbiology is typically a scientific field where well-organized databases are in high demand. For example:

1. The inaccuracy of measurements can be compensated by numerous measurements.
2. The variability of responses is a main focus of investigation, and quantification and modeling of variability require replicate measurements.
3. International data exchange cannot be done without defining compatible database formats.

Accordingly, the first question when building a database is: What to record? In this respect, the construction of a database is similar to that of a mathematical model; both are, in some way, the art of omitting the unnecessary. One cannot record all available information on microbiology experiments; some simplification and categorization is inevitable. The way these simplifications are carried out is sometimes arguable, even subjective, but undoubtedly necessary, just as mathematical models are necessarily simplified descriptions of nature, to understand and predict complex phenomena. The distribution of information among database categories (fields) is parallel to assigning mathematical variables to certain quantities; the relationship between those fields, for fast search and interrogation, is parallel to relationships and equations between mathematical variables for deriving conclusions and predictions in a mathematical way.

In general, producing software of microbiology data commonly involves the following stages of development:

Level 1 — Raw data: these are as recorded by the person generating the data, usually in spreadsheet format.

Level 2 — Database: these are sets of data, systematically categorized and recorded, following a predefined structure (syntax). To create a database, one needs computational skill, understanding of the data, and the ability to make expert judgments when necessary.

Level 3 — Browser: a computer program to navigate in the database.

Level 4 — Simple predictions: these are usually interpolated values given by mathematical equations, for example, the FMM or the PMP is at this level.

Level 5 — Complex predictions (dynamic conditions, growth/no growth boundary, probability of growth, Bayesian methods, expert systems): at the moment, no predictive software package is available at level 5.

A new predictive microbiology initiative called ComBase is described by Baranyi and Tamplin (2002). It aims at pooling data in a common database for predictive microbiology purposes. The database will be a source of publicly available information on bacterial responses to food environments.

One of the most important features of ComBase is that it can record not only a single value in a field but also a "pointer" to a table, representing the variable changing with time. A dynamic profile can be recorded this way, as a list of (time, value) pairs, for either an environmental factor or a response variable. ComBase is available via the Internet.

6.4 EXPERT SYSTEMS

The increased use of computer technology in food production, the widespread use of the Internet in business, and the increasing ease and speed of development of user-friendly software opens up a number of possibilities for software solutions in supporting decisions that are based on predictive microbiology. This section describes a number of ways in which decision-support software can contribute. We focus on the employment of a particular class of software programs called expert systems. These tools formally encapsulate knowledge and data and, given input through a user-interface, generate conclusions and analyses to inform the user.

6.4.1 Types of Computer-Based Decision Support

The simplest way in which computers can support decisions is to provide for intelligent storage and retrieval of data. This storage and retrieval can be easily extended to add documentation retrieval. More advanced decision-support systems can perform complex calculations given user input and then present results in appropriate formats.

Beyond data storage, data retrieval and the implementation of calculations is a class of tools that can bring a higher level of functionality and knowledge to decision support (Level 5 described above). These tools can bring more domain-specific knowledge to the decision-support function by including more complex knowledge, relationships among pieces of knowledge, and can impose a knowledge-based structure on the decision process.

Earlier in this chapter we discussed large predictive microbiology databases (e.g., ComBase) and model interfaces (e.g., PMP) that are derived from these data. Decision-support tools can be used to navigate across vast quantities of predictive microbiology information, retrieve specific information, and then to build and validate microbial models. Furthermore, decision-support tools can potentially direct the user to alternative sources of information when appropriate models and data are lacking.

An example of more complex decision support is the use of expert systems. While a number of software tools whose functionality is derived from expert knowledge may be called *expert systems*, the term is most appropriately applied to software that provides advisory information based on a set of rules and algorithms for inference based on the relationships underlying these rules.

In these systems, the core knowledge is stored as a series of IF–THEN rules that connect diverse evidence such as user input, data from databases, and the formalized opinions of experts into a web of knowledge. Software tools are available to facilitate the development of these rule-based systems, including tools to handle the user interface and creation of an executable, stand-alone software product. Recently, the capacity to create Internet-ready expert systems has been developed and integrated into these development tools to greatly facilitate the dissemination of expert knowledge in terms of both the development time and the ease of its delivery to users.

There are various examples of decision-support systems which encapsulate knowledge that is based in predictive microbiology. Examples include decision-support tools to be applied in predicting food safety and shelf life (Wijtes et al., 1998; Zwietering et al., 1992), and a step-wise system structured as a standard risk assessment process to assist in decisions regarding microbiological food safety (van Gerwen et al., 2000). Other examples of decision-support systems for microbial processes are described by Voyer and McKellar (1993) and Schellekens et al. (1994).

Expert systems provide two main forms of decision support. They impose a structure on the inference such that there is a consistent and deterministic pathway between user input, data, and conclusions. This is useful where the decision domain benefits from the assurance of consistency and where there is a desire to remove more subjective elements of decision-making from the overall decision process. For organizations dealing with recurring and complex food safety issues, consistency and reliability of decision-making may provide assurance to buyers, regulators, and the organization itself that safety will not be compromised by unstructured or ill-informed decision-making. In addition to concerns for subjective and variable decision-making, the proper structuring of the decision (in terms of delineating and weighing the necessary considerations and accessing the required knowledge to support the decision) may itself be the most difficult task for nonexpert users to carry out.

A second form of decision support provided by expert systems is to capture and make available the expert knowledge itself. The result of the conversion of expert knowledge into a set of rules that can generate advice in the absence of direct expert interaction could be a valuable asset for many institutions. Such tools may be particularly useful when provided to smaller organizations that do not have ready access to scientific expertise. The expertise that may be held by relatively few persons can be made available to many users who might otherwise compete for the attention of a few experts. In addition, time spent addressing repetitive enquiries can detract from the expert's own efforts in maintaining and enhancing his knowledge base, thus diminishing the overall generation of knowledge in an organization. With respect to food safety, a key ingredient in the process is the knowledge and decision processes that must be applied to assure a safe product. An expert system could provide assurance of the supply and the quality of this knowledge ingredient.

6.4.2 POSSIBLE APPLICATIONS OF EXPERT SYSTEMS

A number of possibilities exist for the development of decision-support systems based on rule-based expert systems. Examples (Paoli, 2001) include:

Virtual inspection: Simulating interaction with an inspector to allow establishments to self-assess their facility or a food processing operation. This may be of interest to large companies and regulators who find their quality control or inspection resources inadequate for the number of establishments that they are required to assess.

Process deviation assessment: Tools which incorporate expert knowledge regarding the best actions to take in case of process deviations, in terms of assessing the seriousness of the deviation and in recommending corrective actions for the implicated product and the process.

Problem-oriented education: Apart from providing expert advice, an expert system could be used to assist in education regarding the importance of key variables and to foster careful reasoning among operational decision-makers.

In-line real-time expert systems: Examples exist (though not in predictive microbiology) of expert systems that receive real-time data and provide continuous assessments of the status of systems, based on the combination of these data and embedded knowledge-based rules that interpret the data for display to operators. This could be applied to food production systems where a complex set of variables requires monitoring combined with complex reasoning to assure safety and quality.

6.4.3 APPLICATION TO QUALITATIVE RISK ASSESSMENT

Of international interest may be applications in the area of qualitative risk assessment. While most of the recent attention in the field of microbiological risk assessment has been paid to quantitative risk assessment (see Chapter 8), there is increasing interest in the relative merits and drawbacks of a more qualitative form of risk assessment. The merits of such an approach include the speed with which such an analysis could be performed and the decreased reliance on quantitative data which is often unavailable. Some drawbacks include the potential for qualitative risk assessment to become little more than a literature search with loose and unstructured reasoning and untraceable conclusions.

Currently, there are no formal guidelines specifying what type of analysis would constitute a properly performed qualitative risk assessment. Were such guidelines to be developed, an expert system could be produced which encapsulated and imposed an appropriate risk assessment process, as well as some of the requisite knowledge that would be required to assure an adequate qualitative risk assessment. Recent work by Crawford-Brown (2001) is an example of a decision-support tool to assist in the development of a highly structured and well-documented qualitative risk assessment. Further examples of this type of tool might be of great benefit to those who would like to capture the potential value of systematic risk assessment while avoiding the computational and data burden of quantitative approaches.

6.5 CONCLUSIONS

On the basis of the trends seen for the past ten years it can be said that the demand for predictive microbiology software programs will expand with increasing application of microbial models to food systems. This includes the use of predictive models for research and for the development of Hazard Analysis and Critical Control Points systems, product formulation, and risk assessment. Concomitantly, continued advances in database and decision-support software, as well as Internet technologies, will provide risk managers with better tools for seeking relevant information and making informed decisions. However, to sustain this effort, multiinstitutional collaborations will be critical for managing the input, organization, quality control, and dissemination of large quantities of predictive microbiology data. Ultimately, these advances will provide researchers, students, and educators with greater access to information for improving the safety and quality of the food supply.

REFERENCES

Baranyi, J. and Roberts, T. A. 1994. A dynamic approach to predicting bacterial growth in food. *Int. J. Food Microbiol.* 23:277–294.

Baranyi, J. and Tamplin, M.L. 2002. ComBase: a combined database on microbial responses to food environments. 1st International Conference on Microbial Risk Assessment: Foodborne Hazards. College Park, MD, p. 23.

Buchanan, R.L. 1993. Developing and distributing user-friendly application software. *J. Ind. Microbiol.* 12:251–255.

Crawford-Brown, D. 2001. Expert system for waterborne pathogens. Presented at workshop, "Decision-Support Tools for Microbial Risk Assessment" at the Society for Risk Analysis Annual Meeting, Seattle, WA.

Juneja, V.K., Novak, J.S., Marks, H.M., and Gombas, D.E. 2001. Growth of *Clostridium perfringens* from spore inocula in cooked cured beef: development of a predictive model. *Innovative Food Sci. Emerging Technol.* 2:289–301.

Juneja, V.K. and Marks, H.M. 2002. Predictive model for growth of *Clostridium perfringens* during cooling of cooked cured chicken. *Food Microbiol.* 19:313–3297.

McClure, P.J., Blackburn, C. de W., Cole, M.B., Curtis, P.S., Jones, J.E., Legan, J.D., Ogden, I.D., Peck, M.W., Roberts, T.A., Sutherland, J.P., and Walker, S.J. 1994. Modelling the growth, survival and death of microorganisms in foods: the UK Food Micromodel approach. *Int. J. Food Microbiol.* 23:265–275.

Oscar, T.P. 1999. Response surface models for effects of temperature, pH, and previous growth pH on growth kinetics of *Salmonella typhimurium* in brain heart infusion broth. *J. Food Prot.* 62:106–111.

Paoli, G. 2001. Prospects for expert system support in the application of predictive microbiology. Presented at workshop, "Decision-Support Tools for Microbial Risk Assessment" at the Society for Risk Analysis Annual Meeting, Seattle, WA.

Schellekens, M., Martens, T., Roberts, T.A., Mackey, B.M., Nicolai, M.B., Van Impe, J.F., and De Baerdemaeker, J. 1994. Computer-aided microbial safety design of food processes. *Int. J. Food Microbiol.* 24:1–9.

van Gerwen, S.J.C., te Giffel, M.C., Van't Riet, K., Beumer, R.R, and Zwietering, M.H. 2000. Stepwise quantitative risk assessment as a tool for characterization of microbiological safety. *J. Appl. Microbiol.* 88:938–951.

Voyer, R. and McKellar, R.C. 1993. MKES tools: a microbial kinetics expert system for developing and assessing food production systems. *J. Ind. Microbiol.* 12:256–262.

Wijtes, T., Van't Riet, K., in't Veld , J.H.J., and Zwietering, M.H. 1998. A decision support system for the prediction of microbial food safety and food quality. *Int. J. Food Microbiol.* 42:79–90.

Zwietering, M.H., Wijtzes, T., Wit, J.C. de, and Van't Riet, K. 1992. A decision support system for prediction of the microbial spoilage in foods. *J. Food Prot.* 55:973–979.

7 Modeling Microbial Dynamics under Time-Varying Conditions

*Kristel Bernaerts, Els Dens, Karen Vereecken,
Annemie Geeraerd, Frank Devlieghere,
Johan Debevere, and Jan F. Van Impe*

CONTENTS

7.1 INTRODUCTION

Predictive food microbiology essentially aims at the quantification of the microbial ecology in foods by means of mathematical models.[1] These models can then be used to predict food safety and shelf life, to develop and assist in safety assurance systems in the food industry (e.g., Hazard Analysis of Critical Control Points), and to establish exposure studies in the framework of risk assessment (see, e.g., References 2 to 4). Though challenge testing tends to be the common policy in the food industry, information on microbial kinetics — in food products — is increasingly consolidated into mathematical models, which may significantly reduce the number of challenge tests required to determine, for example, shelf life. In combination with predictive models for, e.g., heat transfer, and other process variables, and the initial contamination level, these models are essential building blocks in time-saving simulation studies to optimize and design processing, distribution, and storage conditions (e.g., temperature–time regimes) that guard food safety and spoilage (e.g., Reference 5).

In the early years of predictive microbiology, strong preference has been expressed towards sigmoidal functions that gave a good description of growth curves obtained under nonvarying environmental conditions. The most commonly used growth model was probably the modified Gompertz model.[6] Microbial inactivation at high temperatures — exhibiting a log-linear behavior — could be described as a first-order decay reaction (see, e.g., Reference 7). Effects of environmental conditions on these *primary* models (i.e., evolution of cell number as function of time) are embedded into *secondary* models (see Chapter 2 and Chapter 3 for more details). *Dynamic primary models* capable of (1) dealing with realistic *time-varying conditions* and (2) including the previous *history* of the food product in a natural way have been introduced since the early nineties.[8,9]

Besides the need for such dynamic models, it is also clear that real food product conditions should be taken into account during modeling (e.g., Reference 10). More (mechanistic) knowledge needs to be built into existing models such that the physiological response of microorganisms and the associated microbial dynamics can be accurately explained under fluctuating conditions. For example, reliable predictions for microbial lag phenomena and interaction are lacking nowadays.

In this chapter, the elementary building block for dynamic mathematical models describing microbial evolution is presented (see Section 7.2). Given this general expression, (mechanistic) knowledge on the microbial behavior in foods can be gradually built in to yield a generic model structure describing the microbial dynamics of interest. During this model development process, a continuous trade-off needs to be made between *model complexity* and *manageability*. On the one hand, the mathematical model should incorporate sufficient (mechanistic) knowledge in order to generate *accurate* predictions. Reliable predictions are indispensable to advocate confidence in predictive microbiology within the food industry. On the other hand, these mathematical models must remain user-friendly and computationally manageable in view of their industrial applicability.

The chapter is organized as follows. Section 7.2 introduces the general dynamic model building approach. First, this strategy is illustrated for modeling simple growth and inactivation behavior. However, accurate modeling of microbial dynamics in foods usually requires more complex model structures. In this respect, (1) the modeling of microbial lag under time-varying temperature conditions via an individual-based approach (see Section 7.3) and (2) the modeling of interspecies microbial interactions mediated by product inhibition (see Section 7.4) are discussed. At the same time, the fundamentals of microscopic (*individual-based*) and macroscopic (*population level*) modeling are revisited. Section 7.5 summarizes the general conclusions.

7.2 GENERAL DYNAMIC MODELING METHODOLOGY

The elementary dynamic model building block describing microbial dynamics under batch cultivation within a homogeneous environment consists of the following differential equation:

$$\frac{dN_i(t)}{dt} = \mu_i(N_i(t), <N_j(t)>_{i \neq j}, <env(t)>, <P(t)>, <phys(t)>, \ldots) \cdot N_i(t) \quad (7.1)$$

with $i,j = 1,2,\ldots, n$ the number of microbial species involved (analogous with Reference 11). $N_i(t)$ represents the cell density of species i and $\mu_i(\cdot)$ [h^{-1}] defines its *overall specific evolution rate* depending on interactions within and/or between microbial populations (N_i and/or N_j, respectively), physicochemical environmental conditions ($<env>$), microbial metabolite concentrations ($<P>$), the physiological state of the cells ($<phys>$), among others. Microbial proliferation is generated when $\mu_i(\cdot) > 0$ and microbial decay results from $\mu_i(\cdot) < 0$.

Observe that all influencing factors may depend on time themselves. For example, temperature may change dynamically with time, and thus acts as an *input* when solving the system of differential equations. To describe the time-dependent evolution of metabolite production and the physiological state of the cells, for example, additional *coupled* differential equations are added to the set of differential equations in 7.1. This is abundantly illustrated throughout the paper.

Within structured food systems, Expression 7.1 describes the *local* dynamic behavior of microorganisms. In such case, local inputs are needed. For example, local temperatures can be computed using heat transfer models. Furthermore, microbial dynamics shall be influenced by spatially varying substrate and nutrient concentrations (which may become restricted because of diffusion limitations). Diffusion limitations also cause spatial gradients of metabolic products. In addition, the need for a valid *transport* model for microbial growth (i.e., describing spatial colony dynamics) rises (e.g., Reference 12).

7.2.1 BASIC ELEMENTS FOR MODELING GROWTH

If environmental conditions are constant, the microbial growth curve — the (natural) logarithm of the cell density as function of time — typically exhibits a sigmoidal shape consisting of three phases: the lag phase, the exponential phase and the stationary phase (see Figure 7.1). First, the population needs to adjust to its new environment. Second, the population attains its maximum specific growth rate characteristic for the specific environment. Third, growth ceases because of, e.g., inhibitory effects of metabolites. Eventually, this leads to inactivation.

The overall specific growth rate in Expression 7.1 can be represented by three factors describing these three phases*:

---

* The dynamics of a single species are considered and the subscript i can thus be omitted.

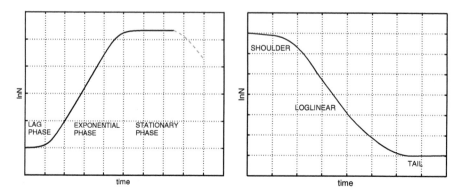

FIGURE 7.1 Left plot: Typical growth curve (full line) at constant environmental conditions. Right plot: Typical inactivation curve under mild constant processing conditions.

$$\frac{dN(t)}{dt} = \mu_{lag}(\cdot) \cdot \mu_{max}(\cdot) \cdot \mu_{stat}(\cdot) \cdot N(t) \tag{7.2}$$

During the exponential phase, the specific growth rate remains constant at μ_{max}, which is the maximum specific growth rate that can be realized within the actual environment. The dependence on environmental factors such as temperature is typically incorporated into secondary models (e.g., Reference 13). The first factor $\mu_{lag}(\cdot)$ is introduced to describe the lag behavior and thus needs to embed the gradual increase of the overall specific growth rate from 0 to μ_{max}. The third factor $\mu_{stat}(\cdot)$ induces the gradual decrease in the specific growth rate towards 0, resulting in the stationary phase.

Dynamic models in predictive microbiology are reported in, e.g., Baranyi and Roberts,[14] Baranyi et al.,[8] Hills and Mackey,[15] Hills and Wright,[16] McKellar,[17] and Van Impe et al.[9,18] A well-known dynamic model is the growth model by Baranyi and Roberts:[14]

$$\frac{dN(t)}{dt} = \left[\frac{Q(t)}{1+Q(t)}\right] \cdot \mu_{max} \cdot \left[1 - \frac{N(t)}{N_{max}}\right] \cdot N(t)$$

$$\frac{dQ(t)}{dt} = \mu_{max} \cdot Q(t) \tag{7.3}$$

Recognize the three factors in the right-hand side of the first equation as presented in Equation 7.2. The first factor, i.e., the so-called adjustment function, describes the gradual adaptation of the population to attain μ_{max}. Hereto, an additional state variable $Q(t)$ is introduced into the model [thus $\mu_{lag}(Q(t))$]. This variable denotes the physiological state of the cells that should augment until the adjustment function reaches (approximately) its maximum value, namely, 1. At that point, the exponential phase starts. The initial value of $Q(t)$ together with the maximum specific growth rate determines the lag-phase duration. Graphically, μ_{max} corresponds with the slope of the log-linear part of the growth curve.* The third factor, i.e., the so-called inhibition function, causes the growth rate to decrease asymptotically to 0 when the population density reaches its maximum level N_{max} [thus $\mu_{stat}(N(t), N_{max})$].

Environmental conditions affecting the outgrowth of microorganisms in food products are often time-varying. In such case, predictions of the food safety and the shelf life can be generated by combining a dynamic primary model with a secondary model relating the typical primary parameters with environmental conditions (e.g., $\mu_{max}(<env(t)>)$). Doing so, it is implicitly assumed that the primary parameters, e.g., the maximum specific growth rate, immediately change according to the changing environmental factors and the secondary model. Consequently, delayed responses (lag) induced by (sudden) fluctuations of the surrounding environment cannot be predicted.[19] Furthermore, the cessation of growth is a response to starvation following exhaustion of nutrients and/or the inhibition by metabolic products.[20] Description of the inhibition within mixed cultures by, e.g., product formation, cannot be consistently described when using the single model parameter N_{max} (see below).

Section 7.3 and Section 7.4 illustrate how such dynamic growth models (7.2) can be *fine-tuned* towards the modeling of microbial lag and growth inhibition. Eventually, we aim at robust mechanistically inspired models.

7.2.2 BASIC ELEMENTS FOR MODELING INACTIVATION

During mild heat treatment (at constant temperature) microbial inactivation often shows a non-log-linear behavior characterized by a delayed response (*shoulder*) and a resistant population (*tailing*) (see Figure 7.1, right plot). According to Expression 7.1, a general model structure reads as follows.**

$$\frac{dN(t)}{dt} = -k_{shoulder}(\cdot) \cdot k_{max}(\cdot) \cdot k_{tail}(\cdot) \cdot N(t) \tag{7.4}$$

To express the specific microbial inactivation rate the symbol k is commonly used.

On the basis of the mechanistic insight on the occurrence of the shoulder and tailing phenomenon,[21–24] Geeraerd et al.[25] established the following functions modeling the shoulder and tailing behavior.

$$\frac{dN(t)}{dt} = -\left[\frac{1}{1+C_c(t)}\right] \cdot k_{max} \cdot \left[1 - \frac{N_{res}}{N(t)}\right] \cdot N(t)$$

$$\frac{dC_c(t)}{dt} = -k_{max} \cdot C_c(t) \tag{7.5}$$

---

* From a mathematical point of view, the adjustment function is exactly equal to 1 only at infinity, whereas the inhibition function approximates 1 when $N(t) \ll N_{max}$. However, from a numerical point of view, both factors are 1 during a considerable part of the growth curve. Hence, it can be reasonably said that during the log-linear part μ_{max} is reached.

** Here too the dynamics of a single species are considered and the subscript i can thus be omitted.

The first factor in the right-hand side of the first equation models the shoulder of the inactivation curve. Before first-order inactivation of the population takes place (at a specific inactivation rate k_{max}), some critical protective component C_c needs to be inactivated. It is assumed that this occurs according to a first-order relationship (i.e., second differential equation in 7.5). The shoulder is obtained by applying a Michaelis–Menten-based adjustment function, namely, $(1 + C_c(t))^{-1}$ [thus $k_{shoulder}(C_c(t))$]. Starting at a low value, the adjustment function increases towards unity and, at that point, log-linear inactivation is observed. Analogous with the physiological state $Q(t)$ in the dynamic growth model 7.3, $C_c(t)$ can be interpreted as the physiological state of the population in the context of inactivation. The tailing phenomenon can be explained by some resistant subpopulation N_{res} that is unaffected during the (heat) treatment. This tailing at a residual population N_{res} is here modeled by $(1-N_{res}/N(t))$ [thus $k_{tail}(N(t), N_{res})$]. Note that this residual subpopulation is not necessarily a constant value but may vary when modeling nonthermal inactivation,[26,27] or when subjecting the microbial population to sequences of inactivation treatments.[28]

To conclude, observe that the general model structure 7.4 and model 7.5 also encompass classical log-linear inactivation. In Equation 7.5, log-linear inactivation is generated by selecting (after identification on experimental data) a very low value for $C_c(0)$ and N_{res}, implying the absence of a shoulder and a tail, respectively.

7.3 EXAMPLE I: INDIVIDUAL-BASED MODELING OF MICROBIAL LAG

Factors affecting the occurrence and extent of the commonly observed *initial* (population) lag phase (i.e., a period after inoculation during which cells adapt themselves to the new environment, see Figure 7.1) can be attributed to the past environment, the new environment, the magnitude of the environmental change, the rate of the environmental change, the growth status (e.g., exponential, stationary) of the inoculated cell culture, and the variability between individual cell lag phases. These environmental changes may involve nutritional and chemical, as well as physical changes. Obviously, environmental fluctuations during exponential growth can also cause lag (i.e., intermediate lag). Large temperature gradients, for example, applied during the exponential growth phase shall induce an intermediate lag phase observed as a transient adaptation of the growth rate.[19,29]

Secondary models describing the relation between the (population) lag-phase duration and the physicochemical environment are usually based on highly standardized experiments during which cells are grown to their stationary phase under optimal growth conditions before being transferred to the new environment, which is not – deliberately — varied upon the subsequent growth. Such mathematical models perform well under the conditions that they have been developed for. However, any deviation within the prehistory of the contaminating population may seriously alter the lag behavior.[30,31] Especially, huge deviations between model prediction and actual microbial dynamics are observed under time-varying environmental conditions.[32]

Overall, lag phenomena induced by (sudden) environmental changes are insufficiently explored, and suitable (generic) predictive models are not available. In this section, a first attempt towards a general model structure that is valid for various microorganisms and various dynamic temperature conditions — as presented in Dens[33] — is described. Opposed to the more traditional population modeling approach, basic *mechanistic knowledge* concerning the mechanism causing lag at *cell level* is embedded into the model. In particular, the theory of cell division is implemented within an *individual-based modeling* approach to enable the description of lag phases that can be induced by sudden temperature rises.

7.3.1 PRINCIPLES OF INDIVIDUAL-BASED MODELING

The fundamental unit of bacterial life is the cell, encapsulating action, information storage and processing as well as variability. It can therefore be appropriate to construct microbial models in terms of the *individual cells*.[34] This is the domain of *individual-based modeling*. The basic idea behind this approach is that, if it is possible to specify the *rules* governing the behavior of the cells, then the global multicellular behavior can be explained by the interactions between the individual cell activities. The rules constituting the model reflect the (presumed) behavior of the individual cells, such as nutrient consumption, biomass growth, cell division, movement, differentiation, communication, maintenance, and death. Since a change in microscopic (individual-based) rules may lead to significantly different macroscopic (population) behavior, it might be possible to rule out impossible mechanisms and to learn about the true mechanisms. A very important property of individual-based models is the fact that they easily allow for differences between the individuals. This is accomplished by using random variables, drawn from a certain statistical distribution. The introduction of a range of randomness and the consideration of a high number of individuals interacting independently with the environment leads to a good representation of reality and leads to a better understanding of the cellular metabolism (see, e.g., Reference 35). Spatial effects can be relatively easily translated into a set of rules. Kreft et al.[34] introduced the spatial aspect in their model to reproduce the growth of *Escherichia coli* cells in a colony.

In general, individual-based models incorporating underlying mechanistic knowledge of microbial dynamics are widely spread, but are relatively unexplored in the field of predictive microbiology. The more general modeling approach in predictive microbiology considers the microbial population as such, i.e., the population is described by a single-state variable, namely, $N(t)$. Furthermore, model parameters are usually assumed to be deterministic, i.e., have one typical value. When incorporating cell-to-cell variability into population-based models, population-related model parameters are considered as random or distributed variables (e.g., Reference 36). Individual-based models have the advantage that the cell-to-cell variability can be incorporated at cell level, i.e., the level from which variability actually originates. The general concepts of individual-based models and their applicability in the context of predictive microbiology are discussed in Dens.[33]

7.3.2 Implementation of Mechanistic Insight into an Individual-Based Model

7.3.2.1 Modeling Mechanistic Insight on the Temperature Dependency of Cell Growth

The mechanistic insight into the *theory of cell division* has been built into an individual-based model BacSim, originally developed by Kreft et al.[34] In contrast to the general expression 7.1 describing the evolution of a bacterial population $N(t)$, biomass growth of the individual cells $m(t)$ is considered and is assumed to occur exponentially at any time*:

$$\frac{dm(t)}{dt} = \mu_{max} \cdot m(t) \tag{7.6}$$

This expression forms the elementary building block of the proposed individual-based model.

Concerning the cell cycle of an individual cell, Cooper and Helmstetter[37] observed that, for a constant temperature, a constant time C is needed for the replication of DNA and a constant time D for cell division. In combination with the fact that DNA replication is always initiated when the cell attains a certain amount of biomass $2m_c$, Donachie[38] derived the following relationship for the amount of biomass at cell division m_d:

$$m_d = 2m_c \exp(\mu \cdot (C + D)) \tag{7.7}$$

with μ the specific growth rate of the cell biomass (in combination with 7.6, μ represents μ_{max}). Following this equation, the cell mass at division (and thus also the average cell mass of the population) is proportional to the exponent of the product $\mu \cdot (C + D)$. With respect to this equation and based on literature, a number of hypotheses on the effect of dynamic temperatures on the cell division process (and thus the overall specific cell-number growth rate) can be formulated:

i. The product $\mu \cdot (C + D)$ stays constant for different temperature conditions. This means that temperature variations do not alter the size and the chemical composition of the cells, as postulated by Cooper.[39] In other words, the biomass growth rate as well as the population growth rate will immediately change when imposing temperature changes and no lag will be observed.

ii. Trueba et al.[40] reported that the average cell volume of E. coli decreases with decreasing temperatures. Consequently, for these observations, the product $\mu \cdot (C + D)$ depends on temperature as the average cell volume is proportional to m_d. In case of a temperature increase, for example, the

---

* The dynamics of a single species are considered and the subscript i can thus be omitted.

biomass growth rate will instantaneously change but the population number will lag behind as the average volume for division has increased.

iii. A lag in biomass growth of E. coli induced by sudden temperature shifts
from low to high temperatures has been reported by Ng et al.[41] The authors
assume that cells growing at low temperatures express some *damaged*
status that needs to be *repaired* before active growth at high temperatures
can be achieved. This damaged state can be reflected by a limiting concentration of one or more enzymes. When passing from a low to a high
temperature, cells first need to increase the concentration of these limiting
enzymes, before they can increase their biomass growth rate.

A (simplified) mathematical translation of this hypothesis reads as
follows:

$$\frac{dm(t)}{dt} = \frac{E(t)}{m(t)} \cdot \frac{\mu_{max}(T(t))}{L} \cdot m(t) = \mu(\cdot) \cdot m(t)$$

$$\text{with} \quad \frac{dE(t)}{dm(t)} = L \quad (L = L_l \text{ or } L_h)$$

with $E(t)$ some critical growth factor, and L the rate at which E is synthesized (after Reference 41). This production rate changes according to
temperature (in a discrete way), i.e., L_l and L_h are the typical production
rates for low and high temperatures, respectively. For E. coli populations,
the high temperature zone ranges from 20 to 37°C and is also known as
the normal physiological range of E. coli.[41]

In conclusion, this hypothesis will predict a lag phase when temperature variations cross the (lower) boundary of the normal physiological
range.

The temperature dependence in the suboptimal growth temperature range can
be modeled by the square root model of Ratkowsky et al.[42]:

$$\sqrt{\mu_{max}(T(t))} = b \cdot (T(t) - T_{min}) \tag{7.8}$$

For more details on the exact implementation (i.e., parameter values, initial conditions, etc.) of these hypotheses, reference is made to Dens.[33]

7.3.2.2 Simulation Results

As a case study, the effect of abrupt shift-up temperatures on the growth of E. coli
is described. The experimental data in Figure 7.2 and Figure 7.3 depict the effect
of a small (i.e., 5°C) and a large (i.e., 20°C) positive temperature shift on the growth
of E. coli, respectively. Full details on the experimental data generation can again
be found in Bernaerts et al.[19] and Dens.[33]

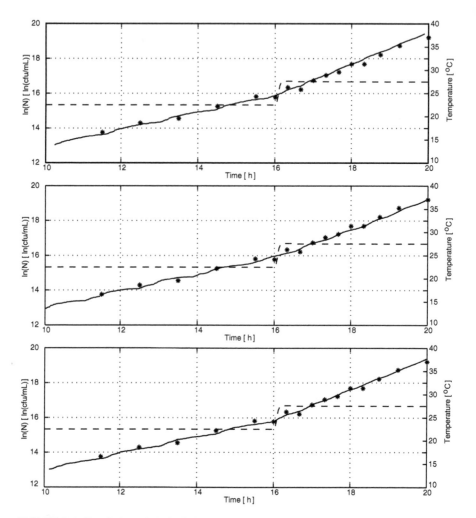

FIGURE 7.2 Simulation of the individual-based models proposed in Section 7.3 on experimental data of *E. coli* (*) submitted to a sudden temperature shift from 22.5 to 27.5°C during exponential growth (Adapted from Dens, E.J., *Predictive Microbiology of Complex Bacterial/Food Systems: Analysis of New Modelling Approaches,* Katholieke Universiteit Leuven, Belgium, 2001). The solid line represents the model prediction using the measured temperature profile (dashed line). Top: hypothesis (i), middle: hypothesis (ii), bottom: hypothesis (iii).

For each of the temperature shifts, the three hypotheses described in the previous paragraph have been implemented. It appears from Figure 7.2 that the small temperature increase from 22.5 to 27.5°C does not alter the balanced growth dynamics of the microorganisms and is properly described in all three cases. On the contrary, cell density data generated during the larger temperature shift from 15°C (low temperature range) to 35°C (high temperature range) induces a lagged growth response that can be predicted by only hypotheses (ii) and (iii) (see Figure 7.3). In hypothesis (ii), the lag phase is due to the time needed to increase the cell volume up to the new critical

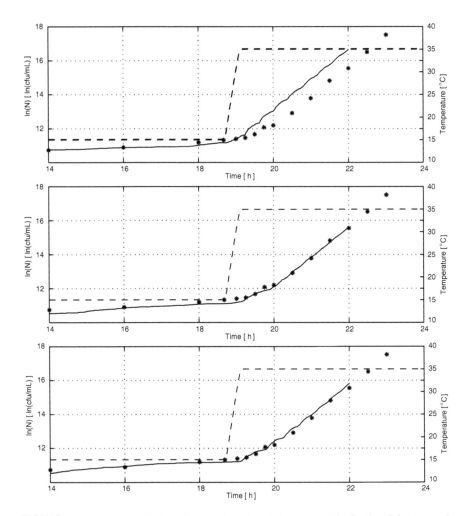

FIGURE 7.3 Simulation of the individual-based models proposed in Section 7.3 on experimental data of *E. coli* (*) submitted to a sudden temperature shift from 15 to 35°C during exponential growth (Adapted from Dens, E.J., *Predictive Microbiology of Complex Bacterial/Food Systems: Analysis of New Modelling Approaches,* Katholieke Universiteit Leuven, Belgium, 2001). The solid line represents the model prediction using the measured temperature profile (dashed line). Top: hypothesis (i), middle: hypothesis (ii), bottom: hypothesis (iii).

mass at division. Biomass growth exhibits an immediate rate adjustment whereas cell number shows lag behavior. In hypothesis (iii), the lag phase is reproduced at the level of biomass growth and propagates into the cell number evolution.

7.3.2.3 Discussion of Results

Individual-based modeling yields an excellent tool to integrate mechanistic knowledge at the level of the individual cell behavior into a model structure. Simulations with the individual-based model can then explain the population dynamics.

In this example, three cell mechanisms describing the effect of dynamic temperatures could be extracted from literature references. Two of the three hypotheses could describe both a small and a large temperature shift equally well. Given only population density measurements, it is therefore impossible to *discriminate* between the established models. At this point, additional (more advanced) measurements are needed to further establish the model structure. Such more advanced measurements can be biomass weight, DNA concentration, RNA concentration, protein concentration, etc. In other words, the revised modeling example clearly points out the *two-way interaction* between *model building* and *data generation*. Besides the selection of essential measurements, this two-way interaction embraces the design of *informative* experiments, i.e., the selection of appropriate (*dynamic*) input conditions (see, e.g., References 19 and 43) or (*static*) treatment combinations (see, e.g., References 44 and 45).

A disadvantage of the individual-based modeling approach is that the models may become relatively complex and computationally tedious. However, the obtained mechanistic knowledge can eventually form a sound basis for population-based models (which are more easily manageable).

7.4 EXAMPLE II. MODELING MICROBIAL INTERACTION WITH PRODUCT INHIBITION

In this section, the interaction of lactic acid bacteria (antagonist) with pathogenic bacteria (target) is discussed and modeled. Information given has been extracted from Vereecken et al.[46–47] and Vereecken and Van Impe.[48]

7.4.1 DESCRIPTION OF THE INHIBITION PHENOMENA

During the fermentation process of lactic acid bacteria, lactic acid is produced (biological process). This lactic acid released into the medium will dissociate and lower the medium pH (chemical process). Both the undissociated lactic acid concentration ($[LaH]$) and the decreased pH ($\sim[H^+]$) have an inhibitory effect on microorganisms. In the first place, the lactic acid production will cause the inhibition of the bacterium growth itself. The cessation of growth observed as the stationary phase can thus be attributed to a self-induced inhibitory effect. In addition, this lactic acid production affects neighboring microorganisms. Pathogenic bacteria, like *Yersinia enterocolitica* (see Figure 7.4), can be very sensitive to this inhibitory compound.[24] The increasing lactic acid concentration will cause an early termination of the growth process. For this reason, lactic acid bacteria can be exploited as a natural antimicrobial agent within (fermented) food products or as a protective culture.

7.4.2 MODELING MICROBIAL GROWTH WITH LACTIC ACID PRODUCTION AND INHIBITION

In contrast to the classical approach, Equation 7.3, where the stationary phase is modeled as function of $N_i(t)$ and N_{max}, growth inhibition emerges from lactic acid production, which is therefore explicitly incorporated into the model structure:

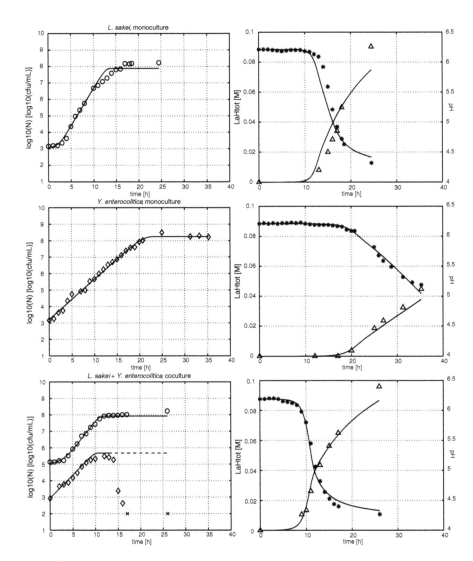

FIGURE 7.4 Description of experimental data of *Lactobacillus sakei* (o) and *Yersinia entero-colitica* (◊) grown in mono- and coculture with the dynamic model structures (Equation 7.9 and Equation 7.10 in combination with 7.11 and 7.12) presented in Section 7.4 (Adapted from Vereecken, K.M. and Van Impe, J.F., *Int. J. Food Microbiol.*, 73(2/3), 239, 2002 [× refers to cell numbers below detection limit]). The total lactic acid concentration [LaH]$_{tot}$ (Δ) and pH (*) are depicted in the right-hand plots. The dissociation kinetics of the applied medium have been computed according to Wilson et al.[49] (Observe that the inactivation of *Y. entero-colitica* cannot be predicted by the model structure [dashed line].)

$$\frac{dN_i(t)}{dt} = \mu_{lag}(\cdot) \cdot \mu_{max}(\cdot) \cdot \mu_{LaH,H^+}([LaH],[H^+]) \cdot N_i(t) \qquad (7.9)$$

This general expression describes the growth characteristics of both target and antagonist.* The growth-related lactic acid production — particularly by the antagonist — requires an additional coupled differential equation:

$$\frac{d[LaH]_{tot,i}(t)}{dt} = \pi(\cdot) \cdot N_i(t) \qquad (7.10)$$

with $\pi(\cdot)$ the specific lactic acid production rate of the antagonistic bacterium (i). Note that $[LaH]_{tot}$ refers to the total lactic acid concentration, i.e., the sum of the undissociated and dissociated lactic acid concentration. In case multiple lactic acid producing strains are present, the overall growth rate of each strain will be affected by the sum of all $[LaH]_{tot,i}$ concentrations.

To describe the chemical process of lactic acid dissociation in complex media, several methods inspired by traditional chemical laws are available (e.g., References 48 and 49). Given the medium, the process of lactic acid dissociation can be fully identified irrespective of the microbial model. Observe that [LaH] and [H⁺] vary with time and are determined by the lactic acid producing strain and the dissociation properties of $[LaH]_{tot}$ in the growth medium.

Several inhibitory functions can be proposed for $\mu_{LaH,H^+}([LaH],[H^+])$. On the basis of a rigorous model structure evaluation, Vereecken et al.[48] translated the inhibitory effect of undissociated lactic acid and the proton concentration (pH) into the following equation:

$$\mu_{LaH,H^+} = \left(1 - \frac{[LaH]}{[LaH]_{max}}\right)^{\alpha} \cdot \left(1 - \frac{[H^+]}{[H^+]_{max}}\right)^{\beta} \quad \begin{array}{l} \text{when } [LaH] \le [LaH]_{max} \\ \text{and } [H^+] \le [H^+]_{max} \end{array}$$

$$= 0 \quad \begin{array}{l} \text{when } [LaH] > [LaH]_{max} \\ \text{or } [H^+] > [H^+]_{max} \end{array} \qquad (7.11)$$

with $[LaH]_{max}$ the lactic acid concentration at which growth ceases, $[H^+]_{max}$ the proton concentration associated with the minimum pH for growth, and α and β some small positive values. The inhibition terms have no effect on the microbial dynamics as long as the undissociated lactic acid concentration and proton concentration remain well below their inhibitory value. In such cases, both functions are approximately equal to 1. When [LaH] and [H⁺] become significant as time proceeds, either function evolves towards 0 and growth stagnates.

To complete the model structure, the specific lactic acid production rate needs to be mathematically modeled. Combining the traditional linear law with the concept

---

* The subscript i thus refers to either the antagonist or the target.

of metabolism inhibitory concentrations, $\pi(\cdot)$ in Equation 7.10 can be represented as follows:

$$\pi(\cdot) = \underbrace{Y_{LaH/N_i} \cdot \mu_i(\cdot)}_{growth} + \underbrace{Y_{m,i}([LaH],[H^+])}_{maintenance} \qquad (7.12)$$

with Y_{LaH/N_i} the yield coefficient [mmol cfu^{-1}], $\mu_i(\cdot)$ the overall specific growth rate (embracing the terms μ_{lag}, μ_{max}, and μ_{LaH,H^+} within Equation 7.9), and $Y_{m,i}$ ([LaH], [H$^+$]) the maintenance coefficient [mmol cfu^{-1} h^{-1}]. The first factors present the growth-related production. The maintenance coefficient assures the observed production of [LaH]$_{tot}$ during the first hours of the stationary phase. This maintenance-related production also ceases when some inhibitory proton or undissociated lactic acid concentration is reached.[50]

The general model structure consisting of the coupled differential Equation 7.9 and Equation 7.10 yields accurate prediction for monocultures as well as mixed-culture growth. This is illustrated for experimental data of *Lactobacillus sakei* and *Y. enterocolitica* in Figure 7.4. More details on parameter values and the practical model implementation are available in References 46 to 48.

7.4.3 DISCUSSION OF RESULTS

The model building strategy described in this example starts from the identification of main phenomena determining the dynamics of the microbial system. The derived general model structure allows the stationary phase to be described in a natural (mechanistically sound) and consistent way. Moreover, the mechanistically inspired model structure can easily describe both single species and multiple species dynamics (with interaction).

To conclude, the present example illustrates how microbial growth on itself may cause a dynamic change of the environmental conditions, e.g., by the production of metabolites.

7.5 CONCLUSIONS

Dynamic mathematical models allow for a consistent computation of the impact of different steps associated with the production, distribution, and retailing of a food (characterized by time-varying conditions) on microbial dynamics. Moreover, the intrinsic properties of microbial evolution such as growth-related product formation and inhibition can be easily integrated and predicted.

Examples given in this paper illustrate how we can *learn* from predictive modeling based on biological and physical ideas. The individual-based modeling approach, for instance, serves as an excellent tool to test generic cell mechanisms with respect to the observed population behavior. However, such a modeling approach with an increased level of detail demands more advanced measurements at the cell or population level or both.

In view of expanding the applicability of predictive models, researchers must be encouraged to aim at an increased *generality* — and thus *transferability* — of model structures. For example, the complete model structure established in Section 7.4 can describe the individual behavior of lactic acid bacteria as well as the inhibitory mechanism in the presence of pathogenic or spoilage bacteria. This example also illustrates that cell density measurements are not always sufficient to establish complex model structures. Components interfering with the microbial dynamics, such as metabolite formation, should be identified, measured, and built into the model structure. Given this increased (experimental) knowledge on the microbial dynamics, we can aim at more *robust* mechanistically inspired models yielding a high predictive quality.

In this respect, it ought to be stressed that model builders can learn (more) from *dynamic experimental data*. Microbial dynamics under realistically time-varying conditions are not necessarily observable from (commonly available) static data. In the first example (see Section 7.3), the application of time-varying temperature profiles revealed the induction of an intermediate lag phase during the exponential growth of *E. coli*.

When extrapolating model structures established on static experimental data to more realistic dynamic conditions, e.g., combination of processing steps, model predictions may fail to describe the microbial evolution accurately. Stephens et al.,[51] for example, observed that slow heating rates applied during inactivation of *Listeria monocytogenes* induced thermotolerance. Predictions using an inactivation model developed on static experiments (not taking into account the magnitude of heating rate) systematically overestimate the effect of the applied heat treatment. Future research should thus pay attention to dynamic model development using dynamic experimental data. In such cases only, complementary effects of dynamic conditions or subsequent treatments can be properly incorporated within the model structure. Observe that synergetic effects form the basic principles within the hurdle technology (see, e.g., Reference 52), which is often addressed in the food industry.

Overall, model improvement aims at an increased predictive accuracy. However, striving for this increased modeling accuracy, one must always keep an eye on the model structure *complexity*. In this respect, it must always be clearly specified for which purpose the model is being developed. An important challenge for the future is therefore the search for a satisfactory trade-off between predictive power and manageability of mathematical models: *When is simple good enough?* (after Reference 53).

ACKNOWLEDGMENTS

This research was supported by the Research Council of the Katholieke Universiteit Leuven, the Fund for Scientific Research – Flanders (FWO), the European Commission, the Belgian Program on Interuniversity Poles of Attraction, and the Second Multi-Annual Scientific Support Plan for a Sustainable Development Policy, initiated by the Belgian State, Prime Minister's Office for Science, Technology and Culture.

REFERENCES

1. Ross, T., McMeekin, T.A., and Baranyi, J., Predictive microbiology and food safety, in *Encyclopedia of Food Microbiology,* Volume 3, Robinson, R.K., Batt, C.A., and Patel, P.D., Eds., Academic Press, San Diego, 2000, p. 1699.
2. Cassin, M.H., Lammerding, A.M., Todd, E.C.D., Ross, W., and McColl, R.S., Quantitative risk assessment for *Escherichia coli* O157:H7 in ground beef hamburgers, *Int. J. Food Microbiol.*, 4, 21, 1998.
3. Coleman, M.E. and Marks, H.M., Qualitative and quantitative risk assessment, *Food Control*, 10, 289, 1999.
4. Whiting, R.C. and Buchanan, R.L., Microbial modeling, *Food Technol.*, 48(6), 113, 1994.
5. Dalgaard, P., Buch, P., and Silberg, S., Seafood spoilage predictor — development and distribution of a product-specific application software, *Int. J. Food Microbiol.*, 73, 343, 2002.
6. Zwietering, M.H., Jongenburger, I., Rombouts, F.M., and van't Riet, K., Modeling of the bacterial growth curve, *Appl. Environ. Microbiol.*, 56(6), 1875, 1990.
7. Anonymous, Kinetics of microbial inactivation for alternative food processing techniques, *J. Food Sci.*, 4(Suppl.), 108, 2000.
8. Baranyi, J., Roberts, T.A., and McClure, P., A non-autonomous differential equation to model bacterial growth, *Food Microbiol.*, 10, 43, 1993.
9. Van Impe, J.F., Nicolaï, B.M., Martens, T., De Baerdemaker, J., and Vandewalle, J., Dynamic mathematical model to predict microbial growth and inactivation during food processing, *Appl. Environ. Microbiol.*, 58(9), 2901, 1992.
10. McMeekin, T.A., Brown, J., Krist, K., Miles, D., Neumeyer, K., Nichols, D.S., Olley, J., Presser, K., Ratkowsky, D.A., Ross, T., Salter, M., and Soontranon, S., Quantitative microbiology: a basis for food safety, *Emerg. Infect. Dis.*, 3(4), 541, 1997.
11. Vereecken, K., Geeraerd, A., Bernaerts, K., Dens, E., Poschet, F., and Van Impe, J., Predicting microbial evolution in foods: general aspects of modelling approaches and practical implementation, *Journal A, Special issue on modelling and control in bioprocesses*, 41(3), 45, 2002.
12. Grimson, M.J. and Barker, G.C., A continuum model for growth of bacterial colonies on a surface, *J. Phys. A Math. Gen.*, 26, 5645, 1993.
13. Zwietering, M.H., de Koos, J.T., Hasenack, B.E., de Wit, J.C., and van't Riet, K., Modeling bacterial growth as a function of temperature, *Appl. Environ. Microbiol.*, 57, 1091, 1991.
14. Baranyi, J. and Roberts, T.A., A dynamic approach to predicting bacterial growth in food, *Int. J. Food Microbiol.*, 23, 277, 1994.
15. Hills, B.P. and Mackey, B.M., Multi-compartment kinetic models for injury, resuscitation, induced lag and growth in bacterial cell populations, *Food Microbiol.*, 12, 333, 1995.
16. Hills, B.P. and Wright, K.M., A new model for bacterial growth in heterogeneous systems, *J. Theor. Biol.*, 168, 31, 1994.
17. McKellar, R.C., Development of a dynamic continuous–discrete–continuous model describing the lag phase of individual bacterial cells, *J. Appl. Microbiol.*, 90, 407, 2001.
18. Van Impe, J.F., Nicolaï, B.M., Schellekens, M., Martens, T., and De Baerdemaeker, J., Predictive microbiology in a dynamic environment: a system theory approach, *Int. J. Food Microbiol.*, 25, 227, 1995.

19. Bernaerts, K., Servaes, R.D., Kooyman, S., Versyck, K.J., and Van Impe, J.F., Optimal temperature input design for estimation of the square root model parameters: parameter accuracy and model validity restrictions, *Int. J. Food Microbiol.*, 73(2/3), 145, 2002.

20. Bailey, J.E. and Ollis, D.F., *Biochemical Engineering Fundamentals*, 2nd ed., McGraw-Hill, New York, 1986.

21. Adams, M.R. and Moss, M.O., *Food Microbiology*, Royal Society of Chemistry, Cambridge, UK, 1995.

22. Cerf, O., A review: tailing of survival curves of bacterial spores, *J. Appl. Bacteriol.*, 42(1), 1, 1977.

23. Moats, W.A., Dabbah, R., and Edwards, V.M., Interpretation of nonlogarithmic survivor curves of heated bacteria, *J. Food Sci.*, 36, 523, 1971.

24. Mossel, D.A.A., Corry, J.E.L., Struijk, C.B., and Baird, R.M., *Essentials of the Microbiology of Foods: A Textbook for Advanced Studies*, Wiley, Chichester, UK, 1995.

25. Geeraerd, A.H., Herremans, C.H., and Van Impe, J.F., Structural model requirements to describe microbial inactivation during mild heat treatment, *Int. J. Food Microbiol.*, 59, 185, 2000.

26. Shadbolt, C.T., Ross, T., and McMeekin, T.A., Nonthermal death of *Escherichia coli*, *Int. J. Food Microbiol.*, 49, 129, 1999.

27. Whiting, R.C., Modeling bacterial survival in unfavorable environments, *J. Ind. Microbiol.*, 12, 240, 1993.

28. Shadbolt, C.T., Ross, T., and McMeekin, T.A., Differentiation of the effects of lethal pH and water activity: food safety implications, *Lett. Appl. Microbiol.*, 32, 99, 2001.

29. Zwietering, M.H., de Wit, J.C., Cuppers, H.G.A.M., and van't Riet, K., Modeling of bacterial growth with shifts in temperature, *Appl. Environ. Microbiol.*, 60(1), 204, 1994.

30. Hudson, J.A., Effect of pre-incubation temperature on the lag time of *Aeromonas hydrophila*,. *Lett. Appl. Microbiol.*, 16, 274, 1993.

31. Whiting, R.C. and Bagi, L.K., Modeling the lag phase of *Listeria monocytogenes*, *Int. J. Food Microbiol.*, 73(2/3), 291, 2002.

32. Alavi, S.H., Puri, V.M, Knabel, S.J., Mohtar, R.H., and Whiting, R.C., Development and validation of a dynamic growth model for *Listeria monocytogenes* in fluid whole milk. *J. Food Prot.*, 62(2), 170, 1999.

33. Dens, E.J., *Predictive Microbiology of Complex Bacterial/Food Systems: Analysis of New Modelling Approaches*, Department of Food and Microbial Technology, Katholieke Universiteit Leuven, Belgium, No. 166, 2001. Promoters: Prof. J.F. Van Impe and Prof. B Nicolaï.

34. Kreft, J.U., Booth, G., and Wimpenny, J.W.T., Bacsim, a simulator for individual-based modelling of bacterial colony growth, *Microbiology*, 144, 3275, 1998.

35. Bermudez, J., Lopez, D., Valls, J., and Wagensberg, J., On the analysis of microbiological processes by Monte Carlo simulation techniques, *Comput. Appl. Biosci.*, 5(4), 305, 1989.

36. Nicolaï, B.M. and Van Impe, J.F., Predictive food microbiology: a probabilistic approach, *Math. Comp. Simul.*, 42(2/3), 287, 1996.

37. Cooper, S. and Helmstetter, C.E., Chromosome replication and the division cycle of *Escherichia coli* b/r, *J. Mol. Biol.*, 31, 519, 1968.

38. Donachie, W.D., Relationship between cell size and time of initiation of DNA replication, *Nature*, 219, 1077, 1968.

39. Cooper, S., *Bacterial Growth and Division*, Academic Press, San Diego, CA, 1991.

40. Trueba, F.J., van Spronsen, E.A., Traas, J., and Woldringh, C.L., Effects of temperature on the size and shape of *Escherichia coli* cells, *Arch. Microbiol.*, 131, 235, 1982.

41. Ng, H., Ingraham, J.L., and Marr, A., Damage and derepression in *Escherichia coli* resulting from growth at low temperatures, *J. Bacteriol.*, 84, 331, 1962.

42. Ratkowsky, D.A., Olley, J., McMeekin, T.A., and Ball, A., Relationship between temperature and growth rate of bacterial cultures, *J. Bacteriol.*, 149, 1, 1982.

43. Ljung, L., *System Identification: Theory for the User*, 2nd Ed., Prentice-Hall, Upper Saddle River, NJ, 1999.

44. Davies, K.W., Design of experiments for predictive microbial modeling, *J. Ind. Microbiol.*, 12, 296, 1993.

45. Poschet, F., Geeraerd, A.H., Versyck, K.J., Van Loey, A.M., Ly Nguyen, B., Hendrickx, M.E., and Van Impe, J.F., Comparison of Monte Carlo analysis and linear regression techniques: application to enzyme inactivation, in *Proceedings of the Seventh European Conference Food-Industry and Statistics*, Duby, C. and Cassar, J.Ph., Eds., Agro-industrie et méthodes statistiques, Lille, France, 2002, p. 43.

46. Vereecken, K., Antwi, M., Janssen, M., Holvoet, A., Devlieghere, F., Debevere, J., and Van Impe, J.F., Biocontrol of microbial pathogens with lactic acid bacteria: evaluation through predictive modelling, in *Proceedings and Abstracts of the 18th Symposium of the International Committee on Food Microbiology and Hygiene (ICFMH)*, Axelsson, L., Tronrud, E.S., and Merok, K.J., Eds., MATFORSK, Norwegian Food Research Institute, Oslo, 2002, p. 163.

47. Vereecken, K.M., Devlieghere, F., Bockstaele, A., Debevere, J., and Van Impe, J.F., A model for lactic acid induced inhibition of *Yersinia enterocolitica* in mono- and coculture with *Lactobacillus sakei, Food Microbiol.*, 20, 701–713, 2003.

48. Vereecken, K.M. and Van Impe, J.F., Analysis and practical implementation of a model for combined growth and metabolite production of lactic acid bacteria, *Int. J. Food Microbiol.*, 73(2/3), 239, 2002.

49. Wilson, P.D.G., Wilson, D.R., Waspe, C., Hibberd, D., and Brocklehurst, T.F., Application of buffering theory to food microbiology, in *Predictive Modelling in Foods — Conference Proceedings*, Van Impe, J.F. and Bernaerts, K., Eds., KULeuven/BioTeC, Belgium, 2000, p. 52.

50. Breidt, F. and Fleming, H.P., Modeling of the competitive growth of *Listeria monocytogenes* and *Lactococcus lactis* in vegetable broth, *Appl. Environ. Microbiol.*, 64(9), 3159, 1998.

51. Stephens, P.J., Cole, M.B., and Jones, M.V., Effect of heating rate on the thermal inactivation of *Listeria monocytogenes, J. Appl. Bacteriol.*, 77, 702, 1994.

52. Leistner, L., *Principles and Applications of Hurdle Technology*, Blackie Academic and Professional, Bedford, TX, 1995.

53. Buchanan, R.L., Whiting, R.C., and Damert, W.C., When is simple good enough: a comparison of the Gompertz, Baranyi and three-phase linear models for fitting bacterial growth curves, *Food Microbiol.*, 14, 313, 1997.

8 Predictive Microbiology in Quantitative Risk Assessment

Anna M. Lammerding and Robin C. McKellar

CONTENTS

8.1 INTRODUCTION

Food-borne disease arises from the consumption of microbial pathogens, microbial toxins, or both, by a susceptible individual. The risk of food-borne disease is a combination of the likelihood of exposure to the pathogen, the likelihood of infection or intoxication resulting in illness, and the severity of the illness. In a system as complex as the production and consumption of food, many factors affect both the likelihood and the severity of the occurrence of food-borne disease. Many of these factors are variable and often there are aspects for which little information is currently available. To manage food safety effectively, a systematic means of examining these factors is necessary.

Historically, the production of safe food has been based on numerous codes of practice and regulations enforced by various governing bodies worldwide. With the increased concern regarding the existence of microbial hazards in foods, a more objective approach is warranted, which has led to the introduction of the Hazard

Analysis Critical Control Point (HACCP) system. HACCP as a tool for safety management consists of two processes: building safety into the product and exerting strict process control.[1] The principles of HACCP have been set out by the Codex Alimentarius Commission[2] and consist of seven steps: hazard analysis; determination of critical control points (CCP); specification of criteria; implementation of monitoring system; corrective action; verification; and documentation.[3] HACCP processes as defined for various food products are often based on qualitative information and expert opinion. Moreover, the microbiological criteria underlying HACCP are often poorly understood or defined.[4]

The concept of risk assessment as defined by the Food and Agriculture Organization (FAO) and the World Health Organization (WHO)[5] provides a more quantitative approach to food-borne hazards. Quantitative risk assessment (QRA) is the scientific evaluation of known or potential adverse health effects resulting from human exposure to food-borne hazards.[1,6] Risk assessment is a systematic framework and process that provides an estimate of the probability and impact of food-borne disease. In doing so, exposures to food-borne pathogens are translated into actual human health outcomes.

Quantifying the human health risks associated with the ingestion of specific pathogens in specific foods has been considered feasible only within the last decade. Historically, until the mid-90s, risks associated with foods were estimated, at best, qualitatively, largely with reliance on epidemiological evidence and expert opinion to determine "high risk" vs. "low risk." Evaluations of the risks associated with food-borne hazards, in general or attributable to specific foods, have been predominantly qualitative descriptions of the hazard, routes of exposure, handling practices, consequences of exposure, or all of these. Quantifying any of these elements is challenging, since many factors influence the risk of food-borne disease, complicate interpretations of data about the prevalence, numbers, and behavior of microorganisms, and confound the interpretations of human health statistics. Consequently, policies, regulations, and other types of decisions concerning food safety hazards have been largely based on subjective and speculative information.

Today, advances in our knowledge, analytical techniques, and public health reporting, combined with increased consumer awareness, global trade considerations, and realization of the real economic and social impacts of microbial food-borne illness, have moved us toward the threshold of using QRA to support better prioritizing and decision-making.

Developments in the field of microbial risk assessment have some resemblance to the growth characteristics of a microbial population (see Chapter 2). During what might be termed the lag phase, few researchers attempted to define and model the food chain quantitatively. Today, the field can be described as entering the log phase, as efforts increase internationally to develop sophisticated models in response to risk managers' needs in decision-making.

The recent ratification of the World Trade Organization (WTO) agreement is having a major impact on the development of new approaches for the regulation of food. Countries are encouraged to base their procedures on Codex standards and guidelines to maintain and enhance safety standards.[7] This will lead to the development of harmonized risk assessment and risk management frameworks, providing input into HACCP, which is the primary vehicle for achieving enhanced food safety goals.[7] As the use of HACCP increases, there will be a need for a clear understanding of the relationship among HACCP, microbiological criteria, and risk assessment.[4] Regulators will be called upon to participate in all aspects of HACCP development, in particular to establish public health-based targets, elucidate microbiological criteria, develop improved techniques in microbiological risk assessment, and develop the means for evaluating the relative performance of HACCP systems.[4] Harmonization of international rules will clearly require standardized approaches.[8]

For microbial pathogens in foods, formal risk assessment has evolved from the traditional fields of application such as toxicology and environmental health risks, but with distinct differences. In particular, survival and/or inactivation of pathogens, and the growth of bacteria, must be accounted for, and assessors require predictive models to estimate these parameters. Second, human responses to microbial pathogens can vary significantly, depending on characteristics of the host's immunity and other defense factors; the pathogen's characteristics and survival and virulence mechanisms; and the characteristics of the food matrix in supporting growth, or in protecting the microorganism from inactivation by processing, or in the human after ingestion. However, the focus of this paper will consider the parameters that are driven in part by our ability to predict exposures.

8.2 ASSESSING MICROBIAL RISKS

There are different approaches that can be taken in microbial risk assessment; however, the basic sections of formal risk assessment are hazard identification, hazard characterization, exposure assessment, and risk characterization.[5,6] These describe, respectively, the nature of the food, the contaminant, and associations with human illness; the characteristics of the disease, the pathogen–host interaction, and if data are available, a mathematical model that quantifies the dose–response relationship; an evaluation of the likely intake of the agent in the food; and an integration of the foregoing information to provide a risk estimate, i.e., the likelihood and severity of the adverse effects in a given population. Risk characterization should also delineate the uncertainties and variability in the data used, and in our understanding of the food system, pathogen behavior and human health response. QRA is also considered to be part of the larger concept of risk analysis, which includes, in addition, risk management and risk communication steps.[9]

QRA and HACCP have some common parameters. Process risk models (see later) may help to identify CCPs and specify where significant risks exist. QRA can have input into specification of criteria for CCPs (step 3 of HACCP), as shown in Figure 8.1.[3] Risk assessment is intended to provide a scientific basis for risk management decisions, while HACCP is a systematic management approach to the control of potential hazards in food operations.[10] Thus, risk assessment concerns the overall product safety, while HACCP enhances overall product safety by assuring day-to-day process control.[10] The view of risk assessment being associated with one step of HACCP may be a limited one; in a contrasting view, both HACCP and risk assessment are encompassed in risk analysis, with HACCP representing one management strategy (Figure 8.1).[10] Nauta[11] has recently clarified this relationship by

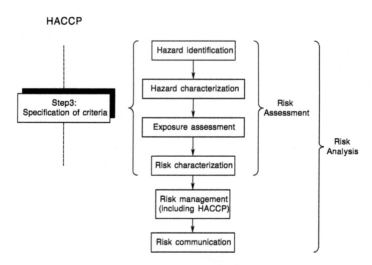

FIGURE 8.1 Relationship among HACCP, quantitative risk assessment, and risk analysis. (Modified from Notermans, S., Gallhoff, G., Zwietering, M.H., and Mead, G.C., *Food Microbiol.*, 12, 81, 1995. With permission.)

stating that while HACCP is typically linked to industrial processes, QRA is used for public health purposes and to help set hazard targets for industry as a whole.

Traditionally, food-borne pathogens and the risk of human illness are often described by descriptive *hazard assessments*, which typically do not actually provide a measure of the risk in terms of likelihood of occurrence and extent of illness (or other endpoint) expected in a population. However, if appropriate, this type of approach can be useful because the information can usually be compiled and summarized quickly if necessary. *Expert knowledge* has also often been relied upon to help decision-makers; however, even experts can misinterpret data, and may be biased towards certain conclusions.

Formal risk assessment based on the four-step framework relies on the basic elements of data, models, and assumptions. The variability and uncertainty in all three elements must be described, either quantitatively or descriptively. The risk assessment must be well documented and transparent; that is, all the data, assumptions, calculations, and technical descriptions should be presented to allow others to understand completely how the conclusions were reached, using what data, and what types of analyses.

We refer to *qualitative* assessments, in which the information used for the assessment is described in general terms, as categories or ratings. For example, ratings of "high," "medium," "low," or "negligible" may be assigned for the various parameters (e.g., pathogen concentration; prevalence, extent of growth/inactivation, or both; amount of food eaten; severity of illness) and for the final risk estimate, based on defined ranges of values for each rating and for each parameter. There are few examples of comprehensive hazard ranking systems. The ICMSF book *Microorganisms in Foods 7: Microbiological Testing in Food Safety Management*[12] gives a good table (in its Chapter 8 appendix) on the ranking of food-borne hazards or

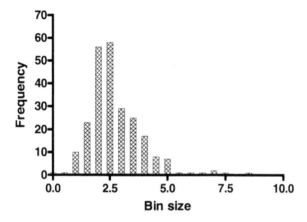

FIGURE 8.2 Example of a log normal distribution.

toxins into hazard groups. In a similar manner, various seafood products have been ranked qualitatively into risk categories.[13]

By contrast, *quantitative risk assessments* require mathematical equations to describe the relationships among all the factors that influence the risk. Quantitative assessments can be point estimates or stochastic (or *probabilistic*). Point-estimate and stochastic models can be differentiated along the lines of their treatment of randomness and probability. Point-estimate models do not include any form of randomness or probability in their characterization of a system, whereas these are fundamental characteristics of probabilistic assessments.

Point-estimate models use a single number for each data set that is used as an input into the model analyzed. For example, the mean concentration (i.e., colony forming units [CFUs] per gram) of *Salmonella* in raw ground beef is a point estimate; the 95th percentile value from a collection of data points would be a "worst-case" point estimate. Probabilistic analyses consider the entire possible range of the numbers of *Salmonella* that may be in the raw product, with the likely frequency at which the various concentrations might occur. Thus, the distribution curve may range from 1.0 to 4.0 logs per gram CFU per gram in product that is positive for the pathogen. Some values will be more likely than others, and this is represented by the height of the distribution curve at those values. This information is derived from one or more sets of laboratory data, or, if few data are available, estimations using sound scientific rationale will be required. The outcome of a point-estimate risk assessment is a single value for the risk estimate, such as 1 in 100,000 probability of illness. A probabilistic risk estimate is a range of values, and how probable each value is likely to be, again depicted by a distribution curve. An example of a lognormal distribution is given in Figure 8.2.

8.3 ROLE OF PREDICTIVE MICROBIOLOGY IN QRA

Most of the risk model development takes place within exposure assessment and dose–response assessment (part of hazard characterization). For some agents, particularly those involving voluntary exposure, such as prescription drugs, exposure assessment is relatively straightforward. But for other agents, such as environmental or food contaminants, an exposure assessment is usually based on considerable uncertainties. It is often not possible to measure exposures directly; rather they must frequently be predicted, for example, by monitoring data, using mathematical modeling, and reconstructing historical exposure patterns. There are two broad types of mathematical models used in exposure assessment: those that predict probable exposure to the agent and those that predict the probable concentration of the agent. Exposure models can be used to estimate population exposures based on small numbers of representative measurements. Models that predict concentrations can be combined with information on human time–activity patterns to estimate exposures.

Key components of assessing exposures may include:

- The microbial ecology in relation to food
- Intrinsic and extrinsic microbial growth requirements
- Prevalence of infection in food animals
- The initial contamination of the raw materials
- The impact of production, processing, cooking, handling, storing, distribution steps, and preparation by the consumer on the microbial agent
- The variability in processes involved and the level of process control
- Slaughter or harvesting practices and the level of sanitation
- The potential for contamination or recontamination
- The conditions for packaging, distribution, and storage of the food, and the food attributes that could influence growth, toxin production, or both

Implicit in the concept of exposure assessment is the influence of processing and environmental factors on the survival and growth of food-borne pathogens. Mathematical models can predict the extent of impact of unit operations on the numbers of microorganisms, which in turn determines the exposure.[14] Specific mathematical functions to quantitate microbial growth and death can be incorporated into risk assessments.[14-17] For example, the Gompertz function is used to evaluate growth parameters:

$$\log x(t) = A + C \exp\{-\exp[-B(t - M)]\} \tag{8.1}$$

where $x(t)$ is the number of cells at time t, A is the asymptotic count as t decreases to 0, C is the difference in value of the upper and lower asymptotes, B is the relative growth rate at M, and M is the time when the absolute growth rate is maximum.[18,19]

Thermal death models can be used to establish the D-value for a microorganism:

$$\log S_t = -\frac{t}{D} \tag{8.2}$$

where S_t is the survival ratio at time t. Much information on microbial growth and survival has been documented, and resulting predictive software such as Food Micro-Model has been used to predict the influence of food composition and environmental conditions on growth and survival of potentially hazardous microorganisms.[20] Models can therefore be used to develop CCPs, and show where data for risk assessments are missing.[21] In addition, models can support regulations and optimize product formulations and support process control.[21] Mathematical modeling can also support quantitation in dose–response assessment. For example, the Beta-Poisson is a commonly used distribution model for dose–response:[14]

$$P_i = 1 - \left(1 + \frac{N}{\beta}\right)^\alpha \tag{8.3}$$

where P_i is the probability of infection, N is the exposure, and α and β are coefficients specific to the pathogen.

In QRA, mathematical models are used to estimate the ultimate risk to the consumer as a function of input values taken from various points along the "farm-to-fork" continuum. Because of heterogeneity of microorganisms, variability around single point estimates of risk can be significant. Thus, point estimates give limited information, describing single instances such as worst-case scenarios without any insight into how likely, or unlikely, this is to occur.[14,22] Improvements in prediction can be made by incorporating uncertainty. Uncertainty is an important factor in risk analysis, since failure to account for it limits our ability to make reliable predictions of risk. Uncertainty may arise from inherent variability in the biological system, or from lack of information or understanding of the mechanisms involved.[15] Uncertainty due to lack of information or understanding vs. uncertainty due to variability can sometimes be minimized by obtaining more, high quality data; however, as this is not always feasible, alternatives must be sought. One approach is to use probability distributions to represent parameter values. These distributions can be built from empirical data, knowledge of underlying biological phenomena, or expert opinion.[22] Using distributions as inputs leads to an output where risk is expressed as a probability distribution. Risk analysis software such as @RISK™, which uses Monte Carlo analysis to simulate output distributions of risk on the basis of variability of input data, can facilitate the risk assessment process.[14,22] In Monte Carlo analysis, the point-estimate relations are replaced with probability distributions. Samples are randomly taken from each distribution in a series of iterations, and the results of each iteration are tallied, usually in the form of a probability density function, or cumulative distribution function. This approach yields an output, the risk estimate, that reflects the uncertainty and variability in the data used for the assessment.

As probabilistic models include components of randomness within their definition, these result in outputs that are in fact estimates of the true system. Probabilistic assessments attempt to capture the variability that is naturally present in a biological system. Such models tend to be a better representation of natural systems, given the randomness inherent in nature itself. Clearly, a point-estimate model to describe a natural system is a significant simplification of a biological system; however, with these caveats, a point-estimate model could be entirely appropriate for the problem at hand, and with given resources. What is important is that assessors and managers alike acknowledge the limitations of the information derived from any such risk model.

Nauta[23] has emphasized the need to separate true biological variability (due to heterogeneity of populations) from uncertainty, the lack of perfect knowledge of the parameter values. This is commonly neglected in risk assessment studies. Working with data on growth of *Bacillus cereus* in pasteurized milk, Nauta[23] showed that prediction of outbreak size may depend on the way that uncertainty and variability are separated. Using a deterministic estimate, the exposure assessment model predicted that there was no risk. A stochastic model without separation of uncertainty and variability predicted individual risk, but no major outbreak. In contrast, when uncertainty and variability were differentiated, a potential major outbreak was predicted.

8.4 SCOPE OF RISK ASSESSMENTS

In addition to different modeling approaches, risk assessments can also differ in their scope. In risk ranking, several foods may be compared within the assessment to determine which pose higher or lower risk. This type of assessment is useful for setting priorities for risk management. Farm-to-fork (production-to-consumption) models describe each stage of the food: growing, harvesting, processing, distribution, retail, and preparation pathway. Alternatively, assessments may focus only on the stages after retail distribution. Typically, risk assessments are constructed in a modular sequence of relevant stages in food harvesting/processing/handling/consumption. Submodels within the individual modules, including ones that integrate predictive equations for growth, inactivation, or both, are defined as appropriate. These may be simple or complex equations, reflecting the precision necessary to estimate significant parameters and changes in pathogen number.

8.5 PROCESS RISK MODELING

Evaluating the microbial safety of a food typically requires consideration of multiple factors that influence the prevalence and numbers of a microbial pathogen in the product. As a tool for strategic decision-making, the scope of a risk assessment should include activities that provide relevant information for the risk manager. This approach has been taken for many microbial risk assessments to describe the production-to-consumption pathway. The main goal of such work is to develop a tool that can be used to analyze the relationship between the factors that affect the presence, behavior, and ultimately concentration of microorganisms and the probability of human illness.

The phrase "Process Risk Model" (PRM) has been introduced to describe such risk assessments.[24,25] The basis of a PRM is the mathematical model that predicts the probability of an adverse impact as a function of multiple process parameters. By manipulating the parameters of the sequential stages of food production, the effect of hypothetical risk-reducing strategies that are based upon changing some component of the system is estimated by the changes in the risk prediction under different scenarios. For example, the probabilistic model allows the prediction of a change in a health effect endpoint, such as the expected number of illnesses within a defined population and time frame, under different HACCP or other types of intervention/control strategies. As a result the model acts as a predictive tool for evaluating future scenarios, rather than presenting a static picture of the present risk to health. Simulation provides this important link between HACCP and QRA.

In addition, intermediate results or outputs from the risk model may be of interest. For example, an intermediate output might be a prediction of the distribution of cell numbers in a package of ground beef, simulated from input parameters of initial carcass contamination, and modeling the changes that occur during slaughter and fabrication. Further analysis of the probabilistic model provides information about key inputs or uncertainties that most significantly influence the risk outcome, thereby identifying potentially effective interventions or research opportunities. Manipulation of the model, by altering input values in "what if" scenarios, can readily provide insight into the effectiveness of proposed risk interventions.[24]

Nauta's[11] modular process risk model (MPRM) constitutes a further improvement on the PRM. In this approach, it is assumed that in any food pathway, all processing steps can be described by six process modules: two microbiological processes (growth or inactivation) and four product handling steps (mixing, partitioning, removal, and cross-contamination). This approach highlighted the importance of including variability in microbial growth models, and provided a tool to identify the most important gaps in knowledge along a food pathway.

8.6 EXAMPLES OF RISK MODELING

There have been very few examples of well-developed QRA for specific microbial hazards in foods. Much of the work published to date is about quantitative models that describe either exposure or dose–response relationships. Schlundt[26] reviewed several microbial risk assessments published between 1996 and 1998 with a view to assessing the state of the art. He noted that few of the studies comprised a full Codex–based risk assessment.[6] Often the purpose of these studies did not relate directly to risk analysis, and the factors that determined the risk were not identified. The driving force for many of these studies was the use and application of mathematical models; thus the focus was largely on exposure assessment. Examples of some recently published food safety assessments are given in Table 8.1.

One of the important developments in QRA is the establishment of risk assessment simulations that can be easily accessed by users. A few examples of these follow. Oscar[27] has developed an interactive Microsoft® Excel-based spreadsheet called Poultry Farm Assess Risk Model (Poultry FARM). This model uses the risk analysis software @RISK™ to provide poultry companies and regulatory agencies with the tools they need to make informed public health decisions. van Gerwen et al.[28] have described a system for microbiological QRA of a cheese spread. Predictive models were incorporated with a decision support expert system called SIEFE: Stepwise and Interactive Evaluation of Food Safety by an Expert System. This approach combined quantitative information on the production processes with qualitative expert knowledge expressed as a series of rules. Ross and Sumner[29] have also developed a Microsoft Excel spreadsheet model for QRA. In this model,

TABLE 8.1
Examples of Currently Published Microbial Food Safety Exposure, Dose–Response, and Risk Assessments

Microorganism	Commodity	Type of Assessment	Reference
Bacillus cereus	Pasteurized milk	Semiquantitative, from retail to consumer	33
	Chinese-style rice	Probabilistic risk estimation, raw product to consumer	34
	Vegetable puree	Exposure assessment, retail to consumer	35
B. cereus, C. perfringens	Cooked chilled vegetables	Probabilistic exposure assessment, product preparation and storage	36
Listeria monocytogenes	Pasteurized milk	Point-estimate hazard/exposure assessment	37
	Pâté and soft cheese	Quantitative (simple), retail to consumer	38
	Ready-to-eat meat and smoked fish	Quantitative dose–response assessment	39
	Raw milk soft cheese	Probabilistic risk estimation, from farm to consumer	38
	Smoked or gravad fish	Quantitative, from retail to consumer	40
	Ready-to-eat foods	Probabilistic risk ranking	41
Salmonella Enteritidis	Cracked eggs	Semiquantitative, from eggs to consumer	42
	Pasteurized liquid egg	Probabilistic risk estimation, from eggs to consumer	43
	Shell eggs	Probabilistic risk estimation, from farm to consumer	31 44, 45
Salmonella spp.	Poultry and products	Probabilistic risk estimation from processing to consumer	27, 31, 46
Mycobacterium paratuberculosis	Pasteurized milk	Probabilistic vs. point-estimate exposure assessments	47
Escherichia coli O157:H7	Ground beef	Probabilistic risk estimation, from cattle to consumer	25, 48
		Probabilistic risk estimation, retail to consumer	49
	Raw fermented sausage	Probabilistic exposure assessment, cattle to retail	50

qualitative inputs are converted into numerical values, and then combined with quantitative inputs in a series of mathematical and logical steps. It was designed as a generic model, to give a quick and simple means of comparing food-borne risks from diverse products.

There are different approaches to risk modeling; thus it is appropriate to discuss here some illustrative examples in more detail. The PRM for *Escherichia coli* O157:H7 in ground beef reported by Cassin et al.[25] is one of the best developed and most detailed models available. The PRM is limited to a particular food production system, and predicts the distribution of probability of illness attributable to *E. coli*

O157:H7 in a particular ground beef scenario.[25] In contrast, the Poultry FARM model described by Oscar[27] is a packaging-to-consumption model that assesses the risk and severity of *Salmonella* spp. and *Campylobacter jejuni* infections from chicken. Outputs include the concentrations of pathogens at each stage of the process, and the health outcome assessment, which takes into account the number of patients who might seek medical treatment, and suffer death or chronic sequelae.

8.6.1 PRM FOR *E. COLI* O157:H7 IN GROUND BEEF

The model of Cassin et al.[25] describes the probability of becoming ill with *E. coli* O157:H7 as a result of consuming undercooked ground beef. It models the production of beef trimmings by a hypothetical abattoir, which are subsequently ground and sold by retailers. Figure 8.3 shows the flow diagram of the process. *E. coli* O157:H7 is the primary microbial hazard identified with ground beef. Cattle are known to be a reservoir for this pathogen, which is shed in feces, and can then subsequently contaminate the carcass during slaughter. *E. coli* O157:H7 is a human pathogen that can cause severe infection, often resulting in death or permanent damage. There have been a number of food-borne outbreaks attributed to undercooked ground beef.

As mentioned earlier, exposure assessment is one of the most important aspects of QRA. In this step, the potential exposure to the pathogen was determined in a single-serving meal. In this model, multiple stages of product handling were described, with appropriate probability distributions assigned to each step, based on available data. The various stages include production; processing and grinding; postprocessing conditions such as microbial growth and thermal inactivation; and consumption.

The production stage concerns the potential concentration of fecal material on the beef carcass. This depends on the level of *E. coli* O157:H7 in feces, which is affected by many factors including season, age of animal, and feeding practices. Prevalence relates to the relative number of animals that shed the pathogen, both within and between herds. Processing includes skinning, evisceration, and trimming. During the skinning process, fecal material from the hide can contaminate the carcass. Previous studies have shown that the various decontamination steps such as trimming of visible contamination and washing using a variety of methods have limited effect on the level of contamination. During the subsequent chilling of the carcass, some microbial growth can occur.

Trimmings collected during the deboning stage are then combined into 5-kg lots, and sent to retailers for grinding. During storage of the ground beef, some microbial growth can occur, and this can be modeled using common functions such as the Gompertz equation (see Chapter 2). The effect of temperature can be modeled using Food MicroModel (see Chapter 6). Finally, cooking is the most effective barrier against *E. coli* O157:H7 exposure, and modeling was based on the cooking preference of the consumer.

A dose–response model based on the Beta-Poisson model was constructed. It was assumed that the virulence of the pathogen is similar to that of *Shigella dysenteriae*, and model parameters were selected based on human feeding studies. The

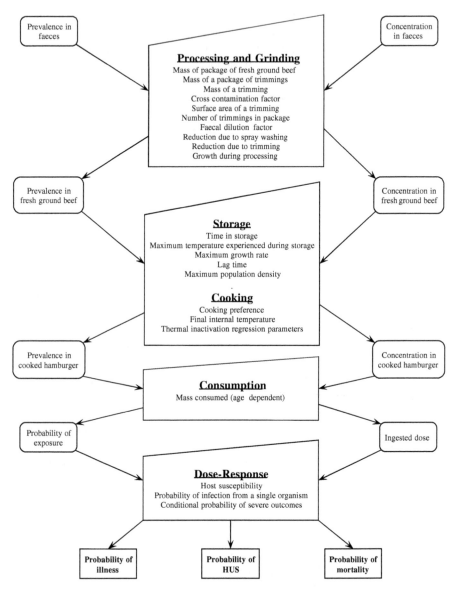

FIGURE 8.3 Flow diagram of the mathematical model of exposure assessment and dose–response for *E. coli* O157:H7 in hamburgers. (From Cassin, M.H., Lammerding, A.M., Todd, E.C.D., Ross, W., and McColl, R.S., *Int. J. Food Microbiol.*, 41, 21, 1998. With permission.)

dose–response curve for an adult population is shown in Figure 8.4. The susceptible population, i.e., young children, was assumed to have a similar vulnerability, but an increased propensity for more severe outcomes. As a final step in development of the model, the probability of illness was the product of the probability of a nonzero exposure and the output of the dose–response model.

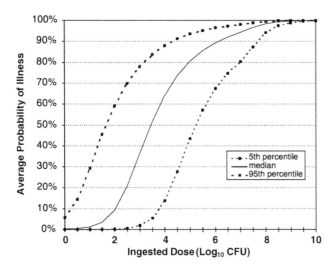

FIGURE 8.4 Beta-Binomial dose–response model — uncertainty in average probability of illness vs. ingested dose of *E. coli* O157:H7. (From Cassin, M.H., Lammerding, A.M., Todd, E.C.D., Ross, W., and McColl, R.S., *Int. J. Food Microbiol.*, 41, 21, 1998. With permission.)

A simulated distribution of probability of illness per meal is shown in Figure 8.5. It is not a simple matter to determine the risk of a specific health outcome, since a wide range of scenarios can exist. The risk for most scenarios is less than 1 in 10,000 (Figure 8.5). The expected value of risk was also calculated, which is a point estimate of the probability of a particular health effect occurring. This value is often used to compare with regulatory objectives to meet standards of acceptable risk; however, the range of risk experienced by the population is lost.

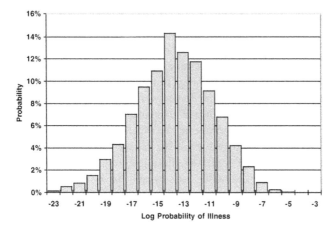

FIGURE 8.5 Probability distribution for probability of illness from a single hamburger meal predicted by the *E. coli* O157:H7 Process Risk Model (PRM). (From Cassin, M.H., Lammerding, A.M., Todd, E.C.D., Ross, W., and McColl, R.S., *Int. J. Food Microbiol.*, 41, 21, 1998. With permission.)

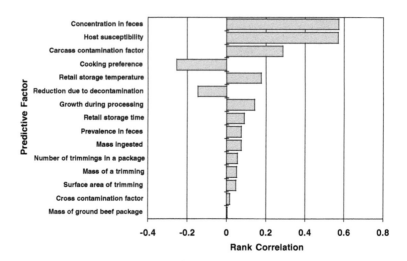

FIGURE 8.6 Spearman rank correlation between the estimated probability of illness and the 15 most important predictive factors of the Process Risk Model (PRM). (From Cassin, M.H., Lammerding, A.M., Todd, E.C.D., Ross, W., and McColl, R.S., *Int. J. Food Microbiol.*, 41, 21, 1998. With permission.)

CCPs can be identified from a PRM using importance analysis. Importance analysis includes the sensitivity of the outcome to a factor, and the uncertainty and variability of that factor. The Spearman rank correlation coefficient was used to measure importance, and a tornado graph showing the 15 predicting factors most highly correlated with risk is shown in Figure 8.6. The concentration of *E. coli* O157:H7 in the feces was the most highly correlated factor, which points out the importance of animal prescreening prior to slaughter, or some intervention that reduces numbers in the feces of the live animal. Host susceptibility (probability of illness from a single organism), carcass contamination factor (relationship between concentration in the feces and on the carcass), and cooking preference were also important risk factors.

The ability to propose appropriate risk mitigation strategies is an important outcome of a PRM. Hypothetical strategies such as improvements in storage temperature, better preslaughter screening, and institution of a consumer information program were simulated using the *E. coli* O157:H7 PRM. These were defined to have an assumed level of compliance with the intervention. It was found that reducing the average temperature of storage at the retail level from 10 to 8°C, with the maximum expected of 13 instead of 15°C, reduced the risk of illness by 80%. In contrast, the effectiveness of consumer education on the importance of fully cooking hamburgers was predicted to reduce risk by only 16%.

8.6.2 POULTRY FARM MODEL

This model predicts the change in concentration of *Salmonella* spp. or *C. jejuni* in a single serving of chicken from packaging at the processing plant, through to consumption by the consumer, as well as adverse health outcomes. It is structured as one simulation model (using @RISK in a Microsoft® Excel spreadsheet) and four

TABLE 8.2
Simulation Conditions for Poultry FARM Model Describing the Fate of *Salmonella* spp. on Chicken from Production to Consumption

Node	Incidence (%)	Extent[a] Minimum	Extent[a] Most Likely	Extent[a] Maximum
Packaging	20	0.0	2.5	4.5
Cold storage	100	−2.0	−0.3	0.0
Distribution	40	0.0	0.3	3.0
Cooking	15	−9.0	−6.0	0.0
Cooling	20	0.0%	0.1%	0.5%
Consumption	20	0.0	0.2	2.0
Infection		1.3	3.3	8.3

[a] Units are: Packaging, log number/serving; Cold Storage, Distribution, Cooling, Consumption (log change/serving); Cooling (% transfer/serving); Infection (log number).

predictive models. In this example, we will address the prediction of only *Salmonella* spp. during the process.

The various nodes described in the model are given in Table 8.2. At each stage, the change in numbers per serving is calculated by drawing a value from the given extent, ranging from the minimum to the maximum expected value, with an intermediate value representing the most likely outcome. The value for incidence is used to specify the number of servings predicted to be contaminated with *Salmonella* e.g., 20% of 100,000, or 20,000 (Table 8.2). During cold storage, the numbers are expected to decrease by a minimum of −2.0, a maximum of 0.0, and a most likely value of −0.3 log cfu per serving in 100% of the servings (Table 8.2).

The other nodes work in a similar fashion. The incidence value in the cooking node represents the 15% of consumers who are expected to undercook their chicken, and during cooling, it is expected that 20% of consumers will expose the cooked chicken to temperature abuse. Changes in the *Salmonella* content of each serving of the chicken were accumulated over the whole process, and infection was expected to occur if the cumulative numbers exceeded the minimum infective dose. This calculation assumed that one *Salmonella* was capable of causing an infection, and that resulted in 100% probability of developing salmonellosis. This calculation is similar to an exponential dose–response model.

Health outcomes were further calculated from the infection incidence. It was assumed that 45.4% of those infected became sick, that 20.7% of infected victims visited a doctor, 4.1% were admitted to hospital, 2.3% experienced chronic sequelae, and 0.1% died.[30]

The @RISK model was simulated 100,000 times to represent the number of servings being considered, and the outputs are given in Figure 8.7 and Figure 8.8. In Figure 8.7, the exposure assessment is presented as the change in log number of

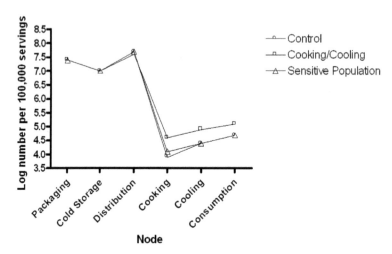

FIGURE 8.7 Exposure assessment for *Salmonella* spp. on chicken for the Poultry FARM model of Oscar.[27]

Salmonella per 100,000 servings over the whole process. In Figure 8.8, the health outcome assessment is given as number of cases per 100,000 consumers. When the simulation was performed using the values given in Table 8.2, the exposure assessment (Figure 8.7; control) showed that the cooking step had the greatest impact on numbers of *Salmonella*. Under these conditions, the number of consumers infected was 43 out of 100,000, with <0.1 deaths (Figure 8.8).

The simulation was repeated with changes made in the initial assumptions. To simulate cases of abuse, it was assumed that the proportion of consumers under-cooking their chicken was 75% rather than 15%, and the proportion of consumers exposing the cooked chicken to temperature abuse during cooling was 75% rather than 20%. In the exposure assessment (Figure 8.7), the combined abuse treatments resulted in an increase of 0.5 log numbers by the end of the cooling step. This translated into an increase in infections to 138 per 100,000 consumers, with a 0.14 death rate (Figure 8.8). In a further example, the incidence of cooking or cooling abuse was kept the same as in the control, and the influence of exposure to a more susceptible population was simulated. This was achieved by decreasing the assumed infection level (minimum, most likely, and maximum) by 1 log. The results of this simulation showed that, as expected, this change did not influence the exposure assessment (Figure 8.7); however, the infection rate increased to 168 per 100,000, and the death rate to 0.17 (Figure 8.8). With other simulations based on predicted changes in processing or consumption patterns, it would be possible to determine those factors that have the greatest impact on health outcomes.

8.7 MODIFYING RISK: CONCENTRATION VS. PREVALENCE

In addition to identifying and quantitating risks, risk assessments can also provide information on the relative impact of intervention strategies. The number of

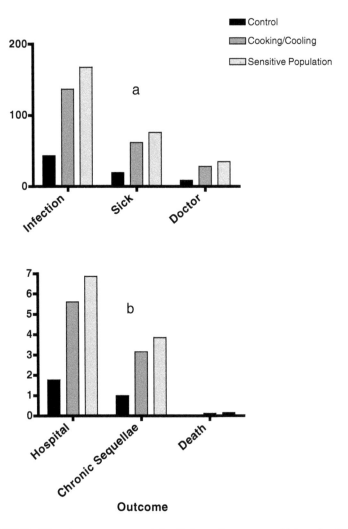

FIGURE 8.8 Health outcome assessment for *Salmonella* spp. on chicken for the Poultry FARM model of Oscar.[27]

microorganisms present in a sample of raw food has a direct impact on the number finally consumed; however, initial numbers may be influenced by either concentration (number per unit weight) or prevalence (proportion of units contaminated). This can be demonstrated by reference to data used for a FAO/WHO risk assessment on *Salmonella* in broiler chickens.[31] In this study, the prevalence of *Salmonella* was determined after immersion in the chill tank with and without chlorine. Data show that carcass cross-contamination was significantly reduced by inclusion of chlorine. Reducing prevalence of *Salmonella*-contaminated carcasses was estimated to have a one-to-one effect on risk reduction.

The effects of reducing the numbers of *Salmonella* on poultry carcasses without changing the prevalence of contaminated carcasses was also assessed using the risk assessment model. A change in concentration does not necessarily have a linear relationship with risk outcome, as is found for prevalence. Assuming a constant prevalence of 20%, and reducing the concentration, gave a reduction in illnesses per million servings from 11.3 to 4.28.[31] However, these observations pertain to individual units that will be prepared by a consumer vs. raw material units that would be comingled during processing and before consumption.

8.8 WHAT IS THE RIGHT MODEL TO USE?

Clearly, there are many options available to microbial risk assessors, from simple descriptive evaluations to highly complex and detailed analyses. In reality, a combination of techniques and analyses will often be incorporated into a single assessment, for example, qualitative information, expert knowledge, and quantitative analyses of available data, when appropriate. The decision of what approach to use, what analytical techniques are needed, and the scope and level of detail of the assessment will be dependent on the nature of the risk management question, and practical issues such as time, expertise, and other resources that are needed. In international trade disputes, the demands are for quantitative microbial risk assessments with some measure of variability and uncertainty. At the national level, the urgency and nature of the risk issue will dictate what approach is needed.

Microbiological models to predict pathogen growth, survival, or inactivation can differ in mathematical complexity, but a complex model may not necessarily be the best choice to answer a particular risk management question.[28] The need for an accurate prediction needs to be offset by consideration of whether the model is easy to use, whether it is robust and precise, and whether it has been validated against independent data. For example, if the objective of a risk assessment is to identify the most significant risk factors in a process, a simple model may have advantages over a complex model. However, if an accurate prediction of bacterial numbers is necessary, a more complex and accurate model may be preferable. In the choice of a suitable model, one must also consider the quality of data that are going to be used to generate a prediction. If the temperature data on a process are poor, it may not be appropriate to use a complex model for the predictions. Often, this can lead to a misinterpretation of the accuracy of the final prediction. The most appropriate model would be the simplest model possible for a given purpose and the given data quality, provided that it is validated and precise. A good model should also be subjected to an analysis that quantifies the accuracy and bias of its predictions.[32] Ideally, a model should be both accurate and unbiased. Models in risk assessment must adequately reflect reality.

8.9 FUTURE DIRECTIONS

At the present time, microbial risk assessors acknowledge many limitations in providing exact estimates of risk, and in the elements of any one risk assessment: the data available, the models developed to describe both the physical and mathematical aspects, and the assumptions necessary to construct these assessments.

Validating the outcomes of a risk assessment also provides a challenge; in many cases, any and all available data for a particular food/pathogen combination are used for assessing exposures and dose–response relations. This leaves a risk estimate, or intermediate outputs, that cannot be validated against independent data.

An additional challenge will be to facilitate the incorporation of existing and new mathematical models into the QRA framework. Many potentially useful models have been developed and published; however, these often exist independent of the specific needs of the regulators and the food industry. There is no definitive process by which these models can be combined with expert opinion and knowledge and other data on (for example) prevalence to clearly and unequivocally define the risk of consumption of a particular food. Some mathematical model databases exist, but these seldom describe the stochastic aspects of the underlying data. It is clear that in future researchers must work closely with regulators and the industry to improve the technology transfer.

8.10 CONCLUSIONS

Currently, there are many aspects of microbial contamination of foods, and the human health responses to pathogens, for which there are few data. However, the development of even preliminary quantitative risk models in a systematic way will help to identify what critical information is lacking, and help to guide future data gathering efforts. Finally, it is worth recognizing that risk assessment models should be considered "dynamic." As modeling tools improve, better data become available, and as we learn more about pathogen–food–host relationships and microbial responses, risk models will be updated and refined to provide better risk estimates. One continuing challenge will be to make sure that QRA models are kept current, so that they continue to be relevant to the changing needs of the food industry. The process of QRA is still in its infancy, however, and standards have yet to be developed. There is a clear advantage to the food industry and consumers to further develop the concepts of QRA and apply them to both common and novel food processes, and it is expected that significant advances will be made in this field over the next decade.

REFERENCES

1. Notermans, S. and Jouve, J.L., Quantitative risk analysis and HACCP: some remarks, *Food Microbiol.*, 12, 425, 1995.
2. CAC (Codex Alimentarius Commission), Guidelines for the Application of the Hazard Analysis Critical Control Point System, Codex Alimentarius (CCFH) Alinorm 93/13A-Appendix II, FAO, Rome, 1993.
3. Notermans, S., Gallhoff, G., Zwietering, M.H., and Mead, G.C., The HACCP concept: specification of criteria using quantitative risk assessment, *Food Microbiol.*, 12, 81, 1995.
4. Buchanan, R.L., The role of microbiological criteria and risk assessment in HACCP, *Food Microbiol.*, 12, 421, 1995.
5. FAO/WHO, Application of Risk Analysis to Food Standards Issues, in Report of the Joint FAO/WHO Expert Consultation, WHO, Geneva, 1995, pp. 13–17.

6. CAC (Codex Alimentarius Commission), Principles and Guidelines for the Conduct of a Microbiological Risk Assessment, CAC/GL-30, FAO, Rome, 1999.

7. Hathaway, S.C. and Cook, R.L., A regulatory perspective on the potential uses of microbial risk assessment in international trade, *Int. J. Food Microbiol.*, 36, 127, 1997.

8. Lammerding, A.M., An overview of microbial food safety risk assessment, *J. Food Prot.*, 60, 1420, 1997.

9. Notermans, S. and Teunis, P., Quantitative risk analysis and the production of microbiologically safe food: an introduction, *Int. J. Food Microbiol.*, 30, 3, 1996.

10. Foegeding, P.M., Driving predictive modelling on a risk assessment path for enhanced food safety, *Int. J. Food Microbiol.*, 36, 87, 1997.

11. Nauta, M.J., Modelling bacterial growth in quantitative microbiological risk assessment: is it possible? *Int. J. Food Microbiol.*, 73, 297, 2002.

12. ICMSF, *Microbiological Testing in Food Safety Management,* Kluwer Academic/Plenum, New York, 2002.

13. Huss, H.H., Reilly, A., and Embarek, P.K.B., Prevention and control of hazards in seafood, *Food Control*, 11, 149, 2000.

14. Buchanan, R.L. and Whiting, R.C., Risk assessment and predictive microbiology, *J. Food Prot.*, 31 (Suppl.), 1996.

15. McNab, W.B., A literature review linking microbial risk assessment, predictive microbiology, and dose–response modeling, *Dairy Food Environ. Sanitation*, 17, 405, 1997.

16. Walls, I. and Scott, V.N., Use of predictive microbiology in microbial food safety risk assessment, *Int. J. Food Microbiol.*, 36, 97, 1997.

17. van Gerwen, S.J.C. and Zwietering, M.H., Growth and inactivation models to be used in quantitative risk assessments, *J. Food Prot*, 61, 1541, 1998.

18. McMeekin, T.A., Olley, J.N., Ross, T., and Ratkowsky, D.A., *Predictive Microbiology: Theory and Application,* John Wiley & Sons, New York, 1993.

19. Skinner, G.E., Larkin, J.W., and Rhodehamel, E.J., Mathematical modeling of microbial growth — a review, *J. Food Safety*, 14, 175, 1994.

20. Panisello, P.J. and Quantick, P.C., Application of Food MicroModel predictive software in the development of hazard analysis critical control point (HACCP) systems, *Food Microbiol.*, 15, 425, 1998.

21. Baker, D.A., Application of modelling in HACCP plan development, *Int. J. Food Microbiol.*, 25, 251, 1995.

22. Lammerding, A.M. and Fazil, A., Hazard identification and exposure assessment for microbial food safety risk assessment, *Int. J. Food Microbiol.*, 58, 147, 2000.

23. Nauta, M.J., Separation of uncertainty and variability in quantitative microbial risk assessment models, *Int. J. Food Microbiol.*, 57, 9, 2000.

24. Cassin, M.H., Paoli, G.M., and Lammerding, A.M., Simulation modeling for microbial risk assessment, *J. Food Prot.*, 61, 1560, 1998.

25. Cassin, M.H., Lammerding, A.M., Todd, E.C.D., Ross, W., and McColl, R.S., Quantitative risk assessment for *Escherichia coli* O157:H7 in ground beef hamburgers, *Int. J. Food Microbiol.*, 41, 21, 1998.

26. Schlundt, J., Comparison of microbiological risk assessment studies published, *Int. J. Food Microbiol.*, 58, 197, 2000.

27. Oscar, T.P., The development of a risk assessment model for use in the poultry industry, *J. Food Safety*, 18, 371, 1998.

28. van Gerwen, S.J.C., te Giffel, M.C., Van't Riet, K., Beumer, R.R., and Zwietering, M.H., Stepwise quantitative risk assessment as a tool for characterization of microbiological food safety, *J. Appl. Microbiol.*, 88, 938, 2000.

29. Ross, T. and Sumner, J., A simple, spreadsheet-based, food safety risk assessment tool, *Int. J. Food Microbiol.*, 77, 39, 2002.
30. Mead, P.S., Slutsker, L., Dietz, V., McCaig, L.F., Bresee, J.S., Shapiro, C., Griffin, P.M., and Tauxe, R.V., Food-Related Illness and Death in the United States, Emerging Infectious Diseases, 5, http://www.cdc.gov/ncidod/eid/vol5no5/mead.htm, 2002.
31. FAO/WHO, Risk *Assessments of Salmonella in Eggs and Broiler Chickens: Interpretative Summary,* World Health Organization, Geneva, 2002.
32. Ross, T., Indices for performance evaluation of predictive models in food microbiology, *J. Appl. Bacteriol.*, 81, 501, 1996.
33. Notermans, S., Dufrenne, J., Teunis, P., Beumer, R., Giffel, M.T., and Weem, P.P., A risk assessment study of *Bacillus cereus* present in pasteurized milk, *Food Microbiol.*, 14, 143, 1997.
34. McElroy, D.M., Jaykus, L.A., and Foegeding, P.M., Validation and analysis of modeled predictions of growth of *Bacillus cereus* spores in boiled rice, *J. Food Prot.*, 63, 268, 2000.
35. Nauta, M.J., Litman, S., Barker, G.C., and Carlin, F., A retail and consumer phase model for exposure assessment of *Bacillus cereus*, *Int. J. Food Microbiol.*, 83, 205, 2003.
36. Carlin, F., Girardin, H., Peck, M.W., Stringer, S.C., Barker, G.C., Martinez, A., Fernandez, A., Fernandez, P., Waites, W.M., Movahedi, S., van Leusden, F., Nauta, M., Moezelaar, R., DelTorre, M., and Litman, S., Research on factors allowing a risk assessment of spore-forming pathogenic bacteria in cooked chilled foods containing vegetables: a FAIR collaborative project, *Int. J. Food Microbiol.*, 60, 117, 2000.
37. Peeler, J.T. and Bunning, V.K., Hazard assessment of *Listeria monocytogenes* in the processing of bovine milk, *J. Food Prot.*, 57, 689, 1994.
38. Bemrah, N., Sanaa, M., Cassin, M.H., Griffiths, M.W., and Cerf, O., Quantitative risk assessment of human listeriosis from consumption of soft cheese made from raw milk, *Prev. Vet. Med.*, 37, 129, 1998.
39. Buchanan, R.L., Damert, W.G., Whiting, R.C., and van Schothorst, M., Use of epidemiologic and food survey data to estimate a purposefully conservative dose–response relationship for *Listeria monocytogenes* levels and incidence of listeriosis, *J. Food Prot.*, 60, 918, 1997.
40. Lindqvist, R. and Westoo, A., Quantitative risk assessment for *Listeria monocytogenes* in smoked or gravad salmon and rainbow trout in Sweden, *Int. J. Food Microbiol.*, 58, 181, 2000.
41. FDA/Center for Food Safety and Applied Nutrition, USDA/Food Safety and Inspection Service, and Centers for Disease Control and Prevention. 2001. Draft assessment of the public health impact of foodborne *Listeria monocytogenes* among selected categories of ready-to-eat foods. www.foodsafety.gov/~dms/lmrisk.html. USDA/FSIS, Washington, DC, 20250.
42. Todd, E.C.D., Risk assessment of use of cracked eggs in Canada, *Int. J. Food Microbiol.*, 30, 125, 1996.
43. Whiting, R.C. and Buchanan, R.L., Development of a quantitative risk assessment model for *Salmonella enteritidis* in pasteurized liquid eggs, *Int. J. Food Microbiol.*, 36, 111, 1997.
44. USDA/FSIS, *Salmonella* Enteritidis Risk Assessment. Shell Eggs and Egg Products. Final report, USDA/FSIS, Washington, DC, 1998. www.fsis.usda.gov/ophs/risk/index.htm.
45. Whiting, R.C., Hogue, A., Schlosser, W.D., Ebel, E.D., Morales, R.A., Baker, A., and McDowell, R.M., A quantitative process model for *Salmonella* Enteritidis in shell eggs, *J. Food Sci.*, 65, 864, 2000.

46. Brown, M.H., Davies, K.W., Billon, C.M.P., Adair, C., and McClure, P.J., Quantitative microbiological risk assessment: principles applied to determining the comparative risk of salmonellosis from chicken products, *J. Food Prot.*, 61, 1446, 1998.
47. Nauta, M.J. and van der Giessen, J.W.B., Human exposure to *Mycobacterium paratuberculosis* via pasteurized milk: a modeling approach, *Vet. Record*, 143, 293, 1998.
48. USDA/FSIS, Draft Risk Assessment of the Public Health Impact of *Escherichia coli* O157:H7 in Ground Beef, USDA/FSIS, Washington, DC, 2002. www.fsis.usda.gov/oppde/rdad/frpubs/00-23nreport.pdf.
49. Marks, H.M., Coleman, M.E., Lin, C.-T.J., and Roberts, T., Topics in microbial risk assessment: dynamic flow tree process, *Risk Anal.*, 18, 309, 1998.
50. Hoornstra, E. and Notermans, S., Quantitative microbiological risk assessment, *Int. J. Food Microbiol.*, 66, 21, 2001.

9 Modeling the History Effect on Microbial Growth and Survival: Deterministic and Stochastic Approaches

József Baranyi and Carmen Pin

CONTENTS

0-8493-1237-X/04/$0.00+$1.50
© 2004

9.1 INTRODUCTION

The objective of food safety microbiology is to eliminate or at least significantly reduce the concentration of pathogenic microbes in food and to prevent their resuscitation and multiplication. Predictive microbiology focuses on the quantitative methods used to achieve this goal.

In a mathematical sense, a predictive model is a mapping between the *environment* of the microbial cells and their *response*. While primary models describe how the microbial cell concentration changes with time in a constant environment (see Chapter 2), a secondary model (such as *D values* vs. temperature; see Chapter 3) expresses a typical parameter of the response as a function of the environmental factors. Both the environment and the response are characterized by one or more, possibly *dynamic*, variables (i.e., they may change with time during observation). Temperature changing with time is an example of a dynamic environment (see Chapter 7); growth/survival curves are examples of dynamic responses, either in constant or dynamic environments.

In accordance with the standard approach to modeling, microbial responses are studied from a distinguished time point — the *zero time* of observation. Naturally, this zero time is when the environment of the cells undergoes a sudden change. Researchers should know, or at least implicitly assume the conditions at this zero time to provide *initial values* for modeling. Such initial values are, for example, the initial composition of natural flora (in the case of food environment) or the inoculum level and initial physiological state of the studied cells. Respectively, we can speak about *actual* (current) and *previous* (historical or preinoculum) environments — the initial values are the result of the previous environment.

A major obstacle in the development of predictive microbiology is that it is difficult to acquire data at low levels of cell concentration. Currently, commonly available counting methods can only measure cell concentrations of higher than *approximately* 10 cells per gram of sample. Therefore, it is frequently problematic to validate predictive microbiology models since most of them focus on situations when the cell concentration is relatively low and the measurements are inaccurate.

Growth modeling concentrates on the early stages of microbial growth in the actual environment. As mentioned, the previous environment affects growth in the actual environment, which effect gradually diminishes after inoculation. This is called the *adjustment* process; the time during which it happens is called the *lag period*.

Death modeling deals with decay in microbial concentration having two main causes: thermal inactivation, when death is relatively fast; and nonthermal death, which is generally a much slower process. Although the physiological mechanisms of the two can be completely different, it is possible to construct mathematical models for them in an analogous way. What is more, to some extent, growth and death modeling can also be discussed analogously. Accordingly, we show some basic techniques to model the effect of history on growth and death in a parallel way.

We also present a comparative review of the applied deterministic and stochastic models. Deterministic models can always be treated as sort of "averaged out" versions of stochastic models. However, "smoothing down" the biological variability means a certain loss of information — the question is just how much. Biology itself always has stochastic elements, but sometimes they are small enough (or we hope they are small enough) to be negligible. In this case, the applied model (and its one or more variables and processes) is reduced to be deterministic, which is simpler and easier to handle. However, such a reduction can be misleading when the random elements are significant. For example, when modeling the lag period of a population of relatively few cells (the "dormant" phase of the "log conc. vs. time curve" before exponential growth), the random effects originating from the variability among the cells cannot be ignored. Similarly, when modeling the number of survivors in an adverse environment, the stochastic variability of the resistance of individual cells can also become significant.

Many authors have studied the effect of history on the lag time before growth (Augustin et al., 2000a; Augustin and Carlier, 2000; Beumer et al., 1996; Bréand et al., 1999; Buchanan and Klavitter, 1991; Dufrenne et al., 1997; Gay et al., 1996; Hudson, 1993; Mackey and Kerridge, 1988; Membré et al., 1999; Stephens et al., 1997; Walker et al., 1990; Wang and Shelef, 1992) and on the shoulder period before exponential death (Cotterill and Glauert, 1969; Mañas et al., 2001; Ng et al., 1962; Sherman and Albus, 1923; Strange and Shon, 1964). In this chapter, we attempt to provide a simple modeling approach to take into account the history effect via the concept of initial work/damage done before the exponential growth/death phase. First, we apply a deterministic model at population level (i.e., when the population size is represented by a single continuous variable), after which we discuss similar questions at single-cell level, using stochastic processes, when the variability among individual cells is also taken into account.

9.2 MODELING THE HISTORY EFFECT AT POPULATION LEVEL (DETERMINISTIC MODELING)

9.2.1 TRADITIONAL PARAMETERS

Throughout this chapter, $x(t)$ denotes the cell concentration at the time t, and $y(t)$ denotes its natural logarithm. The slope of the "$y(t)$ vs. time" curve is the (instantaneous) specific rate, denoted by $\mu(t)$. Its maximum value, μ_{max}, is the maximum specific rate. It is positive for growth and negative for death (Figure 9.1a and Figure 9.1b). The maximum specific rate is measured at the inflexion point of the curve, if it has a sigmoid shape (Figure 9.1a). If the model is biphasic (Figure 9.1b), then this maximum does not necessarily exist. In this case, μ_{max} is meant in an asymptotic sense: $\mu(t) \rightarrow \mu_{max}$ as $t \rightarrow \infty$. In fact, we define the biphasic class of models by the criterion that $\mu(t)$ converges monotonically from a small, $\mu(0)$ initial rate, to a finite value.

In the case of growth, the most frequently modeled parameter is the maximum specific growth rate, or one of its rescaled versions: the r_{max} maximum rate in terms of $\log_{10}$ cell concentration, or the doubling time, T_d. Note the relation between them:

$$r_{max} = \mu_{max}/\ln 10; \qquad T_d = \ln 2/\mu_{max} \qquad (9.1)$$

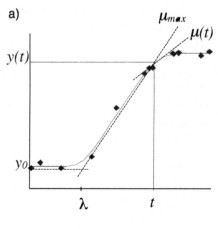

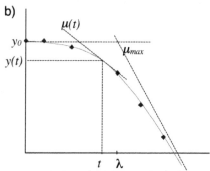

FIGURE 9.1 (a) Main parameters of a sigmoid growth curve. The maximum specific growth rate μ_{max} can be measured at the inflexion point of the curve. (b) Main parameters of a biphasic survival curve. The instantaneous specific death rate converges to a limit value μ_{max}.

In the case of survival (or inactivation), instead of the analogous "halving time," it is more common to use the time to a decimal reduction, called D value:

$$D = -\ln 10/\mu_{max} \tag{9.2}$$

The reason for this slight inconsistency is historical: the first predictive models were created to describe thermal inactivation processes and, at high temperatures, the decimal reduction times were more practical to use than halving times.

Some authors use the maximum specific rate and some the rate of the $\log_{10} x(t)$ function, which can be obtained by an $\ln(10) \approx 2.3$ conversion factor from $\mu(t)$. Most frequently, cell concentration data are given in terms of $\log_{10}$ of the cell number per volume. However, the maximum specific rate has a more universal theoretical meaning as shown by the formulae below:

$$\frac{dy}{dt} = \frac{d(\ln x)}{dt} = \frac{dx\,/\,x}{dt} \mu(t) \tag{9.3}$$

that is,

$$\frac{\Delta x}{x} \approx \mu_{max} \Delta t \qquad (9.4)$$

The left-hand side of the latter equation denotes a relative increase or decrease of the population in Δt time. Therefore, the following interpretation can be given for the maximum specific rate: the probability that, at the time t, a single cell divides (growth) or dies (death), during the small $[t, t + \Delta t]$ interval, is about $\mu(t)\Delta t$; i.e., proportional to the length of Δt, with the factor $\mu(t)$. This is a fundamental link between the specific growth rate measured at population level and the probability of division at single-cell level.

Another important, frequently modeled parameter is the lag time (for growth), or shoulder period (for survival), which is denoted by λ throughout this chapter. For sigmoid curves (Figure 9.1a), it is traditionally defined as the time when the tangent of the $y(t)$ curve with the maximum slope crosses the initial level, $y_0 = \ln x_0$ (Pirt, 1975). (It is irrelevant whether the natural logarithm or $\log_{10}$ is used to define the lag or shoulder.) For biphasic curves (Figure 9.1b), to find a good "geometrical" definition is not so straightforward as pointed out below.

The definition of lag/shoulder does not have the strong basis that would connect the parameter to a single-cell equivalent, as in the case of the specific rate. Baranyi (1998) discussed the mathematical relation between the lag of the population (as defined above) and the lag times of the participating individual cells (see later in this chapter). He showed that the population lag depends on the subsequent specific growth rate and the initial cell number, even if the individual lag times do not. Another anomaly can be shown by means of biphasic models. It is natural to expect that, if a series of tangents drawn to the $y(t)$ curves converges to a limit value (see Figure 9.1b), then the intersections with the initial y_0 level also converge, in which case the lag could be defined by the limit value. Baranyi and Pin (2001) showed a biphasic death model where this was not true; the intersection, denoting the end of the shoulder period, would not converge, as $t \to \infty$, although the specific death rate did converge.

9.2.2 Replacing the Lag Parameter with a Physiological State Parameter

For growth, the "waiting time" of the individual cells is due to an adjustment process before the exponential growth. For survival curves, the shoulder period is due to an initial resistance before exponential death (such as in the case of the multitarget theory, Hermann and Horst, 1970). As shown by Baranyi (2002), it is not easy to relate the distribution of the "waiting times" of individual cells to the apparent lag/shoulder parameter of the population, measured in the traditional way.

Baranyi and Roberts (1994) suggested that the lag should be considered as a derived parameter, a consequence of both the actual environment (via the specific growth rate), and the history of the cells (via a newly introduced parameter, the *initial physiological state* of the cells). Their method can be outlined as follows.

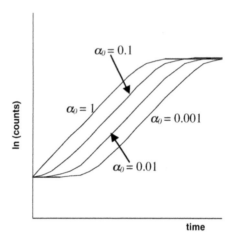

FIGURE 9.2 Interpretation of the α_0 physiological state. $h_0 = -\ln(\alpha_0) = \lambda\mu_{max}$ is the "work to be done," which is the same for different temperatures. h_0 is an initial value, as is $\ln x_0$.

Just as y_0 is an *initial* value, we can introduce another value in order to quantify the "suitability" of the population to the current environment (i.e., the history effect). Let this quantity be denoted by α_0 (which comes from the term *adjustment function,* in the original paper). Let it be a dimensionless number between 0 and 1. For growth, consider it as the fraction of the initial cells, which, without a lag, would be able to produce the same final growth as the original x_0 initial cells with a lag. (Although the real situation is not like this, it is easier to think about α_0 this way. In reality, α_0 is the mean value of the respective parameters of the individual cells. The point is, as we will see later, that it does not matter if we think of it as "α_0, the fraction of the cells that are able to grow," or "each cell is α_0-suitable.")

By simple geometry (see Figure 9.2), it can be shown that

$$\lambda = \frac{-\ln(\alpha_0)}{\mu_{max}} \tag{9.5}$$

This formula expresses our expectation that the lag time depends on both the actual environment, represented by μ_{max}, and the history of the cells, characterized by α_0. The history effect appears as an initial value, and this will be called the initial physiological state. Note that it does not represent simply the "health" of the cells, rather their suitability to the actual environment. Its extreme values are $\alpha_0 = 0$, in which case the lag is infinitely long, and $\alpha_0 = 1$, when all the cells enter the exponential phase immediately and the lag time is 0. The effect of the initial physiological state on the lag is shown in Figure 9.3.

In terms of biological interpretability, the main advantages of using α_0, instead of lag, are:

1. It is an initial value, affecting the lag, and this is in accord with our mechanistic thinking.

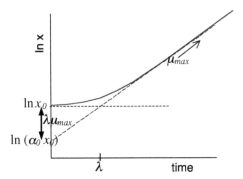

FIGURE 9.3 Depending on the α_0 physiological state, the lag can change while the maximum specific growth rate remains the same.

2. It has a very strong connection with similar parameters of the individual cells: the physiological state of the population is equal to the average of the physiological states of the initial cells, irrespective of their number (Baranyi, 1998).
3. It can be interpreted, via the individual cells, even if no traditional lag can be measured on the bacterial curve (for example, when it does not reach its inflexion point).

9.2.3 A RESCALING OF THE PHYSIOLOGICAL PARAMETER PROVIDES A NEW INTERPRETATION

An interpretation of the history effect is also possible by means of the inverse of the physiological state. Introduce $h_0 = \ln(1/\alpha_0) = -\ln \alpha_0$. It can be conceived as an initial *hurdle*, or the "work to be done" during the lag (Robinson et al., 1998; see Figure 9.2). As we see, it is the product of the lag and the maximum specific growth rate. Observe that, using the doubling time, $T_d = \ln 2/\mu_{max}$ relation, the following formula can be obtained:

$$h_0 = \ln 2 \frac{\lambda}{T_d} \tag{9.6}$$

The quantity λ/T_d is often called "relative lag" (McMeekin et al., 2002) and it is essentially the same as h_0. Some authors have reported that it is unaffected by small changes of temperature (Pin et al., 2002; Robinson et al., 1998) and pH (McKellar et al., 2002a). This can be explained as follows: the more suitable the cells are to the new environment, the less work they have to carry out to adapt. Therefore, the lag is longer at lower temperatures because the "work to be done," after the inoculation, is carried out more slowly, although the amount of work is the same (Figure 9.4). Authors studying the effect of history and actual growth conditions on the work to be done during the lag include Delignette-Muller (1998), Robinson et al. (1998), Pin et al. (2002), and Augustin et al. (2000a).

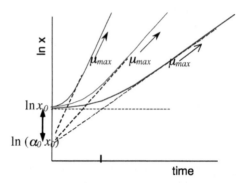

FIGURE 9.4 As the temperature decreases, the work to be done (the difference between ln $x_\bullet$ and $\ln(\alpha_\bullet\, x_\bullet)$) is the same, while the maximum specific growth rate $\mu_{\max}$ becomes smaller and the lag becomes longer. The work to be done during the lag is also carried out at slower rates.

The parameter $h_\bullet = -\ln \alpha_\bullet$, the "work to be done" before the exponential phase, plays the role of a bridge between the history and the actual environment. With an analogy, imagine that cells are crossing this bridge. If its angle is horizontal ($h_\bullet = 0$; no difficulty to overcome), then there is no lag. If it is vertical ($h_\bullet = \infty$), then the work to be done is infinitely big, the lag is infinitely long. On the other hand, if the temperature is higher then every movement is quicker and so is the crossing (lag), although the work to be done is the same (Figure 9.4). It can be expected that this work is smaller when adjusting to a more favorable environment than the opposite way (Delignette-Muller, 1998; Robinson et al., 1998). This theory explains why the environmental effect on the lag has always been much less accurately predicted (Dufrenne et al., 1997; McKellar et al., 2002b; McMeekin et al., 2002; Robinson et al., 1998; Whiting and Bagi, 2002) than the specific growth rate. The lag also depends on the history, the details of which often have not even been recorded, let alone taken into account in the model.

9.3 MODELING THE HISTORY EFFECT AT SINGLE-CELL LEVEL (STOCHASTIC MODELING)

9.3.1 NATURE OF STOCHASTIC MODELING

Finer details of the effect of history on growth/death can be seen through a closer look at the distribution of the division/survival times of individual cells. As mentioned in Section 9.1, if only deterministic models are used to describe responses of microbial populations then the variability among single cells is ignored. This is a problem because predicting microbial variability is increasingly important for Quantitative Microbial Risk Assessment. This variability of the responses increases under stress conditions, which makes predictions even less accurate.

A characteristic feature of stochastic modeling is that it sometimes provides unexpected results. An example of this is the relation between the doubling time of a growing population and the average generation time of the individual cells in the population. Depending on the distribution of the individual generation times, these two quantities can be markedly different. If the generation times follow the classical exponential distribution, then their average is $1/\mu_{max}$, while the doubling time is known to be $\ln 2/\mu_{max}$, where μ_{max} is the (constant) maximum specific growth rate of the population. Baranyi and Pin (2001) described the exponential growth as a Poisson birth process and showed that with this approach the doubling time depends on the number of cells, from $1/\mu_{max}$, when the number of cells is 1, converging to $\ln 2/\mu_{max}$, when the number of cells increases.

9.3.2 PHYSIOLOGICAL STATE AND LAG PARAMETERS FOR SINGLE CELLS

Let $N(t)$ be the *number* of cells in a defined space, at the time t (remember, that $x(t)$ denoted the *concentration* of the cells). Let the lag for the ith cell of the initial population be denoted by τ_i ($i = 1,2,..., N$). We assume that they are identically distributed independent random variables, and their mean value is τ. Note that τ_i is less than the time to the first division of the ith cell, which also includes the first generation time.

Define the physiological states α_i ($i = 1,2,...,N$) for the individual cells by $\alpha_i = \exp(-\mu_{max}\tau_i)$, analogously to the deterministic theory. A basic theorem of Baranyi (1998) is that the average of the individual physiological states is the same as the physiological state of the population, irrespective of the distribution of the individual lag times. Moreover, if $\lambda(N)$ denotes the (traditionally defined) lag of the population consisting of N cells, then

$$\lambda(N) = -\frac{1}{\mu}\ln\frac{\sum_{i=1}^{N}\alpha_i}{N} = -\frac{1}{\mu}\ln\frac{\sum_{i=1}^{N}e^{-\mu\tau_i}}{N} \tag{9.7}$$

This formula is demonstrated in Figure 9.5. The more cells there are in the inoculum, the shorter is the expected population lag, converging to a limit value, λ_{min}. Notice that the correct model for single-cell-generated curve is bi-phasic; the curvature for higher inoculum is caused by the distribution of the individual lag times.

Various authors have reported that at a very low inoculum (approximate to 1 cell per total volume of sample), the lag time is longer than expected (Augustin et al., 2000b; Gay et al., 1996; Stephens et al., 1997). However, as the above formula shows, this finding can also be explained by statistical means without extra biological assumptions. The lag time decreases with the initial cell number because the exponentially growing subgenerations of the cells with the shortest lag times will quickly dominate the whole population.

In particular, Baranyi and Pin (2001) showed the mathematical relation between the respective parameters (mean and variance) of the individual lag times and those of the population consisting of N cells. From these formulae, the distribution of individual lag times is *per se* different from the distribution of the population lag times. This was confirmed by the observation of Stephens et al. (1997) on heat-

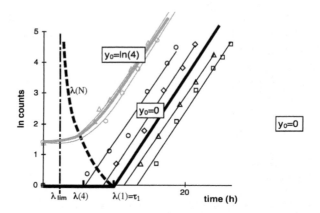

FIGURE 9.5 Simulating four single-cell-generated growth curves ($y_\bullet = 0$) growing separately, and two separate growth curves generated by four cells each ($y_\bullet = \ln 4$). The traditional lag times λ of the subpopulations are estimated by curve fitting. Observe that the $\lambda(N)$ mean population lag converges to a limit value λ_{lim} while its variance (the spread of the growth curves as a function of their inoculum) converges to zero (see Figure 9.6). Symbols: simulated points; continuous thin lines: fitted curves; continuous thick lines: expected curves.

injured *Salmonella* cells. The calculations provided another reason why the lag time should not be considered as a primary growth parameter but as the result of the initial physiological state ($\alpha_\bullet$), the inoculum size and the maximum specific growth rate.

Smelt et al. (2002) estimated the population lag as the minimum of the individual lag times. Our formula shows that the population lag is greater than the minimum of the individual lag times, but closer to it than to the arithmetical mean.

9.3.3 A SIMULATION PROGRAM

To demonstrate the effect of the inoculum on the distribution of the population lag, we developed a simulation program. One hundred growth curves from different inoculum sizes were generated. The distribution functions used by the simulation were based on the theory of Baranyi and Pin (2001).

We assumed that the individual lag times followed the gamma distribution, with the parameters p and v, so the mean individual lag is $\tau = p/v$. The biological interpretation of this scenario is that the cells have to carry out p consecutive tasks, and the times required by the individual tasks are independent, exponentially distributed random variables, with the parameter v (i.e., the average time required for each task was $1/v$).

The parameter p represented the amount of work to be done during the lag time ($h_\bullet = p$). Partly for simplicity, partly for mechanistic reasons, $v = \mu_{max}$ was assumed. The generation times, g_i, of individual cells, after the lag phase, followed the exponential distribution with the parameter μ_{max}.

With the usual terminology for stochastic processes, the initial state of the system is $(N,0)$, meaning that there are N cells in the lag phase and 0 in the exponential phase at the starting time ($t_\bullet$). The time t_1 of the first division in the population is that of the cell with the smallest value for the sum of its lag time, τ_i, and its first generation time, g_i.

$$t_1 = \min(\tau_1 + g_1, \ ... \ \tau_N + g_N)$$

After the first division, $N - 1$ cells remain in lag phase while 2 cells are in exponential growth phase. Thus the state of the system is $(N - 1, 2)$. The time t_2 of the second division will be the minimum of the first division times among the $N - 1$ cells remaining in the lag phase and the division times of the two daughter cells:

$$t_2 = \min(\tau_1 + g_1, \ ... \ \tau_{N-1} + g_{N-1}, \ t_1 + g_{N+1}, \ t_1 + g_{N+2})$$

This second division can happen to a cell in exponential or in lag phase.

1. If a cell in lag phase divides, then the state of the system will be $(N - 2, 4)$ and the time of the third division will be $t_3 = \min(\tau_1 + g_1, \ ..., \ \tau_{N-2} + g_{N-2}, \ t_1 + g_{N+1}, \ t_1 + g_{N+2}, \ t_2 + g_{N+3}, \ t_2 + g_{N+4})$.
2. If a cell in exponential phase divides, then the system jumps into the state $(N - 1, 3)$. Apart from numbering the daughter cells, the time to the third division will be $t_3 = \min(\tau_1 + g_1, \ ..., \ \tau_{N-1} + g_{N-1}, \ t_1 + g_{N+2}, \ t_2 + g_{N+3}, \ t_2 + g_{N+4})$.

The iteration can be continued in a similar manner and this is what our simulation program did, until the population had a given number of cells. One hundred growth curves were generated for each selected inoculum size. The parameters h_0 and μ_{max} were chosen close to practical values: $h_0 = 4$ and $\mu_{max} = 0.5 \ h^{-1}$. Thus the gamma parameters to simulate the individual lag times were $p = 4$ and $v = 0.5 \ h^{-1}$ (mean individual lag $= 8 \ h$) and the parameter of the exponential distribution to simulate the individual generation times was $0.5 \ h^{-1}$ (mean generation time $= 2 \ h$). The traditionally interpreted lag time of each population curve was estimated by fitting a biphasic (no upper asymptote) version of the model of Baranyi and Roberts (1994). Figure 9.5 shows typical examples of the generated population curves (originating from one and four cells). As can be seen, in the case of a single initial cell, when the inoculum is $\ln x = 0$, the population lag (λ, measured by the tangent to the exponential phase) is close to the theoretical individual lag ($\tau_1 = 8 \ h$). If the inoculum size increases then, from the formula

$$\lambda(N) \xrightarrow{\ N \to \infty\ } \lambda_{lim} = \frac{p}{\mu} Ln\left(1 + \frac{\mu\tau}{p}\right) \tag{9.8}$$

the population lag converges to $\lambda_{lim} = 8 \cdot \ln 2 = 5.54$. This convergence is shown in Figure 9.5. Figure 9.6 shows the distribution of the population lag times as a function of the inoculum size.

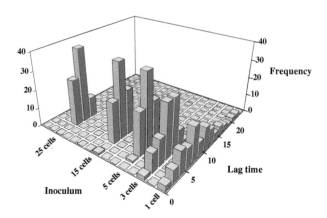

FIGURE 9.6 As the inoculum size increases, the population lag time decreases, converging to a limit value. Its variance converges to zero.

9.3.4 LAG/SURVIVAL DISTRIBUTION OF SINGLE CELLS AND LAG/SHOULDER OF POPULATION CURVES

A general result from Baranyi (2002) connects the expected population growth curve with the distribution of single-cell kinetics:

$$x(t) = N\left(\int_0^t e^{\mu \cdot (t-s)} f(s)ds + \int_t^\infty f(s)ds\right) \qquad (9.9)$$

where $f(t)$ denotes the (common) probability density function (pdf) of the individual lag times. It is analogous to the formula of Körmendy et al. (1998) for death curves:

$$x(t) = N\int_0^\infty e^{-s \cdot t} g(s)ds \qquad (9.10)$$

where $g(s)$ is the probability density function of the individual cells' death rate (i.e., the reciprocal of their survival time). The big difference is that while the probability distribution of the individual survival times determines the whole death curve (actually by a Laplace transformation, as the formula shows) that of the individual lag times does not. In both models, the first compartments are "dormant" states, but the second ones are totally different: for growth, it is the state of exponential growth, characterized by the specific growth rate; for survival curves, it is a "death phase" without any parameter.

The formulae above represent a one-to-one mapping between the $f(t)$ distribution and the population growth/death curve. The problem is that, in practice, this mapping works only in one direction. The expected population curve can be easily derived from the distribution of the individual lag times, but vice versa it would need unrealistically accurate data. This is an example where the performance of stochastic techniques is much better than that of their deterministic counterparts.

9.3.5 HISTORY EFFECT ON THE SHOULDER PERIODS OF SURVIVAL CURVES

For survival curves, with the notations as above, let τ_i be the time for the ith cell to die, and assume that it is distributed exponentially, with the mean value τ (pure Poissonian death process). In this case, $F(t) = 1 - e^{-t/\tau}$, so the survival curve is $y(t) = \ln N - t/\tau$. Therefore, in this case, the survival curve is linear and there is no shoulder parameter: $\lambda = 0$.

Analogously with our approach to modeling the distribution of individual lag times for growth, consider now the situation when τ_i ($i = 1,..., N$) follow the gamma distribution, with the parameters $p \geq 1$ and $v > 0$, where the expected value of the survival time for a cell is $\tau = p/v$. This is interpreted by Baranyi and Pin (2001) in the following way: the cell needs p damaging hits (see the multihit theory in Casolari, 1988), and the times θ_j ($j = 1,..., p$) between the hits are independent, exponentially distributed variables, with a common mean value $\theta = 1/v$. Then the survival time of the ith cell is

$$\tau_i = \sum_{j=1}^{p} \theta_j \qquad (9.11)$$

therefore τ_i ($i = 1,..., N$) are gamma distributed. In this case, as shown by the authors, the derivative of $y(t)$ converges to a constant, namely v, the limit shoulder still does not exist if $p > 1$. In practice, the shoulder would be measured by v and the smallest detectable value of $y(t)$, which means that the shoulder measurement would depend on the precision of the measurement method! This inconsistency makes the shoulder parameter unsuitable for modeling. Instead, the p parameter quantifying the "damage to be done" should be modeled, which is analogous to the h_0 parameter for growth curves.

Another well-known interpretation of the shoulder is the so-called multitarget model (Hermann and Horst, 1970). According to that, a cell has p targets that are being hit synchronously (not consecutively as in the previous case). The times needed to destroy the targets, θ_j ($j = 1,..., p$), are independent, exponentially distributed variables with the common mean value of $1/v$ and the cell is live until all targets are inactivated. Therefore, the survival time of the ith cell is

$$\tau_i = \max_{1 \leq j \leq p} \theta_j \qquad (9.12)$$

and $F(t) = (1 - e^{-vt})^p$, where $p \geq 1$. The population survival curve is now

$$y(t) = \ln N + \ln(1 - (1 - e^{-vt})^p) \qquad (9.13)$$

which converges to the

$$y_a(t) = \ln N - \nu \cdot (t - \ln p/\nu) \qquad (9.14)$$

linear asymptote. Therefore, the limit shoulder parameter can be calculated as

$$\lambda = \ln p/\nu \qquad (9.15)$$

This shows another analogy to the modeling of the history effect on growth: the "ln p" quantity could be used to characterize the initial "damage to be done," and the shoulder period depends on the magnitude of the initial damage and on the rate as the damage is being done.

9.4 CONCLUDING REMARKS

In this chapter, we showed that the history effect can be quantified either by the "initial physiological state" or the "work to be done" parameters; they are simply rescaled versions of each other. The traditional lag parameter has become a derived one, useful for static conditions only. By introducing a variable for the rate at which the work is carried out before the exponential phase, the model can be used for dynamic conditions, too. This concept was successfully applied in a series of papers (Baranyi et al., 1995, 1996) describing the entire bacterial growth profile (not only the exponential phase) under fluctuating conditions.

The priority of α_0 is purely a modeling concept; it does not mean that when fitting a bacterial growth curve, α_0 and not the lag should be estimated. The formula $\lambda = -\ln\alpha_0/\mu_{max}$ to calculate the lag expresses the view that both the history and the current environment influence the lag. This has practical consequences when developing more complex, dynamic models or, for example, when developing predictive software packages. These commonly require the user to provide values for the environmental factors, from which the maximum specific rate (or one of its rescaled versions) and a *default* lag are predicted. If the user gives the inoculum, then the whole growth curve can be constructed, too. What our theory suggests is that it is not only the inoculum that cannot be predicted from the current environment, but also the initial physiological state, α_0. This is shown in Figure 9.2: there are two independent initial points on the y axis:

1. An inoculum value (y_0)
2. A point through which the tangent drawn to the inflexion will pass (which is $h_0 = -\ln\alpha_0$ lower than the inoculum)

Sometimes the value of the initial physiological state should be simply $\alpha_0 = 1$, expressing the fact that the preinoculation conditions are the same as the current environment.

As has been mentioned, many features of the shoulder period before the exponential death phase can be discussed analogously to those of the lag period before the exponential growth phase. Baranyi and Pin (2001) showed that the multitarget theory of inactivation (Hermann and Horst, 1970), modeling the shoulder before the exponential death rate, provides a parallel between the initial "damage to be done" for death and the h_0 "work to be done" for growth. The "bridge" between the history and the current environment is characterized by h_0 either for growth (Figure 9.2) or death (the mirror image of Figure 9.2).

The question can be raised of why the distribution of individual lag times is important once different distributions can result in practically the same population curve. The answer lies in the interest of Quantitative Microbial Risk Assessment in the distribution of the quantity "time to a certain level" (such as time to infective dose or time to legally allowed concentration). The distribution of that time depends heavily on the distribution of individual lag times.

Modeling the effect of history is more efficient with stochastic techniques, but also more data- and maths-demanding. Deterministic models are able to produce "averaged out" solutions only, and studying the variability around those predictions is vital in Quantitative Microbial Risk Assessment.

ACKNOWLEDGMENT

The authors are indebted to Susie George for preparing the manuscript. This work was funded by the EC project QLRTD-2000-01145 and the IFR project CSG 434.1213A.

REFERENCES

Augustin, J.C., L. Rosso, and V. Carlier. 2000a. A model describing the effect of temperature history on lag time for *Listeria monocytogenes*. *Int. J. Food Microbiol.* 57, 169–181.

Augustin, J.C., A. Brouillaud Delattre, L. Rosso, and V. Carlier. 2000b. Significance of inoculum size in the lag time of *Listeria monocytogenes*. *Appl. Environ. Microbiol.* 66, 1706–1710.

Augustin, J.C. and V. Carlier. 2000. Mathematical modelling of the growth rate and lag time for *Listeria monocytogenes*. *Int. J. Food Microbiol.* 56, 29–51.

Baranyi, J. 1998. Comparison of stochastic and deterministic concepts of bacterial lag. *J. Theor. Biol.* 192, 403–408.

Baranyi, J. 2002. Stochastic modelling of bacterial lag phase. *Int. J. Food Microbiol.* 73, 203–206.

Baranyi, J. and C. Pin. 2001. A parallel study on bacterial growth and inactivation. *J. Theor. Biol.* 210, 327–336.

Baranyi, J. and T.A. Roberts. 1994. A dynamic approach to predicting bacterial growth in food. *Int. J. Food Microbiol.* 23, 277–294.

Baranyi, J., A. Jones, C. Walker, A. Kaloti, and B.M. Mackey. 1996. A combined model for growth and thermal inactivation of *Brochothrix thermosphacta*. *Appl. Env. Microbiol.* 62, 1029–1035.

Baranyi, J., T.P. Robinson, A. Kaloti, and B.M. Mackey. 1995. Predicting growth of *Brochothrix thermosphacta* at changing temperature. *Int. J. Food Microbiol.* 27, 61–75.

Beumer, R.R., M.C. te Giffel, E. de Boer, and F.M. Rombouts. 1996. Growth of *Listeria monocytogenes* on sliced cooked meat products. *Food Microbiol.* 13, 333–340.

Bréand, S., G. Fardel, J.P. Flandrois, L. Rosso, and R. Tomassone. 1999. A model describing the relationship between regrowth lag time and mild temperature increase for *Listeria monocytogenes. Int. J. Food Microbiol.* 46, 251–261.

Buchanan, R.L. and L.A. Klavitter. 1991. Effect of temperature history on the growth of *Listeria monocytogenes* at refrigeration temperatures. *Int. J. Food Microbiol.* 12, 235–246.

Casolari, A. (1988). Microbial death. In *Physiological Models in Microbiology. Vol. II,* Bazin, M.J. and Prosser, J.I., Eds. CRC Press, Boca Raton, FL, Ch. 7.

Cotterill, O.J. and J. Glauert. 1969. Thermal resistance of *Salmonellae* in egg yolk products containing sugar or salt. *Poult. Sci.* 48, 1156–1166.

Delignette-Muller, M.L. 1998. Relation between the generation time and the lag time of bacterial growth kinetics. *Int. J. Food Microbiol.* 43, 97–104.

Dufrenne, J., E. Delfgou, W. Ritmeester, and S. Notermans. 1997. The effect of previous growth conditions on the lag phase time of some foodborne pathogenic microorganisms. *Int. J. Food Microbiol.* 34, 89–94.

Gay, M., O. Cerf, and K.R. Davey. 1996. Significance of pre-incubation temperature and inoculum concentration on subsequent growth of *Listeria monocytogenes* at 14°C. *J. Appl. Bacteriol.* 81, 433–438.

Hermann, D. and J. Horst. 1970. *Molecular Radiation Biology.* Springer, Berlin.

Hudson, J.A. 1993. Effect of pre-incubation temperature on the lag time of *Aeromonas hydrophila. Lett. Appl. Microbiol.* 16, 274–276.

Körmendy, I., L. Körmendy, and A. Ferenczy. 1998. Thermal inactivation kinetics of mixed microbial populations. A hypothesis paper. *J. Food Eng.* 38, 439–453.

Mackey, B.M. and A.L. Kerridge. 1988. The effect of incubation temperature and inoculum size on growth of *Salmonellae* in minced beef. *Int. J. Food Microbiol.* 6, 57–65.

Mañas, P., R. Pagan, I. Leguerinel, S. Condon, P. Mafart, and F. Sala. 2001. Effect of sodium chloride concentration on the heat resistance and recovery of *Salmonella typhimurium. Int. J. Food Microbiol.* 63, 209–216.

McKellar, R.C., X. Lu, and K.P. Knight. 2002a. Growth pH does not affect the initial physiological state parameter (p_0) of *Listeria monocytogenes* cells. *Int. J. Food Microbiol.* 73, 137–144.

McKellar, R.C., X. Lu, and K.P. Knight. 2002b. Proposal of a novel parameter to describe the influence of pH on the lag phase of *Listeria monocytogenes. Int. J. Food Microbiol.* 73, 127–135.

McMeekin, T.A., J. Olley, D.A Ratkowsky, and T. Ross. 2002. Predictive microbiology: towards the interface and beyond. *Int. J. Food Microbiol.* 73, 395–407.

Membre, J.M., T. Ross, and T. McMeekin. 1999. Behaviour of *Listeria monocytogenes* under combined chilling processes. *Lett. Appl. Microbiol.* 28, 216–220.

Ng, H., J.L. Ingraham, and A.G. Marr. 1962. Damage and depression in *Escherichia coli* resulting from growth at low temperatures. *J. Bacteriol.* 84, 331–339.

Pin, C., G.D. García de Fernando, J.A. Ordóñez, and J. Baranyi. 2002. Analysing the lag-growth rate relationship of *Yersinia enterocolitica. Int. J. Food Microbiol.* 73, 197–201.

Pirt, S.J. 1975. *Principles of Microbe and Cell Cultivation.* Blackwell, London.

Robinson, T.P., M.J. Ocio, A. Kaloti, and B.M. Mackey. 1998. The effect of the growth environment on the lag phase of *Listeria monocytogenes. Int. J. Food Microbiol.* 44, 83–92.

Sherman, J.M. and W.R. Albus. 1923. Physiological youth in bacteria. *J. Bacteriol.* 8, 27–139.

Smelt, J.P.P.M., G.D. Otten, and A.P. Bos. 2002. Modelling the effect of sublethal injury on the distribution of the lag times of individual cells of *Lactobacillus plantarum*. *Int. J. Food Microbiol.* 73, 207–212.

Stephens, P.J., J.A. Joynson, K.W. Davies, R. Holbrook, H.M. Lappinscott, and T.J. Humphrey. 1997. The use of an automated growth analyser to measure recovery times of single heat-injured *Salmonella* cells. *J. Appl. Microbiol.* 83, 445–455.

Strange, R.E. and M. Shon. 1964. Effects of thermal stress on viability and ribonucleic acid of *Aerobacter aerogenes* in aqueous suspension. *J. Gen. Microbiol.* 34, 99–114.

Walker, S.J., P. Archer, and J.G. Banks. 1990. Growth of *Listeria monocytogenes* at refrigeration temperatures. *J. Appl. Bacteriol.* 687, 157–162.

Wang, C. and L.A. Shelef. 1992. Behaviour of *Listeria monocytogenes* and the spoilage microflora in fresh cod fish treated with lysozyme and EDTA. *Food Microbiol.* 9, 207–213.

Whiting, R.C. and L.K. Bagi. 2002. Modelling the lag phase of *Listeria monocytogenes*. *Int. J. Food Microbiol.* 73, 291–295.

10 Models — What Comes after the Next Generation?

Donald W. Schaffner

CONTENTS

10.1 INTRODUCTION

This chapter will highlight two separate and generally unrelated areas of predictive food microbiology: models for cross-contamination and inoculum size (or models that consider the initial number of organisms present). These two classes of models are being included together here, because they represent some of the newer areas of predictive modeling that are less well developed compared to the more well-known and established research areas such as growth and inactivation modeling. This chapter will summarize the current state of research in these two rapidly evolving areas and will conclude with a short example describing preliminary investigations into the integration of these two fields of study.

10.2 CROSS-CONTAMINATION

Predictive models for microbial behavior have traditionally focused on describing increasing concentrations (as a result of multiplication) and decreasing concentrations (as a result of cell death). A number of lines of inquiry have pointed out the need for a third class of models that may be required in some cases: cross-contamination models. The three lines of inquiry all come from the interaction of risk assessment and epidemiology, and are related to three very different microorganisms: *Listeria monocytogenes*, *Campylobacter jejuni*, and food-borne viruses.

L. monocytogenes is a psychrotrophic pathogen that can cause mild illness in healthy adults and spontaneous abortion in pregnant women. The organism is easily destroyed by heating but readily recontaminates the cooked product prior to packaging.[25] While this recontamination is known to take place, (and can contribute to significant disease outbreaks) very few mathematical models are available for microbial risk assessors to use.

The situation with *Campylobacter* is slightly different. In this case, the organism is known to cause cross-contamination in a significant number of cases[1] but the means by which this occurs is not clear.[11] Risk assessment models for Campylobacteriosis have incorporated cross-contamination events during final preparation in a kitchen environment, but as with *Listeria* risk assessment models, few suitable models are available.

Finally, it is known that a number of foodborne disease agents (primarily viruses) can contaminate foods, and in many cases the source of the agent has been an ill food worker.[12] While quantitative microbial risk assessments have yet to address ill workers, hand-to-food and other cross-contamination rates using a nonpathogenic surrogate have been calculated with sufficient detail to be suitable for risk assessment.[7]

10.2.1 ISSUES OF CONCERN

There are a number of issues of particular concern that are important in modeling cross-contamination — issues that are unique to these sorts of models as compared to the traditional growth and decline models: appropriate statistical treatment of data, transfer from one location to another, factors influencing transfer, and the possibility of multiple transfers.

10.2.1.1 Statistical Treatment

Data on cross-contamination are typically presented as percent transfer, as shown below:

$$\frac{CFU \text{ on target}}{CFU \text{ on source}} * 100 = \text{percent transfer}$$

The problem arises when multiple observations of the same conditions are to be combined and reported as an average. It has been shown that when large numbers of observations are made, the distribution of "percent transfer rates" is distinctly

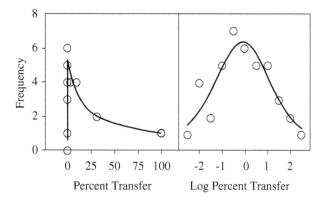

FIGURE 10.1 Distribution of percent transfer data with linear (left panel) and logarithmic transformation (right panel).

nonnormal.[7] This point is made clearer when the data are presented visually. Figure 10.1 shows the same data set plotted as number of observations vs. percent transfer (left panel) and number of observations vs. $\log_{10}$ percent transfer (right panel). This same pattern is borne out for many different types of surface-to-surface transfers.[7,10] Since percent transfer is nonnormally distributed, but $\log_{10}$ percent transfer is approximately normally distributed, this means data should be log transformed before averages are calculated. This apparently subtle distinction has important consequences as illustrated below.

Let us assume we have two observed transfer rates of 5 and 50%. If the mean is calculated arithmetically:

$$(50 + 5)/2 = 27.5\% \text{ transfer}$$

Alternatively, using the statistically appropriate $\log_{10}$ transformed rates leads to a more complex series of calculations:

$$(\log 0.05 + \log 0.50)/2 = (-1.30103 + -0.30105)/2 = -0.80103$$

Then this number should be converted back to the untransformed percent scale:

$$10^{-0.80103} = 15.8\% \text{ transfer}$$

So this simple difference leads to a calculated transfer rate that is over half that obtained when the statistically incorrect method is used.

10.2.1.2 Additive Nature of Cross-Contamination

Another key feature of any cross-contamination model is that some consideration should be made of both the source of the contaminant and the destination. If the source is a contaminated surface, then the number of organisms on that surface must be known, since (at least in theory) a more contaminated source will yield a more contaminated destination. Care must be taken to ensure that the number of organisms to be added to an already contaminated item is added in a numerically correct manner. For example, most calculations in a microbial risk assessment will be $\log_{10}$ CFU increases or decreases. This would not be correct in cross-contamination. If a food contained 10 organisms and 100 were added, this would not be a two $\log_{10}$ increase, $1 + 2 = 3$ or 1000 organisms, but $10 + 100 = 110$ organisms.

10.2.1.3 Factors Influencing Transfer Rate

The next issue that should be addressed would be a consideration of the factors influencing transfer rate. There are a whole host of factors that may influence transfer, like source (air, liquid, or solid), the pressure applied (for solid-to-solid transfers), menstrum effects, contact time, number of organisms present, and surface characteristics. Some of these factors have been investigated for transfer from hands in a healthcare setting, but little data applicable to food systems have been published. What has been published in the food and healthcare literature has not been systematic or comprehensive.[21]

10.2.1.4 Types of Transfers

A recent review[10] describes some of the currently available models for recontamination via air, via processing equipment (i.e., biofilms), or via hand contact. These authors point out that not many available models are directly applicable to the food industry as most models are developed for aquatic or environmental systems. In some cases currently available models are contradictory or incompatible. For example, competing air recontamination models assume that when the concentration in the air increases linearly, the concentration in the product increases either linearly[28] or quadratically.[23] The implications of such assumptions obviously have a critical impact on model predictions.

Models for recontamination that consider the effect of biofilms are quite well developed, not because of their importance in food processing, but because of their application in wastewater treatment. Biofilm models can be one-dimensional, or multidimensional,[16] but the key feature of any biofilm model used for food recontamination is not its dimensionality, but its ability to consider attachment, growth, and detachment averaged over the food contact surface.[10] Some biofilm models appropriate for use in food systems have been developed.[31]

den Antrekker et al.[10] conclude their review by proposing a schematic for a general contamination model that is suitable for modeling recontamination via air, via surfaces, or via hands. This model uses a source, an intermediate phase, and a product, with transfer rates between source and intermediate phases and intermediate and product phases that govern the overall transfer to the product.

10.2.1.5 Multiple Transfers and Complexity

The last issue of concern is the modeling of multiple transfers. The food preparation or handling environment may be such that multiple transfers between many different environments may occur. In a processing plant it may be from air to surface, surface to product, product to clean surface, and finally surface to clean product. In a food service environment it may be from a contaminated surface to workers' hands, and then from hands to food. In a home environment it may be from a contaminated food product to a surface, and then from that surface to another food.

Each of these simple examples would require a series of calculations, using models that generally do not yet exist. At the same time, it should be realized that the real world is considerably more complex than these simple examples show, and that there may be literally dozens of cross-contamination possibilities in even a simple food process or meal preparation.

10.2.2 CROSS-CONTAMINATION SUMMARY

The development of mathematical models suitable for describing cross-contamination events in food production, processing, and preparation is still in its infancy, and might be likened to the general state of predictive food microbiology in the 1980s. The past two decades have seen many improvements in the general state of the art of predictive food microbiology, and there is no reason to believe the next two decades would not experience similar improvements in cross-contamination modeling. This field has attracted the interest of a number of research groups around the world, and the beginnings of a comprehensive body of work are beginning to emerge.

10.3 INOCULUM SIZE MODELING

Predictive models have traditionally been developed using starting bacterial concentration that may be quite high relative to the levels found in some foods. Modelers developing models in this way were quite justified in their choice of this approach. In some cases, inoculum size does not have a significant effect on the response to be modeled (i.e., growth rate).[5] High initial inoculum size also represents a conservative worst-case approach to modeling, and these high starting concentrations helped to assure repeatability and simplified some of the considerations about microbial variability, often called "biovariability."

Despite the logic seen in this worst-case approach, modelers have always sought to improve their models by making them more realistic and representative of real-world conditions. Also, as predictive food microbiologists' modeling tools and abilities have improved, their ability to handle more complex models and modeling techniques have improved concomitantly. As part of this evolutionary improvement in modeling ability, some modelers have sought to address this shortcoming by developing models that take initial microbial concentrations into consideration.

10.3.1 CLOSTRIDIUM BOTULINUM

One of the earliest examples of a predictive model that explicitly acknowledged the influence of inoculum size were models developed for C. botulinum.[13,14] The authors' specific objective in this case was to develop models capable of predicting the probability of toxin formation from a single C. botulinum spore. Later research in this same lab also showed an inoculum size effect in fish[3] and poultry[19] systems inoculated with *C. botulinum*.

The importance of inoculum size for models for *C. botulinum* has since been well documented in the literature.[17,26,27,29,30] These *C. botulinum* models were developed as an aid to the food industry, but also served to point the way towards a more mechanistic understanding of populations of *C. botulinum* spores. The model for nonproteolytic *C. botulinum* developed by Whiting and Oriente,[26] for example, showed that not only did lower population of spores exhibit longer time-to-turbidity, the variability around that time increased markedly with decreasing inoculum size as well. This same effect was also seen with proteolytic *C. botulinum* spores.[27] These results supported the observations made by others conducting microscopy studies that there was a marked variability seen in the germination, outgrowth, and lag time in individual *C. botulinum* spores observed directly.[4] These apparent interactions between spores of *C. botulinum* have also been demonstrated using computer simulation of different inoculum sizes[30] and seem to be caused by the release of a signaling molecule into the culture media by germinating spores.

This inoculum phenomenon does not appear to be unique to *C. botulinum*, as it has also been observed in *Bacillus cereus*[9] and *Bacillus megaterium*[6] using direct microscopic observation and in *Bacillus stearothermophilus*[15] indirectly by time to turbidity.

10.3.2 NONSPORE-FORMING BACTERIA

The effect has also been seen in models for nonspore-forming bacteria, like *Brochothrix thermosphacta*,[18] as well as in nonmodeling research with *Salmonella*[8] and *L. monocytogenes.*[2,22,24] These publications show that the effect is most pronounced when cells are stressed[2,24] or cultured in inhospitable environments.[8] The response in vegetative cells has been interpreted by some as being due to death of a proportion of cells in the inoculum rather than communication, as appears to be the case with *C. botulinum,*[22] although others have shown that addition of spent medium from a stationary-phase culture reduces the variability and length of lag times.[24]

10.3.3 INOCULUM SIZE SUMMARY

Clearly modelers have moved beyond predictive models developed using high initial bacterial concentrations to models using a range of contamination levels. Most of the effort in this area has focused on models for spore-forming organisms, specifically *C. botulinum*. A limited amount of work has also been done with other spore-forming organisms and vegetative cells. It appears that inoculum size has the most dramatic effect on the lag time (for vegetative cells) or germination, outgrowth, and lag time (for spores).

10.4 CROSS-CONTAMINATION AND INOCULUM SIZE

While the two subjects of this chapter do not appear to have much in common, except for both being aspects of modeling on the "cutting edge," there appears to

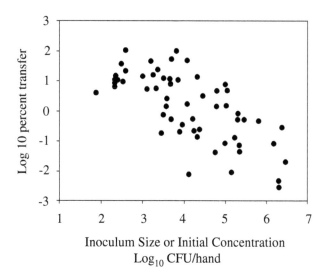

FIGURE 10.2 Percent transfer as a function of inoculum size.

be some evidence that inoculum size is not only important in traditional growth modeling, but that it may also be an important consideration in modeling cross-contamination. Figure 10.2 presents a reanalysis of a portion of the data originally published by Chen et al.[7] for cross-contamination rate from bare hands contaminated with *Enterobacter aerogenes* to lettuce. While the r^2 value (0.49) and a visual inspection of the plot show that the correlation is not ideal, the effect is highly significant ($p > 10^{-10}$). The regression model for the plot indicates that changing the starting concentration by 1.5 $\log_{10}$ CFU will change the $\log_{10}$ percent transfer rate by about 1 (i.e., from 10 to 1%). It is also interesting to note that the relationship between starting concentration and $\log_{10}$ percent transfer rate is an inverse one, so that as the starting concentration decreases, the transfer rate increases. This could have very profound food safety consequences since low levels of pathogens would have a correspondingly greater ability to transfer. This simple example has been used to illustrate the exciting and complex nature of predictive food microbiology at the expanding edges of the discipline.

10.5 SUMMARY

Models for cross-contamination and inoculum size represent areas at the expanding edge of predictive food microbiology. Both areas characterize modelers' attempts to make models more useful and representative of microbial behaviors seen in the real world. Both areas continue to present mathematical, statistical, and methodological challenges to those working in the field.

REFERENCES

1. Preliminary FoodNet data on the incidence of foodborne illnesses — selected sites, US, 2001, *Mor. Mortal. Wkly. Rep.*, 51, 325, 2002.
2. Augustin, J.C., Brouillaud-Delattre, A., Rosso, L., and Carlier, V., Significance of inoculum size in the lag time of *Listeria monocytogenes*, *Appl. Environ. Microbiol.*, 66, 1706, 2000.
3. Baker, D.A. and Genigeorgis, C.A., Predicting the safe storage of fresh fish under modified atmospheres with respect to *Clostridium botulinum* toxigenesis by modeling length of the lag phase, *J. Food Prot.*, 53, 131, 1990.
4. Billon, C.M.P., McKirgan, C.J., McClure, P.J., and Adair, C., The effect of temperature on the germination of single spores of *Clostridium botulinum* 62A, *J. Appl. Microbiol.*, 82, 48, 1997.
5. Buchanan, R.L., Smith, J.L., McColgan, C., Marmer, B.S., Golden, M., and Dell, B., Response surface models for the effects of temperature, pH, sodium chloride, and sodium nitrite on the aerobic and anaerobic growth of *Staphylococcus aureus* 196E, *J. Food Saf.*, 13, 159, 1993.
6. Caipo, M.L., Duffy, S., Zhao, L., and Schaffner, D.W., *Bacillus megaterium* spore germination is influenced by inoculum size, *J. Appl. Microbiol.*, 92, 879, 2002.
7. Chen, Y., Jackson, K.M., Chea, F.P., and Schaffner, D.W., Quantification and variability analysis of bacterial cross contamination rates in common foodservice tasks, *J. Food Prot.*, 64, 72, 2001.
8. Cogan, T.A., Domingue, G., Lappin-Scott, H.M., Benson, C.E., Woodward, M.J., and Humphrey, T.J., Growth of *Salmonella enteritidis* in artificially contaminated eggs: the effects of inoculum size and suspending media, *Int. J. Food Microbiol.*, 70, 131, 2001.
9. Coote, P.J., Billon, C.M.P., Pennell, S., McClure, P.J., Ferdinando, D.P., and Cole, M.B., The use of confocal scanning laser microscopy (CSLM) to study the germination of individual spores of *Bacillus cereus*, *J. Microbiol. Methods*, 21, 193, 1995.
10. den Antrekker, E., Boom, R.M., Zwietering, M.H., and vanSchothorst, M., Quantifying recontamination through factory environments — a review, *Int. J. Food Microbiol.*, 80, 117, 2002.
11. Frost, J.A., Current epidemiological issues in human campylobacteriosis, *J. Appl. Microbiol.*, 90, 85S, 2001.
12. Guzewich, J.J. and Ross, M.P., White paper, Section one: a literature review pertaining to foodborne disease outbreaks caused by food workers, 1975–1998, http://vm.cfsan.fda.gov/~ear/rterisk.html, 1999.
13. Jensen, M.J., Genigeorgis, C.A., and Lindroth, S., Probability of growth of *Clostridium botulinum* as affected by strain, cell and serologic type, inoculum size and temperature and time of incubation in a model system, *J. Food Saf.*, 8, 109, 1987.
14. Lindroth, S.E. and Genigeorgis, C.A., Probability of growth and toxin production by non proteolytic *Clostridium botulinum* in rockfish stored under modified atmospheres, *Int. J. Food Microbiol.*, 3, 167, 1986.
15. Llaudes, M., Zhao, L., Duffy, S., and Schaffner, D.W., Simulation and modeling of the effect of small inoculum size on the time to spoilage by *Bacillus stearothermophilus*, *Food Microbiol.*, 18, 395, 2001.
16. Lu, C., Biswas, P., and Clark, R.M., Simultaneous transport of substrates, disinfectants and microorganisms in water pipes, *Water Res.*, 29, 881, 1995.

17. Lund, B.M., Quantification of factors affecting the probability of development of pathogenic bacteria, in particular *Clostridium botulinum*, in foods, *J. Ind. Microbiol.*, 12, 144, 1993.

18. Masana, M.O. and Baranyi, J., Growth/no growth interface of *Brochothrix thermosphacta* as a function of pH and water activity, *Food Microbiol.*, 17, 485, 2000.

19. Meng, J. and Genigeorgis, C.A., Modeling lag phase of nonproteolytic *Clostridium botulinum* toxigenesis in cooked turkey and chicken breast as affected by temperature, sodium lactate, sodium chloride and spore inoculum, *Int. J. Food Microbiol.*, 19, 109, 1993.

20. Montville, R., Chen, Y., and Schaffner, D.W., Glove barriers to bacterial cross-contamination between hands to food, *J. Food Prot.*, 64, 845, 2001.

21. Montville, R., Chen, Y.H., and Schaffner, D.W., Risk assessment of hand washing efficacy using literature and experimental data, *Int. J. Food Microbiol.*, 73, 305, 2002.

22. Pascual, C., Robinson, T.P., Ocio, M.J., Aboaba, O.O., and Mackey, B.M., The effect of inoculum size and sublethal injury on the ability of *Listeria monocytogenes* to initiate growth under suboptimal conditions, *Lett. Appl. Microbiol.*, 33, 357, 2001.

23. Radmore, K., Holzapfel, W.H., and Luck, H., Proposed guidelines for maximum acceptable air-borne microorganism levels in dairy processing and packaging plants, *Int. J. Food Microbiol.*, 6, 91, 1988.

24. Robinson, T.P., Aboaba, O.O., Kaloti, A., Ocio, M.J., Baranyi, J., and Mackey, B.M., The effect of inoculum size on the lag phase of *Listeria monocytogenes*, *Int. J. Food Microbiol.*, 70, 163, 2001.

25. Tompkin, R.B., Control of *Listeria monocytogenes* in the food-processing environment, *J. Food Prot.*, 65, 709, 2002.

26. Whiting, R.C. and Oriente, J.C., Time-to-turbidity model for non-proteolytic type B *Clostridium botulinum*, *Int. J. Food Microbiol.*, 36, 49, 1997.

27. Whiting, R.C. and Strobaugh, T.P., Expansion of the time-to-turbidity model for proteolytic *Clostridium botulinum* to include spore numbers, *Food Microbiol.*, 15, 449, 1998.

28. Whyte, W., Sterility assurance and models for assessing airborne bacterial contamination, *J. Parenter. Sci. Technol.*, 40, 188, 1986.

29. Zhao, L., Montville, T.J., and Schaffner, D.W., Time-to-detection, percent-growth-positive and maximum growth rate models for *Clostridium botulinum* 56A at multiple temperatures, *Int. J. Food Microbiol.*, 77, 187, 2002.

30. Zhao, L., Montville, T.J., and Schaffner, D.W., Computer simulation of *Clostridium botulinum* 56A behavior at low spore concentrations, *Appl. Environ. Microbiol.*, 69, 845, 2003.

31. Zwietering, M.H. and Hasting, A.P.M., Modelling the hygienic processing of foods — a global process overview, *Trans IChemE part C*, 75, 159, 1997.

11 Predictive Mycology

Philippe Dantigny

CONTENTS

11.1 INTRODUCTION

For over 20 years, predictive microbiology has focused on bacterial food-borne pathogens, and some spoilage bacteria. Few studies have been concerned with modeling fungal development. Predictive modeling is a versatile tool that should not be limited to bacteria, but should be extended to molds. Mathematical modeling of fungal growth was reviewed earlier (Gibson and Hocking, 1997), but at that time very few models were available. The concerns were growth and toxin production, but germination was not examined. On one hand, most of food mycologists are not familiar with modeling techniques, and they tend to use existing models that were developed for describing bacterial growth in foods. On the other hand, people involved in modeling may not be aware of mold specificities. Predictive mycology aims at developing specific tools for describing fungal development.

11.2 CONCERNS

The occurrence of food-borne fungi was described extensively by Northolt et al. (1995). Food raw material and products can be contaminated with spores or conidia and mycelium fragments from the environment. Under favorable conditions, fungal growth occurs. A large number of metabolites are formed during the breakdown of carbohydrates, some of which can accumulate under certain conditions. The main concern is production of mycotoxins, which cannot simply be destroyed by heat.

11.2.1 Mycotoxins Production

It has been reported that 25% of agriculture products are contaminated with myc-otoxins (Mannon and Johnson, 1985). Mycotoxin ingestion by humans, which occurs mainly through plant-based foods and the residues and metabolites present in animal-derived foods, can lead to deterioration of liver or kidney function (Sweeney and Dobson, 1998) and therefore constitutes a risk for human health. The main genera responsible for toxins production are *Fusarium, Aspergillus,* and *Penicillium.* While *Fusarium* species are destructive plant pathogens producing mycotoxins before or immediately post harvesting, *Penicillium* and *Aspergillus* species are more commonly found as contaminants of commodities and foods during drying and subsequent storage (Sweeney and Dobson, 1998). Plant contamination by molds such as *Fusarium* cannot be avoided at the field level because it depends largely on climatic conditions. Predictive mycology would be useful for making predictions on the extent of contamination, growth, and toxin production by these pathogens; however, there are no models currently available. In contrast, controlling the environmental factors during storage of raw materials can prevent the development of *Penicillium* and *Aspergillus.* The prevalence of one species as compared to the other one is related to temperature, *Penicillium* being capable of developing at lower temperatures than is *Aspergillus.*

11.2.2 Economic Losses

Because of the appearance of visible hyphae and production of unpleasant odors, fungal spoilage of food causes economic losses. For example, in the baking industry, these losses vary between 1 and 3% of products, depending on season, type of product, and method of processing (Mälkki and Rauha, 1978). The most widespread and probably most important molds in terms of biodeterioration of bakery products are species of *Eurotium, Aspergillus,* and *Penicillium* (Abellana et al., 1997). But there are many other species responsible for food spoilage. The reason why a particular species dominates in a product is certainly correlated with the species characteristics and the properties of the product (Northolt et al., 1995). Therefore, predictive mycology can well be applied to control fungal development through product formulation, food processing, type of packaging, and conditions of storage.

11.3 MOLD SPECIFICITIES

Fungal growth involves germination and hyphal extension, eventually forming myce-lium. Spores are widely disseminated in the environment, and they are principally responsible for spoilage. Under favorable conditions, spores will swell. Thereafter, when the length of the germ tube is between one half and twice the spore diameter (depending on the source), the spore is considered to have germinated. Germination can be considered as the main step to be focused on, because a product is spoiled as soon as visible hyphae can be observed. However, few studies have concerned germination kinetics. This limitation can be explained in part by the difficulties of acquiring sufficient, reproducible data. In fact, this kind of study requires microscopic observation for evaluating the length of the germ tube. Moreover, observations and measurements should be carried out without opening the dishes (Magan and Lacey, 1984) and experimental devices should be developed for this purpose (Sautour et al., 2001a, 2001c). In contrast, more work was dedicated to the measurement of hyphal extension rate, which is usually reported as radial growth rate (mm d^{-1}).

Because of their ability of dividing, bacteria form single cells and they can be easily enumerated, especially in liquid broth. In such a case, and at high cellular densities, bacterial growth can be estimated automatically, for example, by using the Bioscreen® device, which is based on turbidity measurements. At lower cellular densities and in solid media, colony-forming units per gram (CFU/g) or CFU/ml can be determined.

In contrast, molds form mycelium, and the weight, except at the early stage of growth, does not increase exponentially (Koch, 1975). It is therefore useless to determine the weight of the mycelium for estimating a growth rate parameter. In addition, it is impossible to split the mycelium into individual cells. However, the CFU method can be applied to the enumeration of spores (Vindeløv and Arneborg, 2002).

Temperature (T) is the main factor for controlling bacterial growth, but the effect of water activity (a_w) on mold growth is more important than T (Holmquist et al., 1983). Oxygen is necessary for the growth of food spoilage fungi. Therefore, the use of modified atmospheres to prevent fungal growth and mycotoxin production has been evaluated to extend the shelf life of some kinds of food (El Halouat and Debevere, 1997; Taniwaki et al., 2001).

11.4 MODELS

11.4.1 PRIMARY MODELS

Two aspects of fungal growth can be modeled using primary models: spore germination and radial growth of colonies. The germination of spores of *Fusarium moniliforme* as a function of time was first studied at different a_w (Marín et al., 1996). The percentage of germination vs. time was modeled with the modified Gompertz equation (see Chapter 2) at different water activities (Figure 11.1). In contrast to the case with bacteria where the initial bacterial load (N_0) is a critical parameter to be estimated, the initial percentage of germination was always equal to 0. The asymptotic value where the percentage of germination becomes constant was 100 in most cases. But under harsh environmental conditions some spores are unable to initiate a germ tube, thus leading to a maximum percentage of germination less than 100 (Figure 11.1).

There are two different ways of looking at spore germination: (1) the percentage of germination at a certain time, and (2) the time to obtain a certain germination percentage, or germination time. In the present example, the percent germination after 24 h does not discriminate water activity levels very well (vertical dotted line in Figure 11.1). The response was 100% for a_w in the range 0.94 to 0.98, and 0% in the range 0.88 and 0.92. In contrast, germination time (defined here as half the maximum percent germination) is clearly dependent upon a_w (horizontal dotted line in Figure 11.1). It should be noted, however, that an accurate determination of the germination time requires modeling of the whole germination curve.

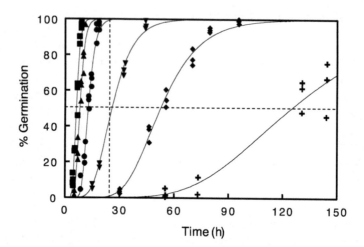

FIGURE 11.1 Effect of water activity and time on germination (%) of spores of *Fusarium moniliforme* (isolate 25N) on MMEA (maize meal extract agar) at 25°C. Water activity levels were 0.98 (■), 0.96 (▲), 0.94 (●), 0.92 (▼), 0.90 (♦), and 0.88 (+). (Redrawn from Marín, S., Sanchis, V., Teixido, A., Saenz, R., Ramos, A.J., Vinas, I., Magan, N. 1996. *Can. J. Microbiol.* 42, 1045–1050.) The vertical dotted line indicates the percent germination at 24 h, and the horizontal dotted line shows the time required for 50% germination at each water activity.

The germination time can also be considered as the probability of a single spore germinating. Accordingly, the logistic function that is usually dedicated to probabilistic models:

$$P = \frac{P_{max}}{(1 - e^{(k\{\tau-t\})})}$$ (11.1)

was used for describing the germination kinetics of *Mucor racemosus* (Dantigny et al., 2002). The parameter P_{max} was substituted with 100% because all spores were capable of germinating. With the objective of designing a secondary model, the parameters of the logistic function were expressed as a function of environmental factors. The rate factor k was constant whatever the temperature, whereas τ (time where $P = P_{max}/2$) was more discriminative.

Shortly after the completion of germination, the mycelium is visible to the naked eye (when the colony diameter reaches approximately 3 mm). Therefore fungal growth can be easily estimated from macroscopic measurements of the radius of the colony. The primary model developed by Baranyi (see Chapter 2) has been adapted to fit colony diameter growth curves of *Penicillium rocqueforti* (Valík et al., 1999), *Aspergillus flavus* (Gibson et al., 1994), and *Penicillium brevicompactum* (Membré and Kubaczka, 2000). In our laboratory we have had considerable success with a simple linear model with breakpoint:

$$r = \mu \cdot (t - \lambda)$$ (11.2)

where r is the colony radius (mm), μ is the radial growth rate (mm d^{-1}), and λ is the lag time (d). The linear section of the graph (with growth rate of μ) is extrapolated to a zero increase in diameter, and the intercept on the time axis is defined as the lag prior to growth (λ). In most cases, the fits are excellent, with the regression coefficients being greater than 0.995. Therefore, under these conditions, it is unlikely that other models such as those of Baranyi or Gompertz would demonstrate superiority over the linear approach. In addition, the parameters μ and λ, can be obtained even when the petri dish is not entirely covered with mycelium. It should also be mentioned that early measurements of diameter of the colony improve the accuracy of the lag period because this parameter is obtained by extrapolation of the straight line.

11.4.2 SECONDARY MODELS

pH, which is usually associated with other environmental factors to prevent bacterial growth, has no marked influence on mold germination or growth. Water activity has a greater effect on mold development than does temperature, whereas an interactive effect between T and a_w is noticed. The effects of temperature and water activity on growth rate of food spoilage molds were compared using normalized variables μ_{dim}, T_{dim}, and $a_{w\ dim}$ within Bělehrádek-type equations : $\mu_{dim} = [T_{dim}]^\alpha$ and $\mu_{dim} = [a_{w\ dim}]^\beta$ (Sautour et al., 2002). It can be observed that for $\alpha = 2$, the equation is equivalent to the square-root model that was originally described by Ratkowsky et al. (1982).

It was reported that the molds studied were characterized by α-values ranging from 0.81 to 1.54 and β-values from 1.50 to 2.44. Because of the lack of specific models for molds there is a tendency to apply models that have been developed for bacteria. For example, the square-root model was used to describe the effect of T on the growth of *Rhizopus microsporus* (Han and Nout, 2000). It is clear that the effect of temperature on molds cannot be modeled by the square-root model because of α-values close to 1. It has been demonstrated that the use of the square-root model when α is less than 2 leads to underestimation of T_{min} (Dantigny and Molin, 2000). Similarly, some doubt can be raised with the use of the cardinal model with inflexion (CMI) described by Rosso et al. (1993) for describing the effect of T on fungal growth rate. For example, a T_{min} value as low as $-12°C$ has been reported for *P. roqueforti* using the CMI model (Cuppers et al., 1997). In contrast, the CMI model was used satisfactorily for describing the effect of a_w on fungal growth rate (Rosso and Robinson, 2001; Sautour et al., 2001b) as suggested by the β-values close to 2. It should also be noted that, in contrast to the square-root model proposed by Gibson et al. (1994) to describe fungal growth, the CMI model allows an estimation of $a_{w\ min}$, which is not easily determinable because fungal growth can well occur after several months of incubation.

11.5 PERSPECTIVES

For the objective of modeling fungal kinetics, the tools that were developed for bacteria can be used, but mold specificities should be taken into account. As a primary step in modeling fungal development, attention should be focused on spore germination. Although there is no widely accepted definition of germination time, this variable provides a pertinent insight into how fast spores are germinating. To determine this variable accurately, spore germination kinetics should be monitored using regular microscopic observations and by developing a specific experimental setup. A significant breakthrough in predictive mycology would be the automation of spore observations. Some attempts were made using image analysis (Paul et al., 1993), but clumping of spores should be avoided. In addition, culture media should be clear enough to allow microscopic observations. Therefore, any model describing germination kinetics could be hardly validated on food products by this technique.

Fungal growth, which is usually reported as radial growth rate, can be easily determined by macroscopic observations. In order to substitute a microscopic observation for a macroscopic one, a relationship between the lag for growth and the germination time was established (Dantigny et al., 2002). However, it was shown that the lag time is very much dependent on the number of spores inoculated at the same spot (Sautour et al., in press). This could be explained, because a large inoculum will form a visible colony more rapidly than a small one, thus decreasing the lag. Some more studies should be conducted to determine the relationship between the lag and the number of spores inoculated. It should also be verified that such a relationship is independent from the environmental factors.

Secondary models concern mainly the influence of environmental factors on fungal growth rate. At present, very few models aimed at assessing the influence of these factors on spore germination have been elucidated. Unfortunately, existing models are polynomial, and cannot be extrapolated to other molds. Secondary models based on parameters with biological significance (e.g., cardinal values) to determine the influence of environmental factors on spore germination and mycotoxins production should be developed. Eventually, other parameters such as preservatives should be included in the list of environmental factors.

REFERENCES

Abellana, M., Torres, L., Sanchis, V., Ramos, A.J. 1997. Caracterización de differentes productos de bollería industrial. II. Estudio de la microflora. *Alimentaria* 287, 51–56.

Cuppers, H.G.A.M., Oomes, S., Brul, S. 1997. A model combined effects of temperature and salt concentration on growth rate of food spoilage molds. *Appl. Environ. Microbiol.* 63, 3764–3769.

Dantigny, P., Molin, P. 2000. Influence of the modelling approach on the estimation of the minimum temperature for growth in Bělehrádek-type models. *Food Microbiol.* 17, 597–604.

Dantigny, P., Soares Mansur, C., Sautour, M., Tchobanov, I., Bensoussan, M. 2002. Relationship between spore germination kinetics and lag time during growth of *Mucor racemosus*. *Lett. Appl. Microbiol.* 35, 395–398.

El Halouat, A., Debevere, J.M. 1997. Effect of water activity, modified atmosphere packaging and storage temperature on spore germination of moulds isolated from prunes. *Int. J. Food Microbiol.* 35, 41–48.

Gibson, A.M., Baranyi, J., Pitt, J.I., Eyles, M.J., Roberts, T.A. 1994. Predicting fungal growth: the effect of water activity on *Aspergillus flavus* and related species. *Int. J. Food Microbiol.* 23, 419–431.

Gibson, A.M., Hocking, A.D. 1997. Advances in the predictive modelling of fungal growth in food. *Trends Food Sci. Technol.* 8, 353–358.

Han, B.-Z., Nout, M.J. 2000. Effects of temperature, water activity and gas atmosphere on mycelial growth of tempe fungi *Rhizopus microsporus* var *microsporus* and *R. microsporus* var *oligosporus*. *World J. Microbiol. Biotechnol.* 16, 853–858.

Holmquist, G.U., Walker, H.W., Stahr, H.M. 1983. Influence of temperature, pH, water activity and antifungal agents on growth of *Aspergillus flavus* and *A. parasiticus*. *J. Food Sci.* 48, 778–782.

Koch A.L. 1975. The kinetics of mycelial growth. *J. Gen. Microbiol.* 89, 209–216.

Magan, N., Lacey, J. 1984. Effect of temperature and pH on water relations of field and storage fungi. *Trans. Br. Mycol. Soc.* 82, 71–81.

Mälkki, Y., Rauha, O. 1978. Mold inhibition by aerosols. *Baker's Digest.* 52, 47–50.

Mannon, J., Johnson, E. 1985. Fungi down on the farm. *New Scientist.* 195, 12–16.

Marín, S., Sanchis, V., Teixido, A., Saenz, R., Ramos, A.J., Vinas, I., Magan, N. 1996. Water and temperature relations and microconidial germination of *Fusarium moniliforme* and *Fusarium proliferatum* from maize. *Can. J. Microbiol.* 42, 1045–1050.

Membré, J.-M., Kubaczka M., 2000. Predictive modelling approach applied to spoilage fungi: growth of *Penicillium brevicompactum* on solid media. *Lett. Appl. Microbiol.* 31, 247–250.

Northolt, M.D., Frisvad, J.C., Samson, R.A. 1995. Occurrence of food-borne fungi and factors for growth. In *Introduction to Food-Borne Fungi*, 4th ed. Samson, R.A., Hoekstra, E.S., Frisvad, J.C., Filtenborg, O. (Eds.), Centraal Bureau voor Schimmelcultures, Delft, The Netherlands, pp. 243–250.

Paul, G.C., Kent, C.A., Thomas, C.R. 1993. Viability testing and characterization of germination of fungal spores by automatic image analysis. *Biotechnol. Bioeng.* 42, 11–23.

Ratkowsky, D.A., Olley, J., McMeekin, T.A., Ball, A. 1982. Relationship between temperature and growth rate of bacterial cultures. *J. Bacteriol.* 149, 1–5.

Rosso, L., Lobry, J.R., Bajard, S., Flandrois, J.P. 1993. An unexpected correlation between cardinal temperatures of microbial growth highlighted by a new model. *J. Theor. Biol.* 162, 447–463.

Rosso, L., Robinson, T.P. 2001. A cardinal model to describe the effect of water activity on the growth of moulds. *Int. J. Food Microbiol.* 63, 265–273.

Sautour, M., Dantigny, P., Divies, C., Bensoussan, M. 2001a. Application of Doehlert design to determine the combined effects of temperature, water activity and pH on conidial germination of *Penicillium chrysogenum*. *J. Appl. Microbiol.* 91, 900–906.

Sautour, M., Dantigny, P., Divies, C., Bensoussan, M. 2001b. A temperature-type model for describing the relationship between fungal growth and water activity. *Int. J. Food Microbiol.* 67, 63–69.

Sautour, M., Dantigny, P., Guilhem, M.-C. Bensoussan, M. In press. Influence of inoculum preparation on the growth of *Penicillium chrysogenum*. *J. Appl. Microbiol.*

Sautour, M., Rouget, A., Dantigny, P., Divies, C., Bensoussan, M. 2001c. Prediction of conidial germination of *Penicillium chrysogenum* as influenced by temperature, water activity and pH. *Lett. Appl. Microbiol.* 32, 131–134.

Sautour, M., Soares Mansur, C., Divies, C., Bensoussan, M., Dantigny, P. 2002. Comparison between the effects of temperature and water activity on growth rate of food spoilage molds. *J. Ind. Microbiol. Biotechnol.* 28, 311–315.

Sweeney, M.J., Dobson, A.D.W. 1998. Mycotoxin production by *Aspergillus, Fusarium* and *Penicillium* species. *Int. J. Food Microbiol.* 43, 141–158.

Taniwaki, M.H., Hocking, A.D., Pitt, J.I., Fleet, G.H. 2001. Growth of fungi and mycotoxin production on cheese under modified atmospheres. *Int. J. Food Microbiol.* 68, 125–133.

Valík, L., Baranyi, J., Görner, F. 1999. Predicting fungal growth: the effect of water activity on *Penicillium rocqueforti*. *Int. J. Food. Microbiol.* 47, 141–146.

Vindeløv, J., Arneborg, N. 2002. Effects of temperature, water activity, and syrup film composition on the growth of *Wallemia sebi*: development and assessment of model predicting growth lags in syrup agar and crystalline sugar. *Appl. Environ. Microbiol.* 68, 1652–1657.

12 An Essay on the Unrealized Potential of Predictive Microbiology

Tom McMeekin

CONTENTS

12.1 INTRODUCTION

This book presents contemporary views of the state of the art of predictive modeling of microbial growth in foods, which, it is worth noting, has been the subject of much research for more than 30 years. Contained herein the reader will sense that the "front end" of the modeling process (data collection, model developing, model fitting) has a scientifically sound basis and that the "middle bit" (tertiary models, applications software, expert systems, etc.) should make the technology readily available.

However, one continues to have the sense that the predictive models and the databases upon which they are based have not nearly reached their potential. This "wrap up" chapter is intended to reinforce the view that predictive microbiology research has addressed, and continues to address, perceived weaknesses of the concept and offers the view that its potential as a food safety management tool remains to be realized. The views expressed are personal, but probably shared by others who have engaged in this type of research for many years. The style adopted is that of an "essay" (i.e., a written composition less elaborate than a treatise) in the hope that a less formal presentation may encourage greater consideration by potential users of the concept.

12.2 A SHORT HISTORY AND THE PHILOSOPHY OF PREDICTIVE MICROBIOLOGY

Esty and Meyer[7] devised a mathematical model to describe the thermal death kinetics of *Clostridium botulinum* type A spores that was used immediately, and since that time as the basis for heat processing of nonacid canned foods. Its use continues, despite subsequent understanding that a log-linear model may be an oversimplified description of the rate at which death occurs. Log-linear kinetics ignore the complications of "shoulders" and "tails," because the performance criterion applied, a 12-log-cycle reduction in spore numbers, introduces a very large safety margin. Practitioners of heat processing use the time/temperature combinations derived with great confidence in the microbiological outcome. Such confidence and immediate and continuing application of the process suggest a single purpose rather than the earliest introduction of the concept of predictive microbiology. This would have implied considering the consequences of selecting less severe processing conditions.

In searching for the origin of predicting microbial behavior in foods, the trail appears to begin with Scott[25] who researched the effect of temperature on microbial growth on meat and wrote:

> A knowledge of the rate of growth of certain micro-organisms at different temperatures is essential to studies of the spoilage of chilled beef. Having these data, it should be possible to predict the relative influence on spoilage exerted by the various organisms at each storage temperature. Further, it would be feasible to predict the possible extent of the changes in populations which various organisms may undergo during the initial cooling of sides of beef in the meatworks when the meat surfaces are frequently at temperatures very favourable to microbial proliferation.

Scott[24] also studied the effect of water availability on microbial growth on meat and, it is interesting to note that while explicit models were not developed, the knowledge accumulated was sufficient to allow shipments of nonfrozen meat from Australia to markets in the U.K. and Europe, based on the combined effects of temperature, water availability, and modified atmosphere. Later in this chapter we will return to the proposition that significant benefit can be obtained from accumulated knowledge or patterns of microbial population responses without transforming that knowledge into an explicit mathematical description.

So, why bother taking the additional steps necessary to convert a response pattern into a mathematical model? The answer is both practical and philosophical. In the practical sense, the additional effort required to construct and validate a model will, if properly carried out, lead to formulation of a general rule describing the effect of the environmental responses studied on the growth, death, or survival of the target population. A response pattern, on the other hand, is more likely to describe the outcome of limited experimental trials, the applicability of which will require testing under even slightly changed conditions. Challenge tests are a well-established means of describing a response pattern, or ensuring that a performance criterion for a particular product/process combination is met.

Like the botulinum cook, the processing criteria or formulations scrutinized by challenge tests will "overcook" the situation to ensure a very high probability that the process will not "fail dangerous." The value of properly constructed and validated models to replace, or significantly reduce, the need for expensive and time-consuming challenge tests by providing a generally applicable solution is well documented. As all microbial population responses are variable and characterized by a distribution,[18] knowledge of that distribution takes the predicted outcome to the next level by allowing calculation of confidence limits for the predictions. The increased precision available in turn allows greater confidence in specifying minimal processing conditions, thus preserving the nutritional and quality characteristics of a product without leading to microbial food safety or shelf-life problems.

The philosophical reasons to develop a predictive model lie in the nature of science itself as expressed by Lord Kelvin: "When you can measure what you are speaking about and express it in numbers you know something about it; but when you cannot measure it, when you cannot express it in numbers your knowledge is of a meagre and unsatisfactory kind."

Thus, quantitative science is inherently more useful than the qualitative description of a phenomenon, in that the latter is embodied in the former. Nevertheless, we should not lose sight of the fact that qualitative information is better than no knowledge, and that semiquantitative response patterns may be appropriate to address a particular question, to serve as a starting point or, as a "check and balance," when moving towards quantitative descriptions of microbial population responses. When using "black box" modeling approaches, such as polynomial models or artificial neural networks, *a priori* knowledge is a valuable adjunct providing reality checks to ensure that the developing model actually describes the observed biological response.[9]

The history and philosophy of science likewise suggest that moving from an empirical or phenomenological description to a mechanistic or deterministic description of a process represents an advance in the "good science" hierarchy. The former descriptions are pragmatic in nature and give rise to stochastic models in which the data are described by useful mathematical relationships. Mechanistic models, on the other hand, have a theoretical basis, allowing interpretation of the observed response on the basis of underlying theory; e.g., in considering microbial growth Monod[17] noted that, "There is little doubt that as further advances are made towards a more integrated picture of cell physiology, the determination of growth constants will have a much greater place in the arsenal of microbiology."

Of the secondary models commonly used in predictive microbiology, none can be viewed as truly mechanistic, although some are regarded as more mechanistic than others. Thus, Arrhenius-type temperature dependence models for microbial growth are often thought to have a greater mechanistic basis than do Belehradek-type (square-root or Ratkowsky-type) models. This contention would not have found support from Belehradek[2] who, because of the origin of Arrhenius models in chemical kinetics, wrote:

The problem of temperature coefficients in biology was initiated by chemists and has suffered from the beginning from this circumstance. Attempts to apply chemical temperature–velocity formulae (the $Q10$ rule and the Van't Hoff–Arrhenius law) to biological processes failed because some of the temperature constants used in chemistry (Q_{10}, μ) can be said not to hold good in biological reactions.

While it may be argued that mechanistic models are more amenable to refinement as knowledge of the system increases, development of fully mechanistic models for microbial growth has been limited by inability, to date, to provide quantitative values for all model parameters. Thermodynamic mechanistic models for the denaturation of proteins based on the enthalpy and entropy convergence temperatures and the heat capacity change[8] appear to apply to a wide range of proteins.[19] The growth of bacterial populations also seems amenable to this approach as foreshadowed by Ross and Olley in Chapter 10 of McMeekin et al.[13] and Ross[22] using *Escherichia coli* as an example. The ever increasing speed of computer simulation has recently enabled Drs. Olley, Ratkowsky, and Ross to apply this thermodynamic model to a very wide range of bacteria, including true psychrophiles.

Heat capacity change, which is the driving force for protein denaturation, may also explain the temperature response for bacterial growth, although it has not been measured for whole cells. This lends support to June Olley's long held view that the master reaction controlling temperature dependence is, in fact, the unfolding of proteins (or other macromolecules), exposing hydrophobic groups to interactions with surrounding water. Recent protein unfolding studies suggest that exposing polar surfaces may also be involved.[19]

When mechanistic models are invoked it is prudent to check the magnitude and sign (+ or −) of parameter estimates. Inappropriate parameter values, e.g., enormous activation energies, indicate that the model does not describe biological reality even though a mechanistic basis may have been inferred. Moving to an even greater level of "malpractice" in predictive model development one encounters situations in which very limited data sets have been fitted to models with a surfeit of parameters, resulting in an apparently perfect fit of the data to the model. This scenario represents an exercise in curve fitting unique to the data set used to develop the model and, in the end, because of false expectations based on erroneous assumptions of a mechanistic, or even a solid, quantitative foundation, is significantly less useful than describing a general response pattern.

12.3 THE BASICS OF PREDICTIVE MODELING

The rules for model selection were laid down by Ratkowsky in Chapter 2 of a previous book on predictive modeling,[13] viz parsimony, parameter estimation properties, range of variables, stochastic assumption, and interpretability of parameters. These continue to apply to the primary and secondary models described in Chapter 2 and Chapter 3 of this book and their use is supported by the model fitting techniques described in Chapter 4.

An important prerequisite to model fitting is the need to plan ahead by selecting an experimental design appropriate to the purpose of the study and to understand the limitations of the method of data collection selected. While "traditional" viable count and turbidimetric methods continue to dominate in modeling studies, several alternatives are given in Chapter 1 of this book. The search, as is the case in developing detection methods, is driven by:

1. The desire to obtain results in a compressed time frame
2. Ease-of-use characteristics, including automation to reduce cost and reduce the physical burden involved in collecting sufficient data at appropriate, usually close, time intervals and over an appropriate, often extended, time period

Regardless of the method chosen, the standard remains the viable count method against which other methods require rigorous validation. Despite the labor-intensive, time-consuming nature of viable count methods, they can, with good laboratory practice, perform with sensitivity, accuracy, precision, reproducibility, and repeatability.

Chapter 5 of this book provides a timely reminder that the test of a predictive model is not how well it performs in well-controlled laboratory conditions, but how well it predicts the behavior of microbial populations in real foods and in environments experienced under practical conditions of food production, processing, and storage. Indeed, some researchers have advanced the opinion that initial development of models in laboratory media represents wasted effort if subsequently the model fails to provide an adequate description of a target organism's behavior in food.[5] In particular, Brocklehurst (see Chapter 5, this book) and colleagues have drawn attention to important effects of food structure, including emulsions and surfaces that may significantly affect microbial behavior.

Similar caveats on the performance and limits of models were advanced by Ross[23] under the headings Model Applicability and Model Accuracy. The necessity that models are applied only to relevant situations requires enunciation of the conditions under which the model performs well and the boundaries beyond which predictions should not be made. If a model is deemed appropriate, its accuracy must also be considered, and this determination must take account of the fact that all microbial responses are variable. Commonly used measures to evaluate model performance are the bias and accuracy factors.[21] Note use of the term "evaluate" by this author, whereas most authors use "validate" to describe the process of ensuring that a model performs well in real foods subjected to anticipated conditions.

12.4 ADDRESSING CONCERNS IN PREDICTIVE MODELING

While variability is recognized as a characteristic feature of response times, such as the generation time or lag phase duration of a microbial population, that variability is characterized by distributions such as the gamma distribution in which the variance is proportional to the square of the response time.[18] This knowledge enables variability to be incorporated in models and confidence limits to be determined.

A more difficult suite of problems arises from the ability of microorganisms to adapt readily to the selective pressures of the food environment. Variability and adaptability were considered by Bridson and Gould[3] in a dichotomy described respectively as classical vs. quantal microbiology, where it was argued that "large populations of microorganisms obey the rules of taxonomy but the individual cells exhibit uncertainties (caused by mutations and fluctuating local environments) which are buried within the macropopulations" and "the functional stability of classical microbiology masks minority subpopulations which, nevertheless, contribute to the complex dynamics of microbial populations."

Conditions under which the adaptability of individual cells will influence the performance of a predictive model include those where small numbers of cells are found in a harsh environment causing decline in the viable population. This situation, often found in food processing environments, will lead to a few viable cells repairing, resolving the lag phase, and becoming the parent stock for the next phase of active growth.[14]

Adaptability and conditions where the rules of quantal microbiology apply give rise to uncertainty and the remaining challenges for predictive modeling. These now are centered on assessment of the initial conditions in a food, which will determine the level of contamination (expressed as concentration or prevalence in a sample), the initial physiological state of the population (or the survivors from an original population), and the complexity of the food system, including the microenvironment in which an individual organism is deposited. Uncertainty may also arise if interactions occur between different components of the microbiota and in fluctuating environments. The impact of the latter will depend on the magnitude of the fluctuations. Approaches to deal with dynamic environments and history effect on microbial growth and survival are discussed in Chapter 7 and Chapter 9, respectively.

Fluctuating environments, particularly with respect to temperature, probably represent a normal situation in food processing and during storage and distribution, and the consensus is that lag times will be affected but that there is little (if any) effect on generation times once the lag phase has been resolved. As an example, in the author's laboratory, cycling *Streptococcus thermophilus* between 30 and 40°C produced immediate changes to the anticipated growth rate. The microbiological outcome of such fluctuations can, therefore, be predicted easily by a growth rate temperature dependence model.

When temperature fluctuations are larger, a transient lag phase may be introduced before exponential growth is resumed. A good example of this behavior was provided by Baranyi et al.[1] working with *Brochothrix thermosphacta*. When the incubation temperature was dropped from 25 to 5°C, the growth rate changed as anticipated, but a shift from 25 to 3°C induced a lag phase in this psychrotolerant spoilage bacterium.

Modeling the effects of severe fluctuations in temperature or other environmental factors will be more difficult in that both lag and growth phases need to be considered and, in some instances, the model will also have to account for death of a proportion of the population. These situations provide examples where patterns of microbial population behavior, without an explicit mathematical description, may indicate a practical control situation. Let us consider a temperature-based example that has been studied at the pilot-plant level, and a laboratory-based water-activity example.

In the former, a pilot-plant-scale cheese-milk pasteurizer was used to study the development and control of thermoduric streptococci biofilms during milk pasteurization. Under normal operating conditions thermoduric streptococci grew on pasteurizer plates where the bulk milk temperature was between 35 and 50°C, and were detected in the product stream after 8 to 10 h. Introducing a temperature step change in the growth region of 55°C for 10 min with a 60-min interval between step changes resulted in a 20-h production run without detectable growth of thermoduric streptococci.[11]

In the latter example, Mellefont et al.,[15] using optical density (OD) methods showed that abrupt osmotic downshifts significantly increased the lag times of Gram-negative organisms, but those of Gram-positive organisms were largely unaffected. Further studies using viable count methods at close time intervals indicated that the apparent increase in lag time in fact comprises a death phase, a true lag period and growth back to the initial population levels (Mellefont, L.A., personal communication).

Returning to temperature shifts, Mellefont and Ross[16] found that downshifts induced significant increases in the lag phase duration of *Escherichia coli* and *Klebsiella oxytoca* that were dependent on the magnitude of the shift. These authors suggested that lags were introduced when the culture was shifted to temperatures beyond the normal physiological temperature range (NPTR) for growth, where cells are required to do additional work to adjust to the new environment and the rate at which that work is done.[20] At temperatures within the NPTR, the energy requirement for cellular functions to proceed is unchanged, the cells are "cruising," and thus shifts within this region are characterized by a simple change of rate without interruption to the growth cycle. This is a plausible hypothesis but much more work is required to firm up the concept and determine the physiological significance of the NPTR. The notion of a normal physiological range for other factors such as water activity and acidity should also be addressed.

Thus, predicting lag phase duration in foods is considered problematic because of the twin uncertainties of the initial physiological state of organisms and the numbers initially contaminating a product. Despite the fact that lag phase models can be developed in the laboratory with reasonable success (as the uncertainties above are minimized), this has remained one of the more intractable problems in predictive modeling.

A potential solution, however, was provided by Ross[23] using the trusted modelers device of moving from a kinetic to a probabilistic modeling approach to confront increasing uncertainty in describing initial conditions. The approach was to concede that while lag times are highly variable, the variability can be reduced by using the concept of relative lag times or "generation time equivalents," i.e., the ratio of lag time to generation time. By this device it was shown that although lag times may take almost any value, there is a common pattern of distribution of relative lag times for a wide range of species across a wide range of conditions. That common distribution of relative lag times has a sharp peak in the range of four to six generation time equivalents accounting for >80% of lag phase duration determinations.[23] This stochastic approach has significance for the application of predictive models in that when an adjustment of ~5 generation time equivalents is added to generation time predictions, good agreement of observed microbial proliferation on carcasses during chilling was observed.

12.5 IDENTIFYING OPPORTUNITIES FROM PREDICTIVE MODELING

In a practical sense it is clear that interrupting the exponential growth phase of a target organism is an effective strategy to buy time before critical limits of spoilage or pathogenic bacteria are reached. In the pasteurizer example, interrupting growth by intermittent temperature changes has the potential to more than double the run time of the heat exchanger with respect to thermoduric streptococci. An alternative biofilm control strategy is to minimize microbial adhesion to the pasteurizer plates, e.g., by a Teflon surface coating. However, this approach is inherently limited in its efficacy, with a 50% reduction in the initial load translating to an extension in run time equivalent to one generation time, for *S. thermophilus* ~20 min, and a 90% reduction to ~3 generations (~1 h) at its optimum temperature.

Buying time may also be useful in the context of a sequence of processing operations if the effect of changes occurring during each operation on microbial population dynamics is understood. Meat processing provides an interesting case study as regulatory authorities contemplate mandating a microbial "kill" step for certain pathogens during the conversion of live animals to meat. This philosophy will require the application of an intervention to achieve a reduction in pathogen numbers and, with current technology, the options are acid or hot water treatments. But could a better understanding of the chilling process reveal a potentially valuable intervention step without the significant cost burden of acid or hot water cabinets or steam pasteurization equipment? Such an opportunity is suggested by the results of Chang et al.[4] who studied reduction of bacteria on pork carcasses associated with chilling, concluding that "the effects of chilling techniques on microbial populations could provide pork processors with an additional intervention for pork slaughter or information to modify and/or improve the chilling process."

While lowering temperature and, in many chilling operations, simultaneously reducing surface water activity may not result in as great a reduction in microbial numbers as a heat treatment, understanding and "tweaking" these operations may prove valuable:

1. To identify and characterize a new critical control point (CCP)
2. As part of the "farm to fork" philosophy where every operation has an impact on the final level of risk to the consumer

Food safety professionals write often about the Hurdle concept in which multiple barriers confront microbial contaminants to delay resolution of the lag phase and the onset of exponential growth. However, mostly we think of applying hurdles via the intrinsic properties of the food (water activity, pH, organic acid concentration, etc.) and the extrinsic conditions of storage (temperature, atmosphere, etc.).

It may also be useful to think of the cumulative effect of hurdles applied in a sequence of processing operations leading to a final product. In the meat chilling situation outlined above, hurdles leading to inactivation and stasis of microbial populations will be supplemented by the effect of downstream processes. A case in point is freezing cartons of meat destined for the hamburger trade, where ice crystal formation might be expected to cause further damage to cells injured during chilling.

12.6 MODELING ATTACHMENT TO AND DETACHMENT FROM SURFACES

Don Schaffner, in Chapter 10 of this book, draws attention to a recent research trend in food microbiology concerned with modeling contamination and decontamination processes. This subject was reviewed recently by den Aantrekker et al.[6] Encompassed within the general area, researchers will need to consider contamination of foods and food contact surfaces (including hands), transfer of organisms to foods from surfaces and vice versa, removal of organisms from surfaces, the potential for and significance of recontamination of foods with small numbers of pathogens, etc.

The field of research, arising from the propensity of organisms to become intimately associated with surfaces as a survival mechanism, has spawned a subdiscipline of microbiology concerned with a continuum from the early events of adhesion to the formation of mature biofilms and their detachment from surfaces. Two aspects of these types of studies will be considered briefly here. The first is modeling attachment and detachment of biofilms, a general treatment of which was reviewed[6] under the heading Recontamination through Equipment. From this readers will be able to identify the relevant original literature. A specific example of this type of study, supporting the step change control strategy for *S. thermophilus* biofilms[11] was reported by Lee.[12] The model developed suggested that increasing the generation times of *S. thermophilus* represents the most effective way of controlling biofilm formation and subsequent detachment into cheese-milk.

The second aspect considers the important role of fluid transport in contamination events, which, in this author's opinion, is an obvious but often overlooked phenomenon in recent times. However, a considerable literature was generated on the role of fluid transfer in poultry processing in the 1970s (a quarter of a century ago), mainly by Dr. S Notermans and colleagues in The Netherlands and Dr. CJ Thomas and colleagues at the University of Tasmania. A model formulated to describe contamination during immersion chilling of poultry carcasses relies on the observation that the number of bacteria transferred from processing water to the skin of a broiler carcass is directly proportional to the number of organisms in the water. Effectively, when the carcass (or any other surface) is removed from a liquid it carries with it a sample of the liquid in which the density of the organisms is the same as that of the bulk liquid. The simple relationship has found utility, e.g., in the design of countercurrent immersion chillers in which the carcasses exit the chiller at the point where the microbial load in the water is lowest.

12.7 MODELING FUNGAL GROWTH

The great majority of predictive modeling studies to date have described the effect
of environmental factors on bacterial growth. Viruses and parasites effectively have
no growth ecology in foods, but patterns of decline have been described, e.g., for
the effect of freezing on *Trichinella* in pork. As we are reminded in Chapter 11,
molds are very important food spoilage organisms and the toxins they produce may
lead, usually in the longer term, to public health problems. Despite their importance
in food microbiology, the study of predictive mycology is very limited when compared with studies on bacteria. Nevertheless, sufficient material is available for
Philippe Dantigny to provide a short review of fungal modeling studies in foods,
including some interesting points for discussion such as the inability of the square-
root model to describe the effects of temperature on the kinetics of mold growth.

Insights might also arise from returning to the plant pathology literature that,
almost 50 years ago when I was a boy in Northern Ireland, was the basis of public
broadcasts predicting the likelihood of potato blight (*Phytophthora infestans*) or
apple scab (*Venturia inequalis*) problems as a result of prevailing climatic conditions
(temperature and relative humidity). In essence, these forecasts might be considered
an early example of risk assessment, predicting the incidence of plant disease rather
than human disease.

12.8 APPLICATION OF PREDICTIVE MICROBIOLOGY

To use predictive models practically in the food industry requires devices that
monitor the environmental conditions in that part of the paddock to plate continuum
of interest and a means to translate that environmental history into an estimate of
the growth, survival, or death of a target organism.

The devices available are chemical or physical monitors where the interpretative
function is built into the device, e.g., as a color change resulting from a chemical
reaction or, in physical mode, the extent of migration of a dye along a wick.[13] Implicit
in the efficacy of monitors with a built-in interpretative function is that the rate at
which the chemical or physical change occurs mimics that of microbial behavior
under the same conditions. Unfortunately, such monitors are based almost exclusively on Arrhenius reaction kinetics, which deviate from biological response rates
as limits for growth are approached. Often these regions are of greatest interest in
predicting the shelf life and safety of foods.

An alternative to built-in interpretation is to construct a tertiary model,[26] usually
a spreadsheet-based program that converts a monitored temperature history into an
estimate of microbial growth potential. Many such programs are available and
prominent ones are described in Chapter 6 of this book. That chapter also describes
a most significant initiative in predictive modeling: the development of COMBASE,
initially by combining the U.S.-developed Pathogen Modeling Program and the
U.K.-developed Food MicroModel, but to which other significant databases could
be added. Further, the authors of Chapter 6 draw attention to the main pillars of
predictive microbiology software packages: databases and mathematical models.
Here they point out that while most chapters in this book deal with aspects of

mathematical modeling, more emphasis needs to be placed on the value of databases and the scientific study termed Bioinformatics. The power of the database, perhaps through the agency of a mathematical model or algorithm, is described via the use of Expert Systems that provide decision support. These include the possibility of in-line real-time systems.

Such developments highlight the interface of predictive microbiology with information technology systems, allowing the application of models in food safety management strategies including Hazard Analysis Critical Control Point (HACCP), Quantitative Risk Assessment, and Food Safety Objectives. Much has been written about specific opportunities for predictive models to underpin these strategies that need not be repeated here, e.g., see Chapter 8 of this book. Further, there is a particular role for predictive models in comparing the microbiological outcomes of various operations in food manufacture. Predictive models based on detailed knowledge of the microbial ecology of any product/pathogen combination enable us to:

1. Quantify the effect of preservation technologies on microbial populations
2. Optimize existing or suggest new processing procedures
3. Indicate risk management options
4. Identify regulations that are unwarranted
5. Support the need for outcome-based regulations
6. Enable equivalence determinations

The last two applications are likely to become prominent in the global debate on regulatory frameworks to assure the safety of food in international trade and attendant market access issues that sit alongside the protection of public health.

However, reticence to realize the potential value of predictive modeling applications continues as exemplified by comments from the Food Safety and Inspection Service (FSIS) of the USDA. In July 2002, FSIS issued a notice entitled "Use of microbial pathogen computer modeling in HACCP plans" (www.usda.fsis.gov) that presented a particularly negative account of the potential to use microbial pathogen computer modeling (MPCM) programs in the development and use of HACCP plans. While acknowledging that MPCM may be useful in "supporting hazard analyses, developing critical limits and evaluating the relative severity caused by process deviations," FSIS states categorically, "It is not possible or appropriate to rely solely upon a predictive modeling program to determine the safety of food and processing systems." Furthermore, "determining pathogen growth and survival, and controlling it in food products often requires complete and thorough analysis by an independent microbiology laboratory, challenge studies and surveys of the literature."

In effect, FSIS listed the elements of a properly conducted and independently evaluated predictive modeling study:

- Thorough analysis of the literature to reveal general patterns of microbial population behavior
- Challenge studies to validate that the proposed model is applicable in practice

- Independent evaluation to verify that an MCPM program consistently allows a process to meet agreed critical levels at identified CCPs

There is, of course, a requirement for those proposing the use of predictive models in HACCP plans to ensure that the model predictions are used conservatively. This is a fact well recognized by predictive modeling researchers who promote caution in the use of models, e.g., McMeekin and Ross.[14] These authors recognized clearly that

> there are two criteria that must be satisfied if any form of application software is to be used effectively. The first is a properly developed and validated model and the second the ability of an operator to interpret correctly the microbiological significance of the results. Poorly performing models coupled with poor interpretation continue to be the greatest threat to widespread use of predictive microbiology as a technology with the potential to assure the microbiological quality and safety of foods.

A somewhat different perspective on the use of predictive models to evaluate food safety was provided by an International Food Technology (IFT) committee in their report "Evaluation and Definition of Potentially Hazardous Foods." This was prepared for the FDA (Dec. 31, 2001; IFT/FDA Contract No 223-98-2333, Task Order No 4).

In this report IFT recognizes that certain foods have combinations of pH, a_w, preservatives, etc., that restrict microbial growth and may, therefore, not require refrigeration to protect public health. Under certain circumstances time alone can be used to control product safety, e.g., "if *S. aureus* is a concern the Pathogen Modeling Program V.5.1 could be used to estimate the time of storage where the pathogen could grow."

The committee further suggests that general growth models such as Pathogen Modeling Program should be used conservatively or in combination with challenge testing. However, if an in-house model has been developed and validated for a particular food, it could be used itself or with challenge testing.

12.9 CONCLUDING REMARKS

This is only the third major book devoted to predictive microbiology in approximately 30 years of scientific endeavor in the field. The first,[13] referred to in the Preface of this book, appeared in 1993; the second,[23] which focused on predictive models for the meat industry, was not widely distributed.

The chapters in this book cover the entire gamut of predictive modeling research and it is appropriate that the "middle" chapter, Chapter 6, describes database development, the fulcrum on which predictive modeling swings.

Here we find a description of how new knowledge is accumulated and stored in databases that, when properly constructed, are dynamic information repositories to which new information can be added at any time and from which that found to be "dodgy" can be deleted. Speaking the same language is an important element of database construction and a prerequisite to merging existing databases. The authors of Chapter 6 have expended considerable effort to make two major databases — Pathogen Modeling Program and Food MicroModel — compatible and to merge these into COMBASE. This initiative will also guide researchers toward the best way to collect and collate new information so that it can be easily integrated into existing databases. Properly supported, the COMBASE initiative will be a watershed in the evolution of predictive modeling and its widespread application.

Much of what underpins a predictive models database is regarded by many as difficult science. However, this need not unduly concern potential users of models if there is confidence that the underpinning science is sound. What comes after this point, in the form of user-friendly interpretative devices, will determine whether or not the potential of predictive microbiology is realized.

Many specific applications have been suggested for predictive models and some are mentioned earlier in this chapter. However, in a more general sense, the value of predictive models and databases lies in their support of food safety management strategies such as HACCP, Quantitative Risk Assessment, and Food Safety Objectives.

In particular, the role of quantitative information in empowering the HACCP concept must not be undersold. HACCP is a simple, but elegant, concept by which safety is built into a process. It works well where critical control points, e.g., a lethal heat process, can be identified. Its value is less obvious where critical control points do not stand out, e.g., in the conversion of muscle to meat. Its value is compromised where the HACCP concept is virtual rather than real, i.e., where generic criteria are applied to satisfy mandatory requirements that a process is HACCP-based rather than using criteria derived from a knowledge of microbial population behavior.

The challenge now for predictive microbiology is to move from the phase of model development and model validation (evaluation) to *process validation* and *verification* studies that will enable models to be used with confidence in HACCP systems as demonstrated by Jericho et al.[10]

The "hard yakka" (Australian vernacular for hard work) has been done in developing the databases and models and now there are tantalizing prospects for in-line control of processes, appropriate corrective actions, and sound food safety management decisions arising from the concept of predictive microbiology.

ACKNOWLEDGMENTS

The author is indebted to Dr. June Olley for stimulating discussions in the preparation of this essay and to Meat and Livestock Australia (MLA) for continuing support of predictive modeling research.

REFERENCES

1. Baranyi, J., Robinson, T.P., Kaloti, A., and Mackey, B.M., Predicting growth of *Brochothrix thermosphacta* at changing temperature. *Int. J. Food Microbiol.* 27, 61, 1995.
2. Belehradek, J., Temperature coefficients in biology. *Biol. Rev. Biol. Proc. Camb. Philos. Soc.* 5, 30, 1930.

3. Bridson, E.Y. and Gould, G.W., Quantal microbiology. *Lett. Appl. Microbiol.* 30, 95, 2000.

4. Chang, V.P., Mills, E.W., and Cutter, C.N., Reduction of bacteria on pork carcasses associated with chilling method. *J. Food Prot.* 66, 1019, 2003.

5. Dalgaard, P., Mejlholm, O., and Huss, H.H., Application of an iterative approach for development of a microbial model predicting the shelf-life of packed fish. *Int. J. Food Microbiol.* 38, 169, 1997.

6. den Aantrekker, E.D., Boon, R.M., Zwietering, M.H., and van Schothorst, M.,Quantifying recontamination through factory environments: a review. *Int. J. Food Microbiol.* 80, 117, 2002.

7. Esty, J.R. and Meyer, K.F., The heat resistance of the spore of *B. botulinum* and allied anaerobes, XI. *J. Infect. Dis.* 31, 650, 1922.

8. Fu, L. and Friere, E., On the origin of the enthalpy and entropy convergence temperatures in protein folding. *Proc. Natl. Acad. Sci.* 89, 9335, 1992.

9. Geeraerd, A.H., Valdramidis, V.P., Devlighere, F., Bernaerts, H., Debevere, J., and Van Impe, J.F., Development of a novel modelling methodology by incorporating *a priori* microbiological knowledge in a black box approach. *Proc. 18th International ICFMH Symposium*, Lillehammer, Norway, 2002, p. 135.

10. Jericho, K.W.F., O'Laney, G., and Kozub, G.C., Verification of the hygienic adequacy of beef carcass cooling processes by microbiological culture and the temperature-function integration technique. *J. Food Prot.* 61, 1347, 1998.

11. Knight, G., Nicol, R.S., and McMeekin, T.A., Temperature step changes: a novel approach to control biofilms of thermoduric streptococci in a pilot plant-scale cheese-milk pasteuriser. *Int. J. Food Microbiol.* In press.

12. Lee, M.W., *Growth Rate and Detachment Rate of a Biofilm of Thermoduric Bacteria*, M.Sc. thesis, Monash University, Clayton, VI, 1998.

13. McMeekin, T.A., Olley, J., Ratkowsky, D.A., and Ross, T. *Predictive Microbiology: Theory and Application.* Research Studies Press Ltd., Taunton, U.K., 1993.

14. McMeekin, T.A. and Ross, T., Predictive microbiology: providing a knowledge-based framework for change management. *Int. J. Food Microbiol.* 78, 133, 2002.

15. Mellefont, L.A., McMeekin, T.A., and Ross, T., The effect of abrupt osmotic shifts on the lag phase duration of food borne bacteria. *Int. J. Food Microbiol.* 83, 281, 2003.

16. Mellefont, L.A. and Ross, T., The effect of abrupt shifts in temperature on the lag phase duration of *Escherichia coli* and *Klebsiella oxytoca. Int. J. Food Microbiol.* 83, 295, 2003.

17. Monod, J., The growth of bacterial cultures. *Ann. Rev. Microbiol.* 3, 371, 1949.

18. Ratkowsky, D.A., Ross, T., Macario, N., Dommett, T.W., and Kamperman, L., Choosing probability distributions for modelling generation time variability. *J. Appl. Bacteriol.* 80, 131, 1996.

19. Robertson, A.D. and Murphy, K.P., Protein structure and the energetics of protein stability. *Chem. Rev.* 97, 1251, 1997.

20. Robinson, T.P., Ocio, M.J., Kaloti, A., and Mackey, B.M., The effect of growth environment on the lag phase of *Listeria monocytogenes. Int. J. Food Microbiol.* 44, 83, 1998.

21. Ross, T., Indices for performance evaluation of predictive models in food microbiology. *J. Appl. Bacteriol.* 81, 501, 1996.

22. Ross, T., Assessment of a theoretical model for the effects of temperature on bacterial growth rate. Predictive microbiology applied to chill food preservation. *Proc. Conf. No. 1997/2, Comm. C2*, Quimper, France, International Institute Refrigeration, Sydney, 1999, p. 64.

23. Ross, T., *Predictive Microbiology for the Meat Industry*, Meat and Livestock Australia, Sydney, 1999, 196 pp.
24. Scott, W.J., The growth of microorganisms on ox muscle. I. The influence of water content of substrate on rate of growth at −1°C. *J. Counc. Sci. Ind. Res. (Aust.)* 9, 177, 1936.
25. Scott, W.J., The growth of microorganisms on ox muscle. II. The influence of temperature. *J. Counc. Sci. Ind. Res. (Aust.)* 10, 338, 1937.
26. Whiting, R.C. and Buchanan, R.L., A classification of models for predictive microbiology. *Food Microbiol.* 10, 175, 1993.

Index